UNIVERSITY OF LIVERPOOL
013530726
WITHDRAWN
FROM STOCK

THE
INTERNATIONAL SERIES
OF
MONOGRAPHS ON PHYSICS

# Introduction to the Theory of Ferromagnetism

AMIKAM AHARONI

*Weizmann Institute of Science*
*Rehovoth, Israel*

CLARENDON PRESS • OXFORD

*Oxford University Press, Great Clarendon Street, Oxford OX2 6DP*

*Oxford New York*
*Athens Auckland Bangkok Bogota Bombay Buenos Aires*
*Calcutta Cape Town Dar es Salaam Delhi Florence Hong Kong*
*Istanbul Karachi Kuala Lumpur Madras Madrid Melbourne*
*Mexico City Nairobi Paris Singapore Taipei Tokyo Toronto Warsaw*

*and associated companies in*
*Berlin Ibadan*

*Oxford is a trade mark of Oxford University Press*

*Published in the United States by*
*Oxford University Press Inc., New York*


*Reprinted 1998*



*A catalogue record for this book is available from the British Library*

*Library of Congress Cataloging in Publication Data*
*(Data applied for)*

*ISBN 0 19 851791 2*

*Printed in Great Britain by*
*Bookcraft Ltd, Midsomer Norton, Avon*

# PREFACE

This book is mainly intended to be used as a textbook by first-year graduate students and advanced seniors in physics and engineering. It is built, however, in such a way that it can also serve as a reference book for professionals who work on advanced topics of magnetism, but want to refresh their previous studies, or look more deeply into the basis of what they are doing. It is based on a course that I used to teach in the Feinberg Graduate School of the Weizmann Institute of Science, which has been widened here, mainly in the part dealing with micromagnetics.

The emphasis is on explaining the basic principles, without going too deeply into any of the special fields which are normally discussed in different parallel sessions in the magnetism conferences. The idea is to give beginners as much coverage as is possible within a reasonable size of one book, *and* to have mature researchers in one field have at least a glimpse into what other fields are all about. Only in the particular field of micromagnetics did I allow myself to go somewhat into the state-of-the-art of some more advanced topics, mainly because there is no comprehensive treatise which covers this subject, or even any part of it. In some ways, this book is also meant to be such an advanced review of micromagnetics. But even for this subject I tried to concentrate on the basis, and avoid most of the technical details which belong in such a treatise. The more advanced parts are usually given as references to the literature, which should help researchers without confusing the students.

This book is theoretical, but it is by no means meant to be read only by those who want to become theorists. I have tried to keep in mind all those practising engineers and experimental physicists who only too often do good experimental work, without understanding the theory behind it. They usually look up, and with too high respect, to the theorists whose papers they are unable to read, mainly because these papers are written in an incomprehensible language. I hope they will be able to read this book, and to find out that in many cases there is nothing behind that obscure presentation, and the theorists only pretend that they know what they are doing. For this purpose, the emphasis is always on the disadvantages and the drawbacks of each theory, more than on its advantages, which in my mind are self-evident. In particular, I keep pointing out the *approximations* which some theorists ignore, or even try to hide, claiming that a particular calculation, or a particular result, is *exact*.

It is assumed that the reader has already taken an undergraduate course on fields, and is familiar with Maxwell's equations, and the way they are derived. Some (but definitely not all) of that subject matter is repeated here, and presented from a different angle, and with a more mature point of view. I hope that this repetition will help establish a better understanding of the magnetostatics than is possible in a typical undergraduate course, which rushes to cover the curriculum, with no time left to understand it. However, even in this most basic magnetostatics, I am mostly trying to give the student a good foundation to build the theory on, rather than to cover a lot of ground, or to go into the fine details of particular problems.

Most of this book uses classical physics only, because it is impossible to do it otherwise. In spite of some claims by some enthusiasts, there is no quantum-mechanical theory of magnetism which covers more than a minor little corner of the subject, and even that is done by using very rough approximations. I made a special point to discuss the Bohr–van Leeuwen theorem, in section 1.3, because it is quoted much too often by quantum-mechanics specialists, who look down on everybody else, and claim that theirs is the only true physics. Those magneticians who do not develop an inferiority complex from these arguments may skip the details of that section. Nobody ever tells the quantum-mechanical experts to avoid certain approximations used in their theory, and if told, they could not care less. I am trying to encourage the classical-mechanics theorists to have the same attitude.

In the Feinberg Graduate School, all students are required to take an advanced course in Quantum Mechanics, on which I could build my course. This case, however, is atypical, and in other places many students, mostly in engineering, reach a graduate school without ever being exposed to any quantum mechanics. For the benefit of these students, I collected all the quantum-mechanical discussion into chapter 3. The rest of the book is written without any essential reference to this chapter, and can be followed even if chapter 3 is omitted. Those who know something about quantum mechanics should be able to benefit from the presentation of the basic principles in that chapter. But those who do not know quantum mechanics can easily do without it.

I have used an unconventional order of the subject matter, starting from exchange, then adding to it the anisotropy, and including time effects, which are usually studied without any reference to the magnetostatics anyway. The magnetostatic interactions come only after the superparamagnetism. I believe that this order is easier to follow by students than the order which most of my colleagues would have recommended, and that it helps to grasp the principles behind the equations.

The references are only meant to indicate where the reader may find more information which is relevant to some points. They are *not* meant to help write the history of the subject, although I sometimes mention who started that line of research. Therefore, the older papers are not mentioned if they are quoted in newer ones, unless the older ones contain certain information which cannot be found in newer ones. I never could understand the point of those who cite the original work of Maxwell, for example, which nobody bothers to look up anyway. I have also restricted all citations to papers in English only. Some years ago all students were required to have at least a working knowledge of other languages, but these days are passed, and I see no reason to put a list which nobody will even look at. It is rather strange to discuss the works of Néel without citing any of his original works, or even to discuss the Döring mass in section 10.4 with no mention of Döring. But I prefer to do it this way, and anybody who wants to read the papers of Néel in French can easily find the necessary references in the papers which I cite. After all, I am not citing all the older papers in English either, which could easily make a list of many thousands. I was tempted to make an exception in the case of the Dietze and Thomas model of eqn (8.1.1) which I discuss in much detail, but then I decided against it.

It may not seem so to those who are used to textbooks with hardly any references at all, but I tried to keep their number as small as possible, and it is, after all, also a guide to researchers. In as much as possible, I also tried to refer to reviews for broader aspects of the topic discussed in the text, or to semi-popular articles in *Physics Today*, which are rather easy to be read by beginners. This rule is not always adhered to, and in many cases there is nothing besides the original, and difficult-to-read, articles. Occasionally I may have also been carried away, and cited some more advanced articles, which are definitely not on the intended level of the book. But it should be borne in mind that the beginner is not expected to read all those articles, and some of them *are* only intended for practising researchers who want to go into some more detail.

I tried to avoid citing conference proceedings which are not part of a journal, and also journals which are not easily available in many libraries, unless nothing more appropriate is available. For example, before Brown published the full account of what I refer to here as [171], he published a short version in 1963 *J. Appl. Phys.*, **34**, 1319–20. There is nothing in the latter which cannot be found in [171], and there is no reason to mention it. Then Brown did the same with [508], a short version of which was first published as [507]. But in this case, the full paper is published in an obscure journal, which not many have. Therefore, in this case I also cited the short version [507], so that those who cannot find [508] can at least read *something*.

תושלב"ע

A. A.

*Rehovoth, Israel*
December 1995

CONTENTS

# 1

# INTRODUCTION

## 1.1 Nomenclature

It is known from experiment that every material which is put in a magnetic field, **H**, acquires a magnetic moment. The dipole moment per unit volume is defined as the magnetization, and will be denoted here by the vector **M**. In *most* materials **M** is proportional to the applied field, **H**. The relation is then written as

$$\mathbf{M} = \chi \mathbf{H}, \tag{1.1.1}$$

and $\chi$ is called the *magnetic susceptibility* of the material.

Maxwell's equations are usually written for the vector

$$\mathbf{B} = \mu_0(\mathbf{H} + \gamma_{\mathrm{B}} \mathbf{M}), \tag{1.1.2}$$

instead of **M**. Here $\gamma_{\mathrm{B}}$ is a notation introduced by Brown [1] to include different systems of units. In particular, $\gamma_{\mathrm{B}} = 1$ for the SI units, which are popular in textbooks, while $\gamma_{\mathrm{B}} = 4\pi$ for the Gaussian, cgs units, which are most popular among magneticians, and for which $\mu_0 = 1$. If eqn (1.1.1) is fulfilled, it is also possible to rewrite eqn (1.1.2) as

$$\mathbf{B} = \mu \mathbf{H}, \tag{1.1.3}$$

where

$$\mu = \mu_0(1 + \gamma_{\mathrm{B}} \chi) = \mu_0 \mu_r \tag{1.1.4}$$

is known as the *magnetic permeability.* The material is classified as 'paramagnetic' if $\chi > 0$, *i.e.* $\mu_r > 1$, and as 'diamagnetic' if $\chi < 0$, *i.e.* $\mu_r < 1$.

There are, however, some materials which do not fit this classification, because in these materials the magnetization **M** is *not* proportional to the applied field, **H**. It may be, for example, non-zero at $\mathbf{H} = 0$. Actually, **M** in these materials is not even a one-valued function of **H**, and its value depends on the history of the applied field. A typical case is shown in Fig. 1.1, which plots the component of **M** in the direction of the applied field, $M_H$, as a function of the magnitude of that field. The outermost loop is know as the *limiting hysteresis curve*, and is obtained by applying a sufficiently large field in one direction, decreasing it to zero, and then increasing it to a large value in the opposite direction. The curve is reproducible in consecutive cycles of the applied field.

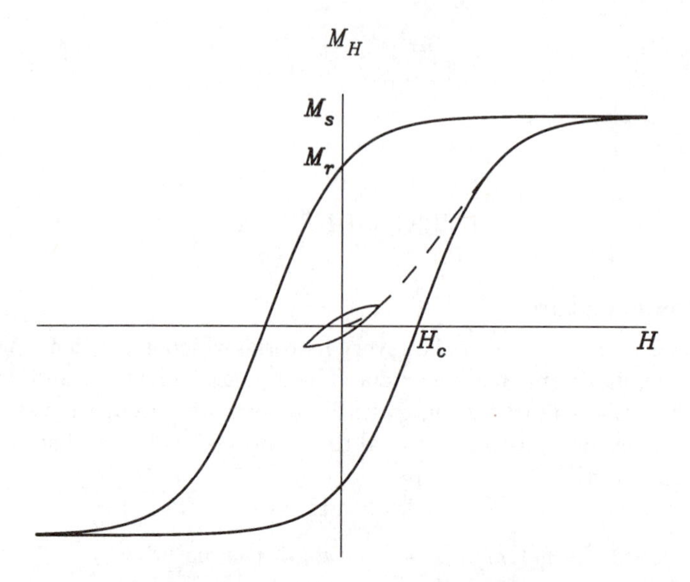


FIG. 1.1. Schematic representation of the *limiting* hysteresis curve (or loop) of a typical ferromagnetic material, displaying also the virgin curve (dashed), and one minor loop. Also shown are the remanence, $M_r$, the saturation magnetization, $M_s$, and the coercivity, $H_c$.

The curve which starts at the origin is known as the *virgin magnetization curve*, and can only be traced once after *demagnetizing* the sample, namely bringing it to a state in which $\mathbf{M} = 0$ at $\mathbf{H} = 0$. This demagnetization may be achieved by heating the sample to a high temperature, and cooling it in zero field, or by cycling the applied field with steadily decreasing amplitudes. If the field is increased, then decreased before the limiting hysteresis curve is reached, and then the field is reversed, a so-called *minor hysteresis loop* is traced. One example of such a curve is shown in Fig. 1.1, but there is actually a whole continuum of them. With an appropriate history of the applied field, one can therefore end at any point inside the limiting hysteresis loop. In particular, it is possible to reach $\mathbf{H} = 0$ with *any* value of $M_H$ between $-M_r$ and $+M_r$, where $M_r$ is the value of $M_H$ on the limiting hysteresis curve, at $H = 0$ (see Fig. 1.1). It is called the *remanence*, or the *remanent magnetization.*

It is possible, although not really necessary, to define permeability for ferromagnetic materials, in order to pretend that they are similar in some way to paramagnetic materials. The definition is not unique, because neither eqn (1.1.1) nor eqn (1.1.2) is fulfilled in ferromagnetic materials. Nevertheless, it is quite customary to introduce some *effective* permeability at a particular value of the applied field, $H$, as

$$\mu_{\text{eff}} = \partial B_H/\partial H, \tag{1.1.5}$$

or over a certain range of values of $H$, as

$$\mu_{\text{eff}} = \Delta B_H/\Delta H. \tag{1.1.6}$$

In either case, it is possible to a certain approximation to apply Maxwell's equations to ferromagnetic materials in the same way they are used for paramagnetic materials, with this effective permeability. However, here $\mu$ is not a constant as it is in paramagnetic materials, and the whole formulation is at best usable at a particular applied field.

Two other important terms are also defined on the limiting hysteresis curve in Fig. 1.1. One is the *coercivity* or *coercive force*, $H_c$, which is the value of $H$ for which $M_H = 0$, and is actually a magnetic field and not a force. The other is the *saturation magnetization*, or *spontaneous magnetization*, $M_s$, which will be defined for the meantime as the value of $M_H$, or the magnitude of $M$, in a very large field. This definition is not accurate, and will be modified in section 4.1, but it should do for now.

This saturation magnetization is an intrinsic property of the material, and is independent of the sample, if properly measured. It is a function of temperature, a typical form of which is plotted in Fig. 1.2. In this figure, $M_s$ is normalized with respect to its value at zero temperature. The temperature is normalized with respect to the so-called *Curie temperature*, $T_c$, of the material, which is the temperature at which $M_s$ becomes 0 *at zero applied field.* When normalized in this way, the curves for different ferromagnets are very nearly the same as in Fig. 1.2. All ferromagnets become regular paramagnets at temperatures above $T_c$, and as such they have a non-zero magnetization in the presence of the field which is used for the measurement. This behaviour is emphasized in Fig. 1.2 which shows the curve as actually measured in a small applied field, at which $M_s$ does not go to zero at the Curie temperature. Specific experimental curves for Ni and Fe can be seen *e.g.* in Fig. 9 of Potter [2].

## 1.2 Weiss Domains

In principle, any theory of ferromagnetism should address *both* of these unusual phenomena, which are not encountered in other materials. It should thus explain the hysteresis displayed in Fig. 1.1 *and* the temperature-dependence of Fig. 1.2, even though most theoreticians work only on one and ignore the other. They do it even when they compare the results with an experiment, which *always* involves both. It is inevitable, because the general *quantitative* problem is too complicated for the present state of knowledge.

Qualitatively, both phenomena are understood to a certain extent due to an explanation already given by Weiss in 1907. Weiss *assumed* that there

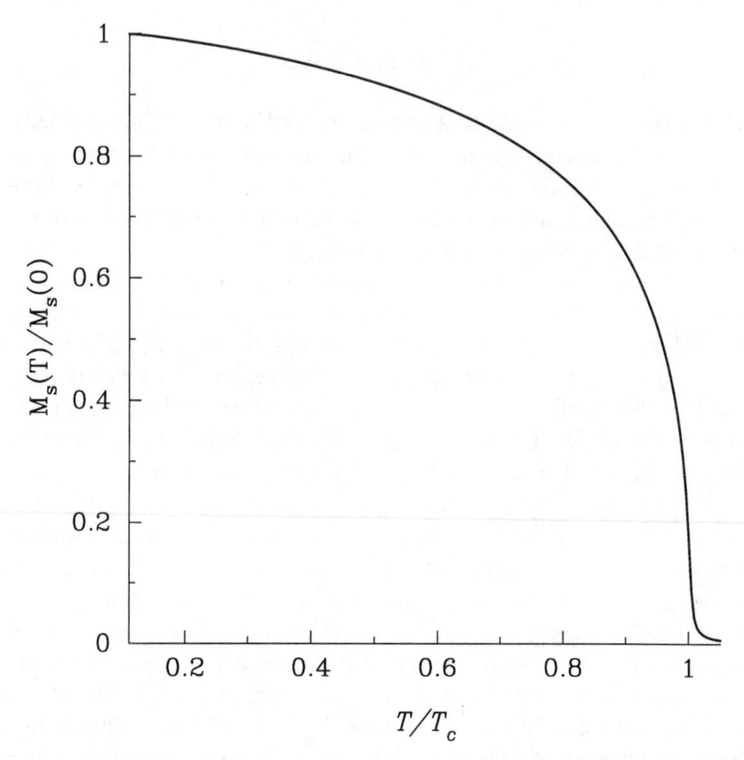


FIG. 1.2. Spontaneous magnetization, $M_s$, of a ferromagnet as a function of the temperature, $T$, normalized to the Curie temperature, $T_c$. The applied field is assumed to be small, but finite, as it is in real measurements.

is a certain internal (or 'molecular') field in ferromagnetic materials, which tries to align the magnetic dipoles of the atoms against thermal fluctuations which prefer a complete disorder of these dipoles. As will be seen in the next chapter, such a *molecular field* is sufficient to explain the temperature-dependence as plotted in Fig. 1.2, and the paramagnetism above the Curie temperature. However, this model leads to a *constant* magnetization $\mathbf{M}$ at any given temperature below $T_c$. In order to explain the unusual field-dependence in Fig. 1.1, Weiss assumed that ferromagnets are made out of many *domains*. Each of these domains is magnetized to the saturation value $M_s(T)$ as in Fig. 1.2, but the *direction* of the magnetization vector varies from one domain to the other. The measured value of the magnetization is the average over these domains, which can be zero in any particular direction when there is an equal number of domains parallel and antiparallel to that direction. It can also have a non-zero value, numerically less than $M_s(T)$, if this number of domains is not equal. The applied magnetic field rotates the magnetization of the individual domains into its own direction, and when this field is sufficiently large to align all domains, the measured average value becomes $M_s(T)$. Without going into fine details, it should be quite clear that this assumption is sufficient to explain the field-dependence in Fig. 1.1, at least qualitatively.

Weiss did not justify any of his two arbitrary assumptions, and could not explain the origin of the molecular field or the existence of the domains. There were also several difficulties in implementing his principles for any quantitative estimations. In particular, using the experimental value of the Curie point in such a theory, the magnitude of the molecular field in iron turns out to be of the order of $10^6$ Oe. It takes a field of the order of 1 Oe to rearrange the domains in iron, and $10^3$ Oe to eliminate them altogether. How come that a field of $10^6$ Oe is not sufficient to align *all* the magnetic moments of iron, and it takes an extra field of only $10^3$ Oe to do it? And how is it that even a 1 Oe field can contribute so very significantly towards a task which a $10^6$ Oe field is not sufficient to accomplish?

In spite of these difficulties (which will be addressed later) the assumptions of Weiss are actually valid and sound, and contain the basic understanding of ferromagnetism. The molecular field is known now to be a certain approximation to a coupling force between spins, which can be derived from more basic principles, as will be shown here on different levels. The existence of domains, magnetized in different directions, is not even an assumption any more. These domains have been observed by several techniques, outlined in section 4.1, and their existence is now an established experimental *fact*. The only difference is that they are known now to be magnetized along certain directions, and are not as *randomly* oriented as Weiss thought. However, being an experimental fact should not stop us from enquiring *why* these domains exist, and an appreciable part of this book will be devoted to answering this question, as well as the question of why a $10^6$ Oe field cannot do what a 1 Oe field can.

Even though the basic properties of ferromagnets are quantum mechanical by nature, most of the treatment here will use classical physics, on a level which can be followed by engineering students who did not take any quantum mechanics in college. It is not only the choice for this book. Most of the development of the theory of ferromagnetism was done using classical concepts only, even in recent years, when everybody knows that a classical theory can at best be only an approximation to the true quantum treatment, *especially* in magnetism. The reason is that pure quantum theory has not advanced yet beyond simple cases which are of very little practical application. However, before adopting this classical approach to ferromagnetism, it is necessary to consider a famous and often-quoted theorem of

Bohr and van Leeuwen, according to which classical physics cannot possibly lead to magnetism, because in pure classical physics the electrons in a material do not interact with an applied magnetic field.

## 1.3 The Bohr–van Leeuwen Theorem

Consider a classical system of $N$ electrons. They have $3N$ degrees of freedom, and are therefore described by their $3N$ coordinates, $q_i$, and their $3N$ momenta, $p_i$. Each electron has a (negative) charge, $e = -|e_e|$. In cgs units, an electron whose velocity is $\mathbf{v}$ creates a current density $\mathbf{j} = e\mathbf{v}$, and a magnetic moment

$$\mathbf{m} = \frac{1}{2c}\mathbf{r} \times \mathbf{j} = \frac{e}{2c}\mathbf{r} \times \mathbf{v}, \tag{1.3.7}$$

at the position $\mathbf{r}$ in space, where $c$ is the velocity of light. The important feature is that this $\mathbf{m}$ is a *linear* function of the velocity $v$ of each particular electron. It means that whatever the pattern of motion of all the electrons is, the *total* magnetic moment is also a *linear* function of all the electron velocities. Therefore, the $z$-component of the *total* magnetic moment of all the electrons must be a function of the form

$$m_z = \sum_{i=1}^{3N} a_i^z(q_1, \ldots, q_{3N})\, \dot{q}_i, \tag{1.3.8}$$

where the dot designates a derivative with respect to time, and where the coefficients, $a_i^z$, are functions of all the coordinates $q_i$, but do *not* depend on $p_i$.

The canonical equations of a classical motion are

$$\dot{q}_i = \frac{\partial \mathcal{H}}{\partial p_i}, \qquad \dot{p}_i = -\frac{\partial \mathcal{H}}{\partial q_i}, \tag{1.3.9}$$

where

$$\mathcal{H} = \sum_{i=1}^{3N} \frac{1}{2m_e}\left(\mathbf{p}_i - \frac{e}{c}\mathbf{A}_i\right)^2 + eV(q_1, \ldots, q_{3N}) \tag{1.3.10}$$

is the Hamiltonian, $m_e$ is the electron mass, $\mathbf{A}$ is the vector potential of the magnetic field, and $eV$ is the potential energy. Substituting eqn (1.3.9) in eqn (1.3.8),

$$m_z = \sum_{i=1}^{3N} a_i^z(q_1, \ldots, q_{3N})\, \frac{\partial \mathcal{H}}{\partial p_i}. \tag{1.3.11}$$

If $k_B$ is Boltzmann's constant, $T$ is the temperature, and

$$\beta = \frac{1}{k_B T}, \tag{1.3.12}$$

the classical statistical average which will be measured is

$$M_z = \frac{\int m_z e^{-\beta\mathcal{H}} dq_1 \ldots dq_{3N} dp_1 \ldots dp_{3N}}{\int e^{-\beta\mathcal{H}} dq_1 \ldots dq_{3N} dp_1 \ldots dp_{3N}} . \tag{1.3.13}$$

According to eqn (1.3.11), the numerator in eqn (1.3.13) is a sum of terms, each of which is proportional to

$$\int_{-\infty}^{\infty} \frac{\partial \mathcal{H}}{\partial p_i} e^{-\beta\mathcal{H}} dp_i = \left[-k_{\mathrm{B}} T e^{-\beta\mathcal{H}}\right]_{p_i=-\infty}^{\infty} \tag{1.3.14}$$

which vanishes, because $\mathcal{H}$ is proportional to $p_i^2$ for large $|p_i|$ according to eqn (1.3.10). Therefore, $M_z = 0$ and there is no magnetic moment at any vector potential (namely for any applied field), no matter what the actual motion of the electrons in the material is. In other words, there is no interaction between an applied magnetic field and the electrons in any material, if these electrons behave according to the laws of classical physics. It means that classical physics cannot account for either diamagnetism or paramagnetism, let alone ferromagnetism.

This theorem is very general, and its proof is rigorous. However, it does *not* eliminate all possibility of using classical physics. All it eliminates is the use of *pure* classical physics, which nobody is doing anyway nowadays. Classical electrons cannot move in a circular orbit around the atomic nucleus without radiating their energy and collapsing into the centre. But many of today's 'classical' theories only use the result of quantum mechanics to force the electron into such orbits, and calculate its radius classically. In fact this circular orbit is all it takes to allow a classical theory of diamagnetism, as will be seen in the next section. Classical electrons do not have a spin either. But by superimposing the quantum-mechanical concept of a spin, a classical theory of ferromagnetism becomes possible. Actually, in the case of ferromagnetism it is not even a real classical theory, like the one which serves as a limit to quantum mechanics in other fields of physics. It is a quasi-classical approach which takes the quantum-mechanical concept of a spin and treats it *as if it were a classical vector.* It essentially uses only a classical *form* to dress up some quantum-mechanical results, which at first sight does not even seem aesthetic.

In principle it would be nicer to treat the whole field of magnetism in general, and ferromagnetism in particular, by pure quantum mechanics. Some books, *e.g.* Wagner [3], and many of the research papers which are published every year, adopt this approach. However, quantum mechanics is only applicable in a very limited part of ferromagnetism. For all the other problems in this field, there is just no other choice. They can be either ignored or studied by the quasi-classical techniques, as used in most of this book. Moreover, classical calculations can give a useful intuitive guidance which is sufficient to prefer them even in some of the cases for which quantum calculations are possible but complicated. Besides, many of the reported quantum calculations use rather rough approximations, as will be seen in chapter 3. It has never been established that these approximations are any better than the use of classical physics.

## 1.4 Diamagnetism

As an illustration, diamagnetism will be studied here both from a quantum-mechanical standpoint, and from a quasi-classical approach on which the Bohr orbit of an electron is superimposed.

Consider first an isolated, quantum-mechanical atom which has $Z$ electrons. Its Hamiltonian is

$$\mathcal{H} = \frac{1}{2m_e} \sum_{i=1}^{Z} \left[ \mathbf{p}_i - \frac{e}{c} \mathbf{A}(\mathbf{r}_i) \right]^2 + \text{other terms}, \tag{1.4.15}$$

where the other terms include interactions between the electrons and the nucleus, and between one electron and another, which do not play a role in the point under discussion. If the applied magnetic field is parallel to the $z$-axis, and is constant in both space and time, the vector potential is

$$\mathbf{A} = \frac{H}{2} (-y, x, 0), \tag{1.4.16}$$

and

$$\mathbf{p}_i \cdot \mathbf{A}(\mathbf{r}_i) = \frac{H}{2} (x p_y - y p_x)_i = \frac{H}{2} \hbar l_{zi}, \tag{1.4.17}$$

where $l_z$ is the $z$-component of the orbital angular momentum.

Substituting eqns (1.4.16) and (1.4.17) in eqn (1.4.15), the expectation value of the Hamiltonian is

$$\overline{\mathcal{H}} = \frac{1}{2m_e} \sum_{i=1}^{Z} \left[ \overline{\mathbf{p}_i^2} - \frac{eH\hbar}{c} \overline{l_{zi}} + \frac{e^2 H^2}{4c^2} \overline{(x^2 + y^2)_i} \right] + \text{other terms}, \tag{1.4.18}$$

which leads to a magnetic moment

$$\overline{m_z} = -\frac{\partial \overline{\mathcal{H}}}{\partial H} = \frac{e}{2m_e c} \sum_{i=1}^{Z} \left[ \hbar \overline{l_{zi}} - \frac{eH}{2c} \overline{(x^2 + y^2)_i} \right]. \tag{1.4.19}$$

For a mole of the material, the susceptibility is

$$\chi = \frac{\partial M}{\partial H} = -\frac{e^2 N}{4 m_e c^2} \sum_{i=1}^{Z} \overline{(x^2 + y^2)_i}, \tag{1.4.20}$$

where $N$ is Avogadro's number, namely the number of atoms in a mole.

In order to obtain the same result by a quasi-classical estimation, consider an electron moving at a constant frequency $\omega_0$ in a circular orbit with a radius $r$, around a nucleus whose electric charge is $Z|e|$. In equilibrium the centrifugal force on the electron is equal to its Coulomb attraction to the nucleus, namely

$$m_e\omega_0^2 r = \frac{Ze^2}{r^2}. \tag{1.4.21}$$

Hence

$$\omega_0 = \sqrt{\frac{Ze^2}{m_e r^3}}. \tag{1.4.22}$$

If a magnetic field, $H$, is applied, a Lorentz force, $(e/c)\,\mathbf{v}\times\mathbf{H}$, is added to the previous forces, the frequency changes to $\omega$, and the new equilibrium equation becomes

$$m_e\omega^2 r = \frac{Ze^2}{r^2} + \frac{|e|\omega r H}{c}. \tag{1.4.23}$$

Using the notation

$$\omega_L = \frac{|e|H}{2m_e c}, \tag{1.4.24}$$

eqn (1.4.23) becomes

$$\omega^2 - 2\omega\omega_L - \omega_0^2 = 0. \tag{1.4.25}$$

Since $\omega_L \ll \omega_0$ even for the largest $H$ which can be technically achieved, the solution of the quadratic equation (1.4.25) may be written as

$$\omega = \omega_L + \sqrt{\omega_L^2 + \omega_0^2} \approx \omega_L + \omega_0. \tag{1.4.26}$$

The application of the magnetic field has, thus, shifted the frequency by the amount $\omega_L$, which is called the *Larmor frequency.*

This change in frequency means that the electron makes extra $\omega_L/(2\pi)$ revolutions per second, thus creating the additional current

$$j = \frac{e\omega_L}{2\pi}. \tag{1.4.27}$$

In cgs units, the magnetic moment is $j/c$ multiplied by the area under the orbit, namely

$$m_z = \frac{j}{c}\pi(x^2+y^2) = -\frac{e^2H}{4m_e c^2}(x^2+y^2), \tag{1.4.28}$$

which leads to the *same* susceptibility as in eqn (1.4.20).

This result does not contradict the general Bohr–van Leeuwen theorem, because a true classical electron cannot sustain the orbit assumed in this 'classical' calculation. It does show, however, that correct results may also be obtained by this quasi-classical approach, when it is superimposed by some artifacts from quantum mechanics, such as the electron maintaining the circular orbit in this case. In a similar way, the quasi-classical study of ferromagnetism is obtained by superimposing the electron spin, with some form for the exchange interaction between these spins.

Before concluding the present case, several comments are in order. In the first place, the picture of an electron completely localized at the atomic site is oversimplified for most solids, and it certainly does *not* apply to metals, in which the degenerate Fermi gas of the electrons occupies certain energy bands that are split by the magnetic field into the so-called Landau levels. It is a completely different problem, which is beyond the scope of this book. The calculation as presented here actually applies only to noble gases, or other gases when they are ionized down to the complete shells. In these cases the material is completely isotropic, and therefore $\overline{x^2} = \overline{y^2} = \frac{1}{3}\overline{r^2}$. For these cases, eqn (1.4.20) is usually found in the literature as

$$\chi = -\frac{e^2 N}{6 m_e c^2} \sum_{i=1}^{Z} \overline{r_i^2}\,, \tag{1.4.29}$$

and the rest of the calculation is the evaluation of $\overline{r_i^2}$, which is rather simple for a hydrogen atom, but less so for other materials.

The second point is that the diamagnetism (*i.e.* negative susceptibility) of eqn (1.4.20) or (1.4.29) exists in *all* materials, including paramagnets. However, in paramagnetic materials the positive susceptibility is much larger than the negative part, which is negligible in comparison. In cgs units, a typical value for the first term in the square brackets of eqn (1.4.19), $\hbar\,\overline{l_z}$, is of the order of $10^{-27}$, while the second term, $|e|H\overline{x^2+y^2}/(2c)$, is of the order of $10^{-36}H$. The largest field $H$ which is physically attainable is $10^6$ Oe, and even at that field the second term of eqn (1.4.19) is negligible compared with the first one. The only case when this second term (namely the diamagnetism) is measurable, is when the first term vanishes, which usually means a closed electronic shell, like in the noble gases. It should be noted that a zero contribution of the *orbital* angular momentum, like in $s$-electrons, is not sufficient, because the electron spin contribution to the first term would normally make the second term negligible anyway. The second term is measurable only when the *total* contribution of orbit *and* spin to the first term vanishes, and the single atom does not have any magnetic moment in zero applied field. This is the case of diamagnetism, when the magnetic moment of each atom, *and* of the ensemble of atoms in the material, is proportional to the applied magnetic field, and is always in the opposite direction to that of the field.

Paramagnetic materials have a non-zero magnetic moment on each of their atoms, which is not caused by, and is independent of, an applied magnetic field. A magnetic field arranges these moments in its own direction, so that the moment increases with an increasing magnitude of the field, and the susceptibility is positive. The diamagnetic, negative susceptibility exists in these materials as well, only it is usually negligible compared with the positive susceptibility part. In the absence of an applied field, the magnetic moments of the individual atoms are randomly oriented, so that their *average* is zero.

In ferromagnetic materials the magnetic moments of the individual atoms interact strongly with each other. As will be seen in the following, this interaction creates a certain degree of order even in the absence of an applied field. This order is the cause of a non-zero average magnetic moment in zero field, which is the basic difference between paramagnetism and ferromagnetism. Thus, the classification of materials can be expressed in terms of the magnetic moment in zero applied field. In diamagnetic materials, this moment is zero for each atom; in paramagnetic materials this moment is non-zero for each atom, but averages to zero over many atoms; and in ferromagnets even the average is not zero.

# 2

# MOLECULAR FIELD APPROXIMATION

## 2.1 Paramagnetism

It is necessary to understand paramagnetism before trying to understand ferromagnetism, and it is best to start from the same quasi-classical approach which will be useful for ferromagnetism. A quantum-mechanical theory of paramagnetism [3] is not particularly difficult to introduce, but it will not serve this purpose as effectively. Therefore, we start by considering an ensemble of atoms, and assume that each of them has a fixed magnetic moment $\mathbf{m}$. In order to be used later for ferromagnets, we just adopt the quantum-mechanical result that the magnitude $m$ of this moment $\mathbf{m}$ is $g\mu_B S$, where $g$ is the so-called 'Landé factor' or 'spectroscopic splitting factor', and

$$\mu_B = \frac{|e|\hbar}{2m_e c} \tag{2.1.1}$$

is known as the *Bohr magneton.* It is another way of expressing eqn (1.4.19), although this magnitude does not even need any justification in the present context of paramagnetism. It may just be taken as a definition of the atomic moment. However, there is another quantum-mechanical property of the spin number, $S$, which is convenient to adopt without considering the details. It is that the component $S_z$ can only assume the $2S+1$ discrete values $-S$, $-S+1$, $\cdots$, $S$, *i.e.* one of the integral or half integral values between $-S$ and $+S$, in integral steps.

These magnetic moments are assumed to interact with an applied magnetic field, $\mathbf{H}$, but not to interact with each other. The energy of interaction of a dipole moment $\mathbf{m}$ with the field $\mathbf{H}$ is known to be $-\mathbf{m}\cdot\mathbf{H}$. If the direction of $\mathbf{H}$ is chosen as the $z$-axis, the average component of $\mathbf{m}$ in that direction at a temperature $T$ is, therefore, according to the classical statistics

$$\langle m_z \rangle = \frac{\Sigma m_z e^{m_z \beta H}}{\Sigma e^{m_z \beta H}}, \tag{2.1.2}$$

where $\beta$ is defined in eqn (1.3.12). Using the above-mentioned magnitude of $\mathbf{m}$ in $m_z$, and noting the allowed values of $S_z$,

$$\langle m_z \rangle = \frac{1}{D}\sum_{n=-S}^{S} g\mu_B n e^{g\mu_B \beta H n} = \frac{1}{D}\sum_{n=-S}^{S} g\mu_B n \xi^n, \tag{2.1.3}$$

$$D = \sum_{n=-S}^{S} e^{g\mu_B \beta H n} = \sum_{n=-S}^{S} \xi^n, \tag{2.1.4}$$

where the notation

$$\xi = e^{\eta}, \qquad \eta = \frac{g\mu_B H}{k_B T}, \tag{2.1.5}$$

is used for short.

The sum in the denominator is the well-known geometric series,

$$\sum_{n=-S}^{S} \xi^n = \frac{1}{\xi^S}\,(1+\xi+\cdots+\xi^{2S}) = \frac{1}{\xi^S}\cdot\frac{1-\xi^{2S+1}}{1-\xi} = \frac{\xi^{-S}-\xi^{S+1}}{1-\xi}. \tag{2.1.6}$$

Differentiating this equation with respect to $\xi$ and rearranging,

$$\sum_{n=-S}^{S} n\xi^{n-1} = \frac{S(\xi^{S+1}-\xi^{-S-1})-(S+1)(\xi^S-\xi^{-S})}{(1-\xi)^2}. \tag{2.1.7}$$

Substituting eqns (2.1.6) and (2.1.7) in eqn (2.1.3),

$$\begin{aligned}\langle S_z\rangle &= \frac{\langle m_z\rangle}{g\mu_B} = \frac{S(\xi^{S+1}-\xi^{-S-1})-(S+1)(\xi^S-\xi^{-S})}{(\frac{1}{\xi}-1)(\xi^{-S}-\xi^{S+1})} \\ &= \frac{2S\sinh\left[(S+1)\eta\right]-2(S+1)\sinh(S\eta)}{\xi^{S+1}+\xi^{-S-1}-(\xi^S+\xi^{-S})},\end{aligned} \tag{2.1.8}$$

where $\eta$ is defined in eqn (2.1.5). Therefore,

$$\langle S_z\rangle = \frac{S\left\{\sinh\left[(S+1)\eta\right]-\sinh(S\eta)\right\}-\sinh(S\eta)}{\cosh\left[(S+1)\eta\right]-\cosh(S\eta)}. \tag{2.1.9}$$

This expression can be simplified by using the following, well-known relations between the hyperbolic functions:

$$\sinh\left[(S+1)\eta\right]-\sinh(S\eta) = 2\sinh\left(\frac{\eta}{2}\right)\cosh\left(\frac{2S+1}{2}\eta\right), \tag{2.1.10}$$

$$\cosh\left[(S+1)\eta\right]-\cosh(S\eta) = 2\sinh\left(\frac{\eta}{2}\right)\sinh\left(\frac{2S+1}{2}\eta\right). \tag{2.1.11}$$

Also, the last term in the numerator of eqn (2.1.9) is transformed according to

$$\sinh(S\eta) = \sinh\left(\frac{2S+1}{2}\eta-\frac{\eta}{2}\right) =$$

$$= \sinh\left(\frac{2S+1}{2}\eta\right)\cosh\left(\frac{\eta}{2}\right) - \cosh\left(\frac{2S+1}{2}\eta\right)\sinh\left(\frac{\eta}{2}\right). \qquad (2.1.12)$$

Substituting all these in eqn (2.1.9), and dividing by $S$,

$$\frac{\langle S_z\rangle}{S} = \frac{2S+1}{2S}\coth\left(\frac{2S+1}{2}\eta\right) - \frac{1}{2S}\coth\left(\frac{\eta}{2}\right). \qquad (2.1.13)$$

Recalling the definition of $\eta$ in eqn (2.1.5), this relation can be written in the form

$$\frac{\langle S_z\rangle}{S} = B_S\left(\frac{g\mu_B SH}{k_B T}\right), \qquad (2.1.14)$$

where the function

$$B_S(x) = \frac{2S+1}{2S}\coth\left(\frac{2S+1}{2S}x\right) - \frac{1}{2S}\coth\left(\frac{x}{2S}\right) \qquad (2.1.15)$$

is called the *Brillouin function.* Besides the argument $x$ it also depends on the spin number, $S$.

As an illustration we shall also consider separately the particular case $S = \frac{1}{2}$. The allowed values of $S_z$ in this case are $n = \pm\frac{1}{2}$, and the summations in eqns (2.1.3) and (2.1.4) are over these two values. Therefore,

$$\langle S_z\rangle = \frac{\sum_n n e^{n\eta}}{\sum_n e^{n\eta}} = \frac{-\frac{1}{2}e^{-\eta/2} + \frac{1}{2}e^{\eta/2}}{e^{-\eta/2} + e^{\eta/2}} = \frac{1}{2}\tanh\left(\frac{\eta}{2}\right). \qquad (2.1.16)$$

That means

$$\langle S_z\rangle = S\tanh\left(\frac{g\mu_B SH}{k_B T}\right), \qquad (2.1.17)$$

instead of eqn (2.1.14). There is no mistake in this algebra, and actually the results are the same. It can be proved by using the appropriate relations between the hyperbolic functions that the definition in eqn (2.1.15) indeed leads to

$$B_{\frac{1}{2}}(x) = \tanh x. \qquad (2.1.18)$$

However, this special function does make the case $S = \frac{1}{2}$ rather atypical. The reader is thus warned that a phrase like 'consider for example $S = \frac{1}{2}$' is encountered very often in the literature; but more often than not it refers to a special case, which is different from what happens for any other value of $S$, even though some salts [4] with $S = \frac{1}{2}$ do exist.

For small values of the argument,

$$\coth x = \frac{1}{x} + \frac{x}{3} + O(x^3). \qquad (2.1.19)$$

Substituting in eqn (2.1.15),

$$B_S(x) = \frac{(2S+1)^2-1}{12S^2}x + O(x^3) = \frac{S+1}{3S}\,x + O(x^3). \qquad (2.1.20)$$

Therefore, if $H$ is not too large, eqn (2.1.14) becomes

$$\langle S_z \rangle = \frac{g\mu_B S(S+1)}{3k_B T}\,H\,, \qquad (2.1.21)$$

which is of the form of eqn (1.1.1). For most paramagnets at ordinary temperatures, no deviations from the linear behaviour of eqn (2.1.21) can be detected even for the largest possible field, $H$. In the cases for which the magnetization is not proportional to the field at high applied fields, it is customary to conserve the form of eqn (1.1.1) anyway, but define a field-dependent susceptibility, $\chi(H)$. The value in eqn (2.1.21) is then referred to as the *initial susceptibility*

$$\chi_{\text{initial}} = \lim_{H\to 0}\frac{\partial\langle M_z\rangle}{\partial H} = \lim_{H\to 0}\frac{\partial N g\mu_B\langle S_z\rangle}{\partial H} = \frac{C}{T}, \qquad (2.1.22)$$

where $N$ is the number of spins per unit volume, and

$$C = \frac{NS(S+1)}{3k_B}(g\mu_B)^2. \qquad (2.1.23)$$

The temperature-dependence of the susceptibility in eqn (2.1.22) agrees with experiment for all paramagnets, and is known as the *Curie law.* It may also be worth mentioning that although $\chi$ may depend on $H$ in paramagnets, it does not depend on the *history* of $H$, as is the case for the ferromagnets which will be discussed later. In paramagnets $\chi(H)$ is a well-defined, single-valued function.

For a very large argument, $\coth x \to 1$, and eqn (2.1.15) implies

$$B_S(\pm\infty) = \pm 1. \qquad (2.1.24)$$

It means that at very large applied fields, the magnetization *saturates*, and does not keep increasing with the field. This saturation obviously occurs when all the spins are aligned in the direction of the applied field, because in these paramagnets the effect of the field is only to change the direction of the individual, fixed magnetic moments. It does not change their magnitudes, except for the small, *diamagnetic* contribution, mentioned in section 1.4, which always exists, but is usually negligibly small in paramagnets.

This saturation cannot be observed in most paramagnets, because the available field $H$ is not large enough to reach that region. It is seen, however, that the argument in eqn (2.1.14) is $SH/T$, rather than just $H$. Therefore, this saturation can be attained at very low temperatures, when $T$ is small. It can also be observed if $S$ is large, which can be achieved by a special phenomenon, known as *superparamagnetism*. In normal materials, $S$ is the spin number of a single atom, and is of the order of 1. However, under certain conditions, which will be described in section 5.2, this $S$ applies to very many atoms' spins which are coupled together. In these cases $S$ can be of the order of $10^3$ or $10^4$, and rather small $H$ is sufficient to reach saturation.

## 2.2 Ferromagnetism

Unlike the paramagnetic atoms of the previous section, which interact only with an external magnetic field, the atomic spins in ferromagnetic materials interact with each other, each of them trying to align the others in its own direction. This interaction between them originates from quantum-mechanical properties of spins, which will be discussed in the next chapter. The readers who do not know any quantum mechanics may skip that chapter, which is not essential for following the rest of this book. They may just adopt as an axiom the existence of such a force which tries to align spins, by the so-called *exchange* interaction. The latter can be expressed as an *exchange energy* between spin $\mathbf{S}_i$ and spin $\mathbf{S}_j$, which is proportional to $\mathbf{S}_i \cdot \mathbf{S}_j$.

Including the same energy of interaction with an applied field, $\mathbf{H}$, as in the case of paramagnetic atoms, the total energy of a system is thus

$$E = -{\sum_{ij}}' J_{ij}\mathbf{S}_i \cdot \mathbf{S}_j - \sum_i g\mu_B \mathbf{S}_i \cdot \mathbf{H}, \tag{2.2.25}$$

where the prime over the first sum indicates that the case $i = j$ is excluded, because the spins do not interact with themselves. Except for these values, both summations extend over all the atomic spins in the material. The coefficients $J_{ij}$ are called the *exchange integrals*, and can be evaluated by methods described in the next chapter. It should be noted that the sign of these coefficients is defined so that if $J_{ij}$ is positive, parallel spins have a lower energy than antiparallel ones, which is the case for a ferromagnetic interaction.

Very many bodies are involved, and some approximation is inevitable. In this chapter we introduce a popular technique, which is known by other names in other branches of physics, but in the context of ferromagnetism it is known by the name *molecular field approximation*, although more recently the name *mean field approximation* is becoming more widely used. In this method, one spin is tagged for checking its statistics in more detail, while the others are just replaced by their mean value (or, rather, their quantum-mechanical expectation value). Then, after some manipulations, that particular spin is *untagged*, saying that on the average it is not any different from the other spins, thus obtaining the mean value.

Specifically, we consider the spin $\mathbf{S}_i$ as something special, and collect together the energy terms in which it is involved (or the Hamiltonian which acts on $\mathbf{S}_i$), when all the *other* spins, $\mathbf{S}_j$, are replaced by their mean value, $\langle \mathbf{S}_j \rangle$,

$$E_i = -2\sum_j J_{ij}\mathbf{S}_i \cdot \langle \mathbf{S}_j \rangle - g\mu_B \mathbf{S}_i \cdot \mathbf{H} = -\mathbf{S}_i \cdot \mathbf{H}_i, \qquad (2.2.26)$$

where

$$\mathbf{H}_i = g\mu_B \mathbf{H} + 2\sum_j J_{ij} \langle \mathbf{S}_j \rangle . \qquad (2.2.27)$$

The factor 2 comes from the fact that the double sum in eqn (2.2.25) actually contains the particular spin *twice*: once under its name $\mathbf{S}_i$, and once as one of the possibilities in summing over $\mathbf{S}_j$. Equation (2.2.27) means that the total energy, $E$, is *not* equal to $\sum_i E_i$, and a factor of $\frac{1}{2}$ has to be introduced in the summation over the interactions, so that they are not counted twice.

To the present approximation, the exchange interaction between the spins has thus turned out to be equivalent to an interaction of each spin with an effective field, $\mathbf{H}_i$, which is non-zero even when the real applied field $\mathbf{H}$ vanishes. It is essentially the assumption of Weiss, mentioned in section 1.2, which is why this effective field became known in the literature as the *Weiss molecular field*, or just the *Weiss field*. More recently, the name of Weiss tends to be forgotten, and the name *mean field* seems to be taking over. Under either name, the above treatment can be regarded as a certain level of a *justification* of the 'molecular field', which Weiss just postulated arbitrarily. To reach this level, we have postulated arbitrarily the existence of an exchange interaction of the form described by eqn (2.2.25). A deeper-level justification calls for deriving eqn (2.2.25) from more basic principles, which will be done in the next chapter.

The problem of ferromagnetism has thus been reduced to the problem of isolated spins interacting with an applied field, which is the problem of paramagnetism treated in the previous section. It should only be noted that the energy here is $-\mathbf{S}\cdot\mathbf{H}$, whereas in the previous section it was $-g\mu_B\mathbf{S}\cdot\mathbf{H}$. Therefore, the argument of the Brillouin function in eqn (2.1.14) needs an appropriate normalization. When that is done, the $z$-component of $\mathbf{S}_i$ is seen to become

$$\langle S_{iz} \rangle = SB_S\left(\frac{SH_i}{k_BT}\right), \qquad (2.2.28)$$

where $H_i$ is defined in eqn (2.2.27).

Now the particular spin $\mathbf{S}_i$ is *untagged*: on average there is no difference between that spin and any other which appears in the summation in eqn

(2.2.27). Therefore, both spins may be written without the index, $i$ or $j$, and eqn (2.2.28) becomes, after substituting from eqn (2.2.27),

$$\langle S_z \rangle = SB_S \left( \frac{S}{k_B T} \left[ g\mu_B H + 2\langle S_z \rangle \sum_j J_{ij} \right] \right), \tag{2.2.29}$$

which is a transcendental equation for determining $\langle S_z \rangle$. Actually, this relation is not strictly defined when written in this way, because the summation over $j$ depends on $i$. It still calls for another assumption about this sum, the most usual of which is that $J_{ij}$ is zero for spins which are not nearest neighbours in the crystal, and it has one universal, non-zero value $J$ for nearest neighbours.

It is customary to use the notation

$$\mu = \frac{\langle S_z \rangle}{S}, \tag{2.2.30}$$

$$h = \frac{g\mu_B S H}{k_B T}, \tag{2.2.31}$$

and

$$\alpha(T) = \frac{2S^2}{k_B T} \sum_j J_{ij} = \left[ \frac{2S^2}{k_B T} pJ \right], \tag{2.2.32}$$

where $p$ is the number of nearest neighbours, and the square-bracketed relation is for the above-mentioned assumption that spins which are not nearest neighbours have a zero exchange integral. Other assumptions about some finite values for the next-nearest neighbours (of which real cases exist [4]), or even further-away spins, are also possible with the same algebra. An example will be given in section 2.3. For these notations, eqn (2.2.29) is

$$\mu = B_S(h + \alpha\mu). \tag{2.2.33}$$

Older texts used to elaborate on graphical solutions of this equation, but it is not necessary any more. With a modern computer a numerical solution of eqn (2.2.33) is a trivial matter, and $\mu$ can be plotted as a function of $\alpha$ for any value of $h$. For a rather small $h$, this solution looks more or less like the plot in Fig. 1.2, with slight variations depending on the value of $S$. In the limit of vanishing $h$, the solution looks approximately like the curve plotted in Fig. 2.1. Both these curves are plotted using another theory, which will be discussed in section 4.6, and are not really plots of the molecular field approximation. They are shown in this stage only as a demonstration of the *qualitative* shape of the solution of eqn (2.2.33). However, for the case $h = 0$, which is the most interesting case for theorists, there is an analytic

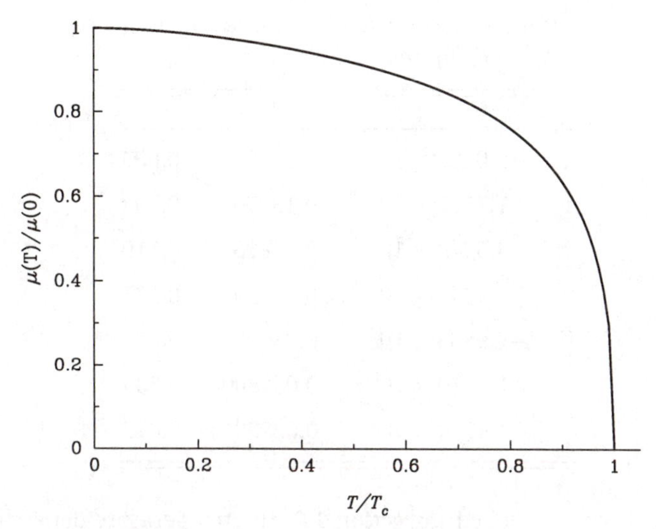


FIG. 2.1. An approximate shape of the solution of eqn (2.2.33) for the case $h = 0$. The temperature, $T$, is normalized to the Curie temperature, $T_c$, above which the only solution of this equation is $\mu = 0$.

approximation [5] to the shape of this solution, which can be particularly useful when the molecular field curve has to be computed many times, as a part of more complex computation, for example in averaging over some parameter. Using the notation

$$m = \mu(T)/\mu(0), \qquad t = T/T_c, \tag{2.2.34}$$

the analytic approximation is

$$(m - a)^2 = a^2 + \frac{1 - 2a}{1 - 2b}(1 - t)(1 - 2b + t^c). \tag{2.2.35}$$

Here $a$ and $b$ are adjustable parameters, the best values for which are tabulated in Table 2.1, and

$$c = 1 + \frac{1}{4S(S + 1)}. \tag{2.2.36}$$

Table 2.1 lists also the maximum deviation of the analytic expression from the exact solution of eqn (2.2.33). This accuracy is adequate for most practical purposes, especially since the maximum of the deviation always occurs for rather small values of $t$, for which the accuracy is usually less important.

**Table 2.1** *The parameters a and b which should be used in eqn (2.2.35), and the maximum relative deviation, D, of this equation from the solution of the molecular field theoretical relation, eqn (2.2.33), for different values of the spin, S.*

| $S$ | $a$ | $b$ | $D$ (%) |
|---|---|---|---|
| $\frac{1}{2}$ | $-1.0182 \times 10^{-2}$ | 0.26166 | 0.695 |
| 1 | $7.7521 \times 10^{-3}$ | 0.19270 | 0.708 |
| $\frac{3}{2}$ | $4.5249 \times 10^{-3}$ | 0.14825 | 0.519 |
| 2 | $1.1241 \times 10^{-3}$ | 0.11229 | 0.277 |
| $\frac{5}{2}$ | $-7.9838 \times 10^{-4}$ | 0.080979 | 0.356 |
| 3 | $-1.5270 \times 10^{-3}$ | 0.052860 | 0.535 |
| $\frac{7}{2}$ | $-1.4780 \times 10^{-3}$ | 0.027221 | 0.677 |

As has been mentioned in section 1.2, the temperature-dependence of the magnetization in zero field, as expressed by eqn (2.2.35) or as plotted in Fig. 2.1, is expected to be valid in the *interior* of the magnetic domains. In practice, measurements are done in sufficiently high fields which are necessary to remove those domains, and are then extrapolated to $H = 0$ in order to compare with the theoretical curves, such as the one plotted in Fig. 2.1. Details of this process will be given in chapter 4.

Actually, if $H = 0$ there is no direction in space which can define the $z$-axis which has been used for deriving eqn (2.2.33) in the first place. In real ferromagnets there is no difficulty, because they are anisotropic, and have a built-in preferred spin direction. However, for the readers who may wonder about it already at this stage, it is sufficient to assume that the case $H = 0$ is the end of a process in which a finite field is applied, then slowly reduced to zero. Such a process is quite close to what is done experimentally anyway.

## 2.3 Antiferromagnetism

The exchange integrals $J_{ij}$ were assumed in the previous section to be positive, so that spins tend to align parallel to each other. This positive value is essential for having a ferromagnetic order, but it is not necessarily so in all materials, and exchange integrals may also be negative. Actually, negative exchange coupling occurs in nature more often than a positive one. When the exchange integral between nearest neighbours is negative, it tends to align the neighbouring spins antiparallel to each other, which can also give rise to a certain order at low temperatures. Such materials do exist, and are called *antiferromagnets*.

The theory of this phenomenon was presented by L. Néel even before it was first observed experimentally; see the history as described in his [6] Nobel lecture. He considered a crystalline material, made out of *two* sublattices, which are constructed in such a way that the nearest neighbour of each spin belongs to the *other* sublattice, with which it interacts by an antiferromagnetic exchange coupling, $-J$, with $J > 0$. Interactions with farther-away spins have also been added in later studies, but here we consider only a relatively simple case which is only a slight generalization of the original Néel assumption of interaction between nearest neighbours only. We assume that besides the $p$ nearest neighbours in the *other* sublattice, each spin also interacts with $p'$ neighbours within the *same* sublattice, by a ferromagnetic coupling, $+J'$. However, in order to maintain a basically antiferromagnetic case, we also assume that $J' \ll J$. Interactions with farther-away spins are taken here as zero.

Other possibilities can also be found in the literature, including cases which cannot be described by such a simple subdivision into two sublattices. An outstanding example is EuSe which does have such two sublattices, but with $J \approx J'$, so that it takes only a small perturbation (*e.g.* some impurities) to change it from a ferromagnet to an antiferromagnet, or vice versa. We ignore these cases here, and do not try to study anything more general than the specified assumptions.

Denoting the $z$-component of the spin in each sublattice by $S_1$ and $S_2$ respectively, the effective fields on each of them are, according to eqn (2.2.27),

$$H_1 = g\mu_B H + 2p'J'\langle S_1\rangle - 2pJ\langle S_2\rangle, \tag{2.3.37}$$

and

$$H_2 = g\mu_B H + 2p'J'\langle S_2\rangle - 2pJ\langle S_1\rangle. \tag{2.3.38}$$

Substituting in eqn (2.2.28)

$$\langle S_1\rangle = SB_S\left[\frac{S}{k_BT}\left(g\mu_B H + 2p'J'\langle S_1\rangle - 2pJ\langle S_2\rangle\right)\right], \tag{2.3.39}$$

and

$$\langle S_2\rangle = SB_S\left[\frac{S}{k_BT}\left(g\mu_B H + 2p'J'\langle S_2\rangle - 2pJ\langle S_1\rangle\right)\right]. \tag{2.3.40}$$

For the case $H = 0$ everything is symmetric, and it is readily seen that the two equations become the same by the substitution $\langle S_1\rangle = -\langle S_2\rangle$. Therefore, the solution is that the magnetization is the same for both sublattices, only in opposite directions, and each of them is a solution of

$$\langle S\rangle = SB_S\left[\frac{2S}{k_BT}\left(p'J' + pJ\right)\langle S\rangle\right], \tag{2.3.41}$$

which is the same as eqn (2.2.29), or (2.2.33), for $H = 0$.

It is thus seen that in zero applied field, the magnetization of each sublattice in an antiferromagnetic material is the same as that of a ferromagnetic material, with the temperature-dependence as in Fig. 2.1. In particular, the order disappears above a certain transition temperature, which is equivalent to the Curie point in ferromagnets, only in the case of antiferromagnets, this transition temperature is called the *Néel point.* It must be emphasized, however, that a measurement of the *total* magnetization, $\langle S_1 \rangle + \langle S_2 \rangle$, gives zero for $H = 0$. Thus these materials could not be discovered from the measurement of the macroscopic magnetization, which is why their existence was not suspected before Néel came up with this concept of two sublattices and an antiferromagnetic exchange coupling. Nowadays, the antiferromagnetic order below the Néel temperature can be seen by neutron diffraction, because neutrons interact with the *local* magnetization when they pass through the crystal. It can also be seen by nuclear magnetic resonance, and by the Mössbauer effect, both of which measure the magnetic moment on the particular atom in which that nucleus is located, and not the macroscopic magnetization, see section 4.1. The existence of antiferromagnetism may also be inferred from measurement of the specific heat, which will not be discussed here, or from the susceptibility *above* the Néel point, to be described in the next section.

## 2.4 The Curie–Weiss Law

The solution of eqn (2.2.33), or (2.3.41), is zero in zero applied field for any temperature above a transition point, namely the Curie point in ferromagnets, and the Néel point in antiferromagnets. In this region of $T > T_c$, it is known from experiment that all ferromagnets and antiferromagnets become regular paramagnets. It is also quite clear from the foregoing, at least qualitatively, that for a sufficiently high temperature the thermal fluctuations overcome the exchange interaction between the spins, thus eliminating the ferromagnetic or antiferromagnetic *order* and making the material as disordered as a paramagnet.

For the quantitative study of the high-temperature region, we will consider together the case of ferromagnets and of antiferromagnets, because it is essentially the same algebra for both, and there is no point in unnecessary repetitions. Actually, we take the ferromagnets to be a particular case of the two-sublattice antiferromagnets, as defined in the previous section. For a true antiferromagnet we assumed there that $J$ and $J'$ are both positive, and that $J' \ll J$. But we can also include the case of a simple ferromagnet as the particular case $J = 0$, $\langle S_1 \rangle = \langle S_2 \rangle$, and $H_1 = H_2$. It may also be obtained as the particular case $J < 0$, $J' = 0$.

In the high-temperature region it is sufficient to approximate the Brillouin function by the first term of a power series expansion, as in eqn

(2.1.20), because the argument of this function is always small. Therefore, eqns (2.3.39) and (2.3.40) may be replaced by

$$\langle S_1 \rangle = \frac{S(S+1)}{3k_BT}\left(g\mu_B H + 2p'J'\langle S_1 \rangle - 2pJ\langle S_2 \rangle\right) \tag{2.4.42}$$

and

$$\langle S_2 \rangle = \frac{S(S+1)}{3k_BT}\left(g\mu_B H + 2p'J'\langle S_2 \rangle - 2pJ\langle S_1 \rangle\right), \tag{2.4.43}$$

which are two linear equations in the magnetizations of the two sublattices.

It is not difficult to solve such a set of two equations, but it is not even necessary to do so, because it is sufficient to *add together* the two equations, and solve for

$$\langle S_{\text{total}} \rangle = \langle S_1 \rangle + \langle S_2 \rangle = \frac{CH}{g\mu_B T} - \frac{\Theta}{T}\langle S_{\text{total}} \rangle, \tag{2.4.44}$$

where

$$C = \frac{2S(S+1)}{3k_B}(g\mu_B)^2, \qquad \Theta = \frac{2S(S+1)}{3k_B}\left(pJ - p'J'\right). \tag{2.4.45}$$

Therefore, the total magnetic moment is

$$\langle M_z \rangle = g\mu_B \langle S_{\text{total}} \rangle = \frac{CH}{T+\Theta}. \tag{2.4.46}$$

This temperature-dependence is known as the *Curie–Weiss law*. It should apply both for ferromagnets and for antiferromagnets, and indeed it fits experiments on both types, with some exceptions [7] which are ignored here. It is more usually expressed in terms of the initial susceptibility,

$$\chi_{\text{initial}} = \lim_{H \to 0} \frac{\langle M_z \rangle}{H} = \frac{C}{T+\Theta}, \tag{2.4.47}$$

which is more general than the relation in eqn (2.4.46). In many cases these expressions are equivalent. But sometimes the first-order approximation to $B_S$ used in eqns (2.4.42) and (2.4.43) is inadequate for large fields, $H$, and it is necessary to apply the Curie–Weiss law only for a small $H$.

In both ferromagnets and antiferromagnets the use of this linear approximation is certainly justified (at least for a small $H$) above the Curie or the Néel temperature, because in both cases, $\langle S_z \rangle = 0$ for $H = 0$. Since we have seen that the ferromagnets are the case $J = 0$, it is clear from eqn (2.4.45) that $\Theta < 0$ in these materials. For antiferromagnets we have assumed that $J' \ll J$, and eqn (2.4.45) yields a *positive* $\Theta$.

It is thus possible to tell the difference between ferromagnets and antiferromagnets from the measurement of the initial susceptibility above the transition temperature, when they are both paramagnets. The Curie–Weiss law in these two types of materials is shown schematically in Fig. 2.2 and the difference is obvious. This property is particularly useful for materials in which the Curie or Néel point is at a very low temperature, which is not easy to access directly. In such a case it is possible to determine from the high-temperature data if the material is going to become ferromagnetic or antiferromagnetic at low temperature. If there is no transition at all, and the material remains paramagnetic down to absolute zero, $J = J' = \Theta = 0$ and the reciprocal of the high-temperature susceptibility should extrapolate to 0 at $T = 0$.

At lower temperatures, the linear approximation to $B_S$ is not adequate. However, it may still be used at the transition temperature itself, where the disorder just begins, because the functions are continuous and a finite $\langle S_z \rangle$ for $H = 0$ must start from small values. Therefore, *at* the transition temperature, the set of linear equations (2.4.42) and (2.4.43) should still have a non-zero solution for $\langle S_1 \rangle$ and $\langle S_2 \rangle$ when $H = 0$. The condition for such a solution is that the determinant of their coefficients vanishes, namely

$$\begin{vmatrix} 1 - C^* p' J' & C^* p J \\ C^* p J & 1 - C^* p' J' \end{vmatrix} = 0 \,, \qquad C^* = \frac{2S(S+1)}{3k_B T_N} . \tag{2.4.48}$$

Hence

$$T_N = \frac{2S(S+1)}{3k_B} \left( pJ + p'J' \right) . \tag{2.4.49}$$

Comparing with eqn (2.4.45), this result means that $T_c = |\Theta|$ for a simple ferromagnet, with interaction between nearest neighbours only. Similarly, $T_N = \Theta$ for a simple antiferromagnet, with interaction between nearest neighbours only, namely for the case $J' = 0$. If $T_N \neq \Theta$, the difference between these experimental values may be taken as a measure of $J'$, provided of course that the difference is rather small, so that it fits the basic assumption of this calculation, that $J' \ll J$. The value of $\Theta$ is always used with eqn (2.4.45) as an experimental evaluation of the exchange integral, $J$. At first sight is seems as if the Curie–Weiss law also contains another parameter, $C$, which can be compared with experiment. However, in practice the number of spins in the crystal sites of ferromagnets is not an integer, as will be discussed in the next chapter, and the susceptibility data must be normalized, so that the parameter $C$ does not contain any useful information, at least for metals.

It should be particularly emphasized that these results apply strictly to cases which fit the assumption of two sublattices, with predominantly nearest-neighbour interaction. Other cases may be studied in a similar way, but do not necessarily lead to similar results. In particular, it is very easy to subdivide a simple cubic, or a bcc, into two such sublattices, and have the nearest neighbour of each lattice point in the *other* sublattice; but it just cannot be done for an fcc lattice. An fcc lattice is more readily subdivided into *four* sublattices, in which the nearest neighbours of each spin are in all other three sublattices. Specifically, one sublattice contains the points $(k, l, m)$, one contains $(k + \frac{1}{2}, l + \frac{1}{2}, m)$, one with $(k + \frac{1}{2}, l, m + \frac{1}{2})$, and one with $(k, l+\frac{1}{2}, m+\frac{1}{2})$, for integral $k$, $l$, and $m$. Numbering the sublattices in this order, counting the nearest neighbours, and using the high-temperature approximation for the Brillouin function, the equations to be solved are

$$\langle S_1 \rangle = \frac{S(S+1)}{3k_BT} \left[g\mu_B H + 8J\left(\langle S_2 \rangle + \langle S_3 \rangle + \langle S_4 \rangle\right)\right], \tag{2.4.50}$$

$$\langle S_2 \rangle = \frac{S(S+1)}{3k_BT} \left[g\mu_B H + 8J\left(\langle S_1 \rangle + \langle S_3 \rangle + \langle S_4 \rangle\right)\right], \tag{2.4.51}$$

$$\langle S_3 \rangle = \frac{S(S+1)}{3k_BT} \left[g\mu_B H + 8J\left(\langle S_1 \rangle + \langle S_2 \rangle + \langle S_4 \rangle\right)\right], \tag{2.4.52}$$

$$\langle S_4 \rangle = \frac{S(S+1)}{3k_BT} \left[g\mu_B H + 8J\left(\langle S_1 \rangle + \langle S_2 \rangle + \langle S_3 \rangle\right)\right]. \tag{2.4.53}$$

By adding together the first two and the last two equations, it is seen that there are actually two sublattices,

$$\langle S_I \rangle = \langle S_1 \rangle + \langle S_2 \rangle, \qquad \langle S_{II} \rangle = \langle S_3 \rangle + \langle S_4 \rangle, \tag{2.4.54}$$

with two equations,

$$\langle S_I \rangle = \frac{2S(S+1)}{3k_BT} \left[g\mu_B H + 4J\left(\langle S_I \rangle + 2\langle S_{II} \rangle\right)\right] \tag{2.4.55}$$

and

$$\langle S_{II} \rangle = \frac{2S(S+1)}{3k_BT} \left[g\mu_B H + 4J\left(2\langle S_I \rangle + \langle S_{II} \rangle\right)\right]. \tag{2.4.56}$$

The fcc lattice has thus been subdivided into two sublattices, but with a number of neighbours in the same sublattice being different from their number in the other sublattice. This slight difference does change the foregoing algebra. Repeating the same calculations for the initial susceptibility leads to the same Curie–Weiss law as in eqn (2.4.47), with

$$\Theta = -\frac{8JS(S+1)}{k_B}. \tag{2.4.57}$$

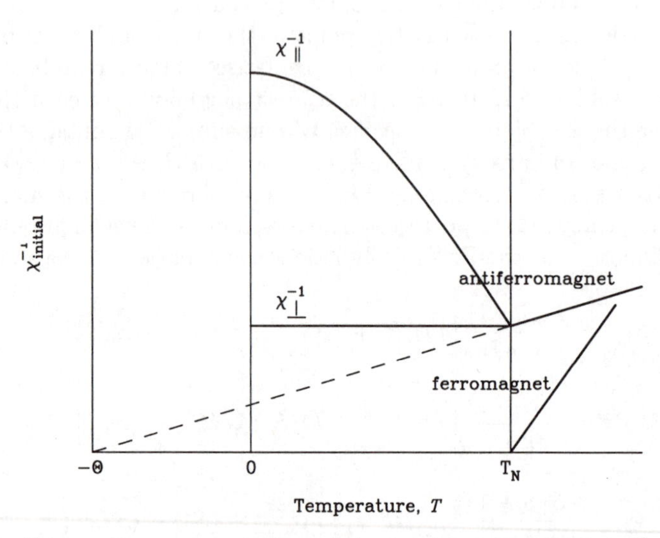


FIG. 2.2. Schematic representation of the initial susceptibility of an antiferromagnet, above and below the Néel temperature, $T_N$, and of a ferromagnet *above* the Curie temperature, $T_c$. For $T < T_N$, a single-crystal antiferromagnet may have different parallel, $\chi_{\|}$, and perpendicular, $\chi_{\perp}$, susceptibilities. For a ferromagnet, $\chi$ has no meaning below $T_c$.

Repeating the same calculation for the transition temperature yields $T_c = -\Theta$, for a ferromagnet, *i.e.* when $J > 0$, but $T_N = \Theta/3$ for an antiferromagnet, *i.e.* when $J < 0$.

It is also possible to use a non-zero field $H$ in eqns (2.3.39) and (2.3.40), solve them numerically, and find the initial susceptibility of an antiferromagnet *below* the Néel temperature. For these two equations as written here, the result is

$$\chi_{\text{initial}} = \frac{(g\mu_B)^2}{2pJ}, \tag{2.4.58}$$

which does not depend on the temperature. However, crystalline materials have an anisotropy, which can be expressed as an internal field that prefers the magnetization to be oriented along certain crystallographic axes. Details will be given in section 5.1, but it may already be realized that the $z$-direction which we have used for the quantization direction may be connected with the crystallographic axes, and is not necessarily identical with the direction of the applied field, $H$. In such cases, the susceptibility in eqn (2.4.58) is the *parallel* susceptibility, $\chi_{\|}$, in which the small applied $H$ is in the same direction as that of the internal field. One may also apply a small field, $H$, perpendicular to the direction of that internal field, and compute the *perpendicular* initial susceptibility, $\chi_\perp$, obtaining the temperature-dependence shown schematically in Fig. 2.2. In principle, both $\chi_\parallel$ and $\chi_\perp$ may be measured in a single-crystal sample.

In practice, this measurement is not that simple, because even single-crystal samples are often subdivided into domains magnetized along different crystallographic axes, and one can only measure some average between the curves for $\chi_\parallel$ and $\chi_\perp$ in Fig. 2.2. For example, in a cubic material with an equal probability of domains in each of the three axes, the measured $\chi$ will be $\frac{1}{3}\chi_\parallel + \frac{2}{3}\chi_\perp$. In a polycrystalline sample one *always* measures just an average of these two curves. Therefore, the details of these curves do not have much use in comparing with experiment. However, because both curves have a discontinuity in the derivative of $\chi$ *vs.* $T$, any average or combination of them will have such a discontinuity. Therefore, the position of this discontinuity, or cusp, in the experimental data is an accurate measurement of the Néel temperature, $T_N$. A more accurate measurement of $T_N$ is obtained from the anomaly in the specific heat, which is beyond the scope of this book.

In a ferromagnet $\Theta < 0$ and according to eqn (2.4.46), the susceptibility diverges when the temperature approaches $T_c$ from above. This infinity is just a manifestation of the possibility to have $M_z \neq 0$ for $H = 0$. It is tempting to define a ferromagnet as the limit of a paramagnet when $\chi \to \infty$, but such a definition does not have any meaning. As has already been mentioned in section 1.1, the relation (1.1.1) is not fulfilled in ferromagnets, and $M_z$ is not even a unique function of $H$ below $T_c$. For the same reason, there is no point in using a small $H$ in eqn (2.2.29) to calculate a susceptibility below $T_c$, in the same way as in an antiferromagnet. It does not prevent theorists from calculating it anyway, but such a susceptibility cannot be measured because a large field is needed to remove the magnetic domains, see section 4.1. In principle, the Mössbauer effect (which measures the magnetization inside the domains, as calculated in this chapter) can see the effect of a small applied field. In practice the accuracy of the Mössbauer effect is not sufficient to see even the effect of quite large fields. Therefore, the susceptibility below the transition temperature has a meaning only in antiferromagnets and not in ferromagnets, and this is the way it is shown in Fig. 2.2.

## 2.5 Ferrimagnetism

The exact cancellation of the opposite magnetization in the two sublattices is possible only for identical magnetic moments in both lattice points. Néel [6] considered also the case of two sublattices in which the magnetic moments are *not* the same. It happens either because they are made of atoms of different materials, or because the *ions* are not the same, for example when there is $Fe^{2+}$ in one sublattice, and $Fe^{3+}$ in the other. In such cases an antiferromagnetic coupling between the two sublattices leads to a *partial* cancellation of the magnetic moment. The resulting, total magnetization at low temperatures is the difference between that of the two sublattices, which is not zero.

Some of the materials to which such a theory applies were actually known before Néel came up with the theory, but they were thought to be of the same class as ferromagnets, which confused the issue. It may be interesting to note that the oldest permanent magnet, known already in ancient times, is magnetite, $Fe_3O_4$, which is a ferrite. As such, it is a ferrimagnet, and not a ferromagnet, according to the classification of Néel. It is also interesting to note that the molecular field theory of ferrimagnets, as presented in this section, fits very well [6] the temperature-dependence of $M_s$ in magnetite.

The name *ferrites* was first given to certain materials made of iron oxides together with some other oxides. Néel used this name as the basis for the class of materials which he called *ferrimagnets.* This name fits only the French pronunciation, but it was also adopted in English, even though it does not fit this language. In spoken English the difference between ferromagnetism and ferrimagnetism can be heard only when the emphasis is put on the wrong syllable.

We start by generalizing the theory of antiferromagnetism in the high-temperature region, for which the Brillouin function can be approximated by the linear term. We consider the same two sublattices, with exchange interaction $-J$ between nearest neighbours *only*, and assume that there are $p$ of them, all in the *other* sublattice. However, here we do not take all atoms to be the same, but assign different quantum numbers, $S_1$ and $S_2$ respectively, to the ions on each sublattice. We also use different $g$-factors, $g_1$ and $g_2$, for the two sublattices. The generalization of the high-temperature equations (2.4.42) and (2.4.43) is straightforward, leading to

$$\langle S_{1z}\rangle = \frac{S_1(S_1+1)}{3k_BT}\left(g_1\mu_B H - 2pJ\langle S_{2z}\rangle\right) \tag{2.5.59}$$

and

$$\langle S_{2z}\rangle = \frac{S_2(S_2+1)}{3k_BT}\left(g_2\mu_B H - 2pJ\langle S_{1z}\rangle\right). \tag{2.5.60}$$

Using the notations

$$C_i = \frac{g_i^2\mu_B^2 S_i(S_i+1)}{3k_B}, \qquad \Theta_i = \frac{2pJ}{3k_B}S_i(S_i+1), \tag{2.5.61}$$

for $i = 1, 2$, these equations become

$$T\langle S_{1z}\rangle + \Theta_1\langle S_{2z}\rangle = \frac{C_1 H}{g_1\mu_B}, \qquad \Theta_2\langle S_{1z}\rangle + T\langle S_{2z}\rangle = \frac{C_2 H}{g_2\mu_B}. \tag{2.5.62}$$

The solution of this pair of equations is

$$g_i\mu_B\langle S_i\rangle = \frac{H}{T^2-\Theta_1\Theta_2}\left(C_iT-\sqrt{C_1C_2\Theta_1\Theta_2}\right), \tag{2.5.63}$$

for $i = 1, 2$, where the index $z$ has been omitted for simplicity, it being understood that the averages are those of the $z$-components. Therefore, the (initial) susceptibility is

$$\chi = \frac{1}{H}\sum_{i=1}^{2} g_i\mu_B\langle S_i\rangle = \frac{(C_1+C_2)T-2\sqrt{C_1C_2\Theta_1\Theta_2}}{T^2-\Theta_1\Theta_2}. \tag{2.5.64}$$

This susceptibility diverges at $T = T_c$, where

$$T_c = \sqrt{\Theta_1\Theta_2}, \tag{2.5.65}$$

as is the case in ferromagnets. Here, as in ferromagnets, $T_c$ is the Curie temperature, above which the magnetization is zero in zero applied field. The latter can be seen as in the previous section, by looking for a non-zero solution of eqns (2.5.62) with $H = 0$, the condition for which is

$$\begin{vmatrix} T_c & \Theta_1 \\ \Theta_2 & T_c \end{vmatrix} = 0, \tag{2.5.66}$$

whose solution is eqn (2.5.65). It is not a coincidence. The divergence of the susceptibility, in ferromagnets or in ferrimagnets, originates from the vanishing of the same determinant in the denominator which appears in the solution of the simultaneous linear equations.

In the case of ferrimagnets, this temperature $T_c$ is called *the ferrimagnetic Curie point.* Above that temperature, $1/\chi$ is the curve given by eqn (2.5.64), which is obviously not the straight line of ferromagnets. However, at much higher temperatures, $T \gg T_c$, eqn (2.5.64) becomes

$$\frac{1}{\chi} \approx \frac{T^2}{(C_1+C_2)T\,(1-\Theta/T)} \approx \frac{T+\Theta}{C_1+C_2}, \tag{2.5.67}$$

where

$$\Theta = \frac{2\sqrt{C_1C_2}T_c}{C_1+C_2}. \tag{2.5.68}$$

The asymptote is thus a straight line which cuts the temperature axis at a negative value, $T = -\Theta$, as is the behaviour of an antiferromagnet, see Fig. 2.3. This $\Theta$ is called the *paramagnetic Curie point*. It is usually different from $T_c$, as can be seen from eqn (2.5.68). This equation actually means that $\Theta = T_c$ only when $C_1 = C_2$.

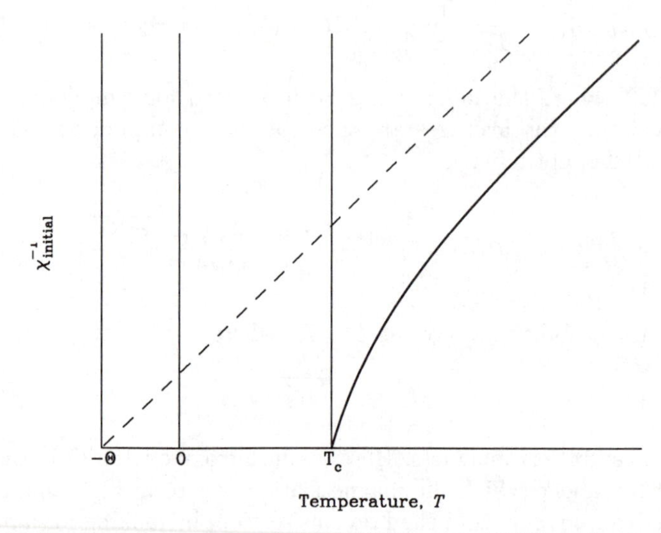


FIG. 2.3. Typical behaviour of the initial susceptibility of a ferrimagnet above the ferrimagnetic Curie temperature, $T_c$, as computed from eqn (2.5.64).

Equation (2.5.67) is the Curie–Weiss law. We have thus seen that this law is obeyed asymptotically in the case of ferrimagnets. In these materials, the high-temperature data for the susceptibility look like those of an antiferromagnet, when they are extrapolated to lower temperatures. However, when the temperature approaches the ferrimagnetic Curie point, $T_c$, the susceptibility of a ferrimagnet looks like that of a ferromagnet. Below $T_c$ there is a spontaneous, non-zero magnetization in a zero applied field which is also similar to a ferromagnet. In that temperature region, a susceptibility can also be calculated, but it does not have any physical meaning because it cannot be measured, as is the case in ferromagnets.

For $T < T_c$ this linear treatment breaks down, and non-linear equations such as those of eqns (2.3.39) and (2.3.40) have to be solved numerically. Since the temperature-dependence of the magnetization is not the same for the two sublattices, the results are usually more complex than in the simple case of antiferromagnets. In particular, it may happen that sublattice $A$ has a much larger spontaneous magnetization, but a smaller Curie temperature than sublattice $B$. At low temperatures the total magnetization is parallel to that of $A$, but at higher temperatures the magnetization of sublattice $A$ vanishes, and the total magnetization is made out of that of sublattice $B$, which means that it turns into the opposite direction. Therefore, there is a temperature in between, known as a *compensation point*, at which the total spontaneous magnetization *passes through zero* before reaching finite values again. This phenomenon has indeed been observed in some ferrites [6] of the general formula $Fe_5M_3O_{12}$ where M is a trivalent rare earth ion. Many experimental data on all sorts of ferrites can be found in books [8] on this subject.

The *linear* study for the temperatures above the Curie point can be easily extended to more than two sublattices. It is readily seen from the foregoing that the more general form for eqns (2.5.59) and (2.5.60) is

$$\langle S_{iz}\rangle = \frac{S_i(S_i+1)}{3k_BT}\left(g_i\mu_BH + 2\sum_j J_{ij}\langle S_{jz}\rangle\right), \tag{2.5.69}$$

for any number of sublattices. Using the notation of $C_i$ as in eqn (2.5.61) and

$$\Theta_{ij} = \frac{2J_{ij}}{3k_B}\sqrt{S_i(S_i+1)S_j(S_j+1)}, \tag{2.5.70}$$

and replacing the average spin component by the variable

$$x_i = \frac{g_i\mu_B}{\sqrt{C_i}}\langle S_{iz}\rangle, \tag{2.5.71}$$

the set of equations (2.5.69) becomes

$$Tx_i - \sum_j \Theta_{ij}x_j = \sqrt{C_i}H. \tag{2.5.72}$$

For the solution of this set of simultaneous linear equations it is convenient to use matrix notation. Let $\theta$ be the matrix whose components are $\Theta_{ij}$, and let $I$ denote the unit matrix. Let $\mathbf{x}$ be the vector whose components are $x_i$, and let $\mathbf{C}^{1/2}$ be the vector whose components are $\sqrt{C_i}$. The set of equations (2.5.72) is then

$$(TI-\theta)\mathbf{x} = \mathbf{C}^{1/2}H, \tag{2.5.73}$$

whose obvious solution is

$$\mathbf{x} = (TI-\theta)^{-1}\mathbf{C}^{1/2}H. \tag{2.5.74}$$

The susceptibility is given by

$$\chi = \frac{1}{H}\sum_i M_{iz} = \frac{1}{H}\sum_i x_i\sqrt{C_i}, \tag{2.5.75}$$

and this sum is the scalar product of the two vectors defined above. Substituting for $\mathbf{x}$ from eqn (2.5.74),

$$\chi = \frac{1}{H}\mathbf{C}^{1/2} \cdot \mathbf{x} = \mathbf{C}^{1/2} \cdot (TI - \theta)^{-1}\mathbf{C}^{1/2}. \tag{2.5.76}$$

This susceptibility diverges when $T$ equals any one of the eigenvalues of the matrix $\theta$. There are materials, such as MnO, which have several different antiferromagnetic patterns, expressed by different eigenvectors $\mathbf{x}$. Transitions from one structure to another occur at the temperatures for which the eigenvalues of $\theta$ cross each other. From the measurements of these transitions it is possible to extract the values of the $J_{ij}$ which appear in the definition (2.5.70) of $\Theta_{ij}$. It is also clear that the *largest* eigenvalue of $\theta$ is the ferrimagnetic Curie temperature, if the material is ferrimagnetic, or the Néel temperature, if the material is antiferromagnetic just below that transition.

Expanding $\chi$ of eqn (2.5.76) in powers of $1/T$, it is seen that at high temperatures,

$$\chi = \frac{\mathbf{C}^{1/2}}{T} \cdot \left(I - \frac{\theta}{T}\right)^{-1}\mathbf{C}^{1/2} \approx \frac{\mathbf{C}^{1/2} \cdot \mathbf{C}^{1/2}}{T - \Theta}, \tag{2.5.77}$$

where

$$\Theta = \frac{\mathbf{C}^{1/2} \cdot \theta\mathbf{C}^{1/2}}{\mathbf{C}^{1/2} \cdot \mathbf{C}^{1/2}} = \frac{\sum_{ij}\Theta_{ij}\sqrt{C_i C_j}}{\sum_i C_i}, \tag{2.5.78}$$

is the paramagnetic Curie point, as defined in the foregoing. Written explicitly,

$$\Theta = \frac{2\sum_{ij} J_{ij} S_i(S_i + 1)S_j(S_j + 1)}{3k_B \sum_i S_i(S_i + 1)}. \tag{2.5.79}$$

The eigenvalues of the matrix $\theta$ are often (but not always) related in some way to those of a similar matrix whose eigenvectors represent the ordered configuration at zero temperature. Details are beyond the scope of this book, and can be found on p. 123 of the treatise [9] of Anderson.

## 2.6 Other Cases

The main advantage of the molecular field approximation is that it is much simpler than any other approximation to the quantum-mechanical Hamiltonian in eqn (2.2.25). It is a very big advantage because a simple theory can be extended to include more complicated additions, which a complex theory cannot.

For those who are only interested in a better approximation to certain parts of Fig. 2.1, obtained from eqn (2.2.25), there are indeed methods to achieve it, as will be seen in chapter 3. However, these methods cannot go *beyond* eqn (2.2.25), which may sometimes be an insufficient approximation to the physical reality. The molecular field approximation is sufficiently simple to allow for *other* energy terms to be added to that equation, which is not usually possible in the more sophisticated methods. In these, even the field **H** is often dropped from eqn (2.2.25) before anything of interest happens. In many cases the additional accuracy of the other methods is more than offset by the reduced accuracy caused by neglecting energy terms which are not really negligible.

One example is the magnetocrystalline anisotropy energy, which will be defined in section 5.1. This term, and the *biquadratic exchange* which is [10] an energy term of the form $(\mathbf{S}_1 \cdot \mathbf{S}_2)^2$, are very easy to add to eqn (2.2.25) under the molecular field approximation, but not so easy with any other method. The latter energy term has become rather popular recently, because it is quite strong in multilayers. It has been shown [11] that it can arise from thickness fluctuations in such films, or from magnetostatic interactions due to [12] surface roughness, and indeed it has been observed [13] in many systems. A more outstanding example is the *anisotropic exchange* which is an energy term [14] of the form $\mathbf{D} \cdot [\mathbf{S}_1 \times \mathbf{S}_2]$. It tries to arrange spins *perpendicular* to each other. Obviously, such a term cannot exist in high-symmetry crystals, because if $\mathbf{S}_1$ and $\mathbf{S}_2$ can be interchanged by some symmetry operation, the vector product changes sign, and **D** must vanish. However, this term does exist in some low-symmetry antiferromagnetic crystals, and by competing with the exchange it causes the direction of neighbouring spins to be slightly off the exact antiferromagnetic direction. Therefore, there is then a net spontaneous magnetization, called *weak ferromagnetism*, in a direction perpendicular to the antiferromagnetic axis. This case is given here just as an example of those which can essentially be treated theoretically by the molecular field approximation only, unless some trivial form [15] is involved. The full details will not be given here, but they can be found in the review of Moriya [16] or in the later literature, such as the study of orthoferrites [17] by Treves.

As a last example for the usefulness of the molecular field, we consider the case of an impurity, non-magnetic atom in a ferromagnetic lattice. For simplicity of the drawing we take a *two-dimensional* square lattice, as shown in Fig. 2.4, where the central atom in the figure does not have a magnetic moment. The problem is essentially described by eqns (2.2.27) and (2.2.28), which we use here with the notations of eqns (2.2.30) and (2.2.32). For the case of interaction between nearest neighbours only, and $H = 0$, these equations are

$$\mu_i = B_S \left[\alpha(T) \sum \mu_j\right], \tag{2.6.80}$$

where the summation is over the nearest neighbours. Specifically, defining 'shells' of distance from the impurity atom, as in Fig. 2.4, and considering the shells next to each atom as in that figure, we can write

$$\mu_1 = B_S \left[(2\mu_2 + \mu_3)\alpha\right], \tag{2.6.81}$$

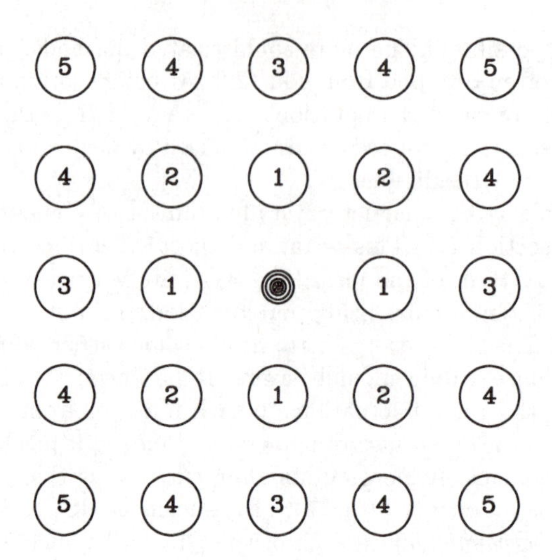


FIG. 2.4. Shell numbering around an impurity atom in a two-dimensional square lattice.

$$\mu_2 = B_S\left[(2\mu_1 + 2\mu_4)\alpha\right], \tag{2.6.82}$$

etc. These equations can be carried up to a shell which is considered far enough from the impurity for its magnetization to be that of a pure ferromagnet. Numerical computations can then try to include an extra shell and see if it makes an appreciable difference, and add another one if it does, until convergence is achieved. This problem was solved [18] as outlined here for a three-dimensional cubic crystal. It is *much* more complicated to approach it by any other method.

The molecular field approximation also has the big disadvantage of being *too* simple. It ignores the actual thermal fluctuations and the possible correlations between fluctuations of neighbouring spins. It can thus be expected to be a good approximation only at rather high temperatures, where the disorder is high. At low temperatures, one can do much better by considering small excitations of the state in which all spins are aligned. This approximation, known as *spin waves*, will be described in the next chapter. Then, when the temperature is still higher, and approaches the Curie temperature, thermal fluctuations become correlated over the whole lattice. In this *critical* region best accuracies are achieved by the so-called *renormalization group* which will be discussed in chapter 4.

# 3

# THE HEISENBERG HAMILTONIAN

## 3.1 Spin and Orbit

In the previous chapter we have always referred to the magnetization of the *spin* at the atomic lattice sites, which is just a manner of speaking. It is not very accurate for some ferromagnetic materials in which there is also a considerable contribution from the electron orbits. However, for the level of this book the difference is mostly semantic, because we consider the total magnetic moment of each atom, for which we have used the spectroscopic splitting factor (also known as the Landé $g$-factor),

$$g = 1 + \frac{J(J+1) + S(S+1) - L(L+1)}{2J(J+1)}, \tag{3.1.1}$$

which can be used in all cases. For a pure spin, $L = 0, J = S$, one obtains $g = 2$, which is twice as much as for the case of a pure orbit, $S = 0, J = L$.

The most direct measurement of this factor $g$ for each material is by a method known as the Einstein–de Haas experiment. In this experiment, the sample is suspended in such a way that its angular displacement can be measured [19] to a high accuracy. A magnetic field is applied to change the sample magnetization, thereby changing the angular momentum of the atoms, thus causing a certain rotation of the whole sample. The value of $g$ is then obtained from the ratio of the change in magnetization to the change in the mechanical angular momentum, and can show which part of the magnetic moment is due to the orbital contribution. It turns out that in most of the common ferromagnets the orbital contribution is negligibly small. The reason is that the electric fields in the lattice turn the plane of the orbits into crystallographic directions, thus making $L_z$ average to zero, or at least to a small number. In some rare earths there is an appreciable orbital contribution, and to include them properly we should have actually used a magnetic moment of $g\mu_B\mathbf{J}$ at the lattice site. However, these cases will be ignored here, and we shall just take the magnetic moment to be $\mu = g\mu_B\mathbf{S}$, where $\mu_B$ is the Bohr magneton, as in chapter 2.

It is interesting to note that the original experiment of Einstein and de Haas was done with a poor accuracy and led to the wrong conclusions. It was repeated by others who dared not publish their (better) results [20] because of the high prestige of Einstein's name, and it took years before it was established that the correct $g$ in Fe is nearly 2, and not 1.

It should be worth noting that the spin responsible for ferromagnetism is that of the d shell in the transition metals, and the f shell in the rare earths. For a filled shell, the total spin is known to be zero, and these shells are unfilled in the above-mentioned materials. The conduction electrons of the outer shell in both cases are not bound to their atoms in the solid state, but are free to move and are actually shared by the whole crystal, as is the case in *all* metals. These itinerant electrons do carry with them some interaction between the localized spins at the lattice sites, as will be discussed in section 3.4.

## 3.2 Exchange Interaction

Besides the indirect interaction carried by the conduction electrons in metals, there is a direct exchange interaction between spins of the ions at the lattice sites, both in metals and in insulators. It has no classical analogue, and is caused by overlap of the electronic wave functions in quantum mechanics. It is this part which is discussed in this section.

Consider a system of $N$ electrons which are bound to $M$ atoms. Let the eigenfunctions of an electron bound to atom No. 1, when the latter is isolated from the rest of the system, be denoted by $\varphi_i(\boldsymbol{\rho}_1)$, where $\boldsymbol{\rho}_1$ are *all* the coordinates of that electron, including the spin. Since all the atoms are identical, if atom No. 2 is separated from the others, an electron bound to it will have the *same* set of eigenfunctions, only at different coordinates, namely $\varphi_i(\boldsymbol{\rho}_2)$, and the same applies to all the other atoms.

Suppose that the $M$ atoms start from a position where they are well separated from each other, and then they are pushed towards each other. When these atoms get closer together, the single-atom levels start to become mixed. However, when this mixing just starts, there must still be some relation between the energy levels of the whole system and those of the separate atoms. In particular, for atoms at a very large distance from each other, the energy levels are

$$\nu = \prod_{i=1}^{M} (2S_i + 1)$$

times degenerate, if $S_i$ is the spin of the $i$-th atom. This degeneracy is removed when the atoms are nearer together, each level being split into $\nu$ ones. We assume, however, that the atoms are not yet very close together, and the interaction between them is sufficiently small, so that this splitting is still small compared with the distance between the different original levels. In such a case the original, atomic levels are still distinguishable in the whole spectrum.

It is felt intuitively that in such a case there should be a way to build the eigenfunctions (or at least a reasonable approximation for them) out of the functions $\varphi_i(\boldsymbol{\rho}_j)$, even though it is not so easy to justify such a feeling mathematically, or even to state the conditions for it by a rigorous, mathematical definition. The simplest combination which can be built up from $\varphi_i(\boldsymbol{\rho}_j)$ is the product

$$\psi = \varphi_1(\boldsymbol{\rho}_1)\varphi_2(\boldsymbol{\rho}_2)\cdots\varphi_N(\boldsymbol{\rho}_N), \tag{3.2.2}$$

and its permutations. However, this function is not allowed, because it does not obey the Pauli exclusion principle, not being antisymmetric to interchanging two of the electrons. We must take a linear combination of functions of the form of eqn (3.2.2) to achieve the necessary antisymmetry, for which it is possible to use the determinant

$$\psi = \frac{\det[\varphi_k]}{\sqrt{N!}}, \tag{3.2.3}$$

where $\det[\varphi_k]$ is a notation for

$$\det[\varphi_k] = \begin{vmatrix} \varphi_1(\boldsymbol{\rho}_1) & \varphi_1(\boldsymbol{\rho}_2) & \cdots & \varphi_1(\boldsymbol{\rho}_N) \\ \varphi_2(\boldsymbol{\rho}_1) & \varphi_2(\boldsymbol{\rho}_2) & \cdots & \varphi_2(\boldsymbol{\rho}_N) \\ \vdots & & & \\ \varphi_N(\boldsymbol{\rho}_1) & \varphi_N(\boldsymbol{\rho}_2) & \cdots & \varphi_N(\boldsymbol{\rho}_N) \end{vmatrix}. \tag{3.2.4}$$

Interchanging any two electrons is equivalent to interchanging the position of two columns in the determinant, which is known to reverse the sign. Therefore, the form of eqns (3.2.3) and (3.2.4) is in accordance with the Pauli exclusion principle.

I will not try to present the most general case, and will just assume here that the set of functions $\varphi_i$ is an *orthonormal* set, even though the conclusions which we are going to draw can also be proved under less restrictive conditions. An orthonormal set is one for which

$$\int \varphi_i^*(\boldsymbol{\rho}_1)\varphi_j(\boldsymbol{\rho}_1)d\tau_1 = \delta_{ij}, \tag{3.2.5}$$

where the asterisk designates the complex conjugate, and $\delta$ is the Kronecker symbol, which is equal to 1 when $i = j$ and to 0 otherwise. The integration in eqn (3.2.5) is over *all* the coordinates in $\boldsymbol{\rho}_1$, namely an integration over the real space and a summation over the two spin coordinates of the electron. In practice, the assumption is that $\varphi_i(\boldsymbol{\rho}_1) = \phi_i(\mathbf{r}_1)\eta_i(\zeta_1)$, where $\mathbf{r}_1$ are the space coordinates, and $\eta_i$ are normalized functions of the $z$-component of the spin, which may be named 'spin up' and 'spin down'. The latter are always orthogonal, so that our assumption means that $\phi_i$ *also* make an orthonormal set of functions.

We also assume that the electrons of the inner shells are tightly bound to their nuclei, and only the wave functions of the electrons in the outer shells are affected by the interaction with electrons of the other atoms. Obviously, the more shells that are taken as the 'outer' ones, the more accurate the calculation is, but usually it is not practical to extend the computations to more than one or two shells. When this technique is used as the so-called Hartree–Fock method for computing wave functions, one rarely extends the second group beyond the valence electrons. In any case, the inner electrons together with the nucleus are taken as an ion, which creates a certain potential at the position of the $i$-th electron. The potentials due to all the $M$ ions add up at the position of the $i$-th electron to a value which we denote by $V_i$. The Hamiltonian of the system of $N$ electrons is then

$$\mathcal{H} = \sum_{i=1}^{N} \mathcal{H}_i + \frac{1}{2} \sum_{i,j=1}^{N}{}' \frac{e^2}{r_{ij}} + \mathcal{H}_c, \tag{3.2.6}$$

where $\mathcal{H}_c$ is the Hamiltonian operating on the ion cores, $r_{ij}$ is the distance between electrons $i$ and $j$, the prime over the second sum eliminates the case $i = j$ from the summation, and

$$\mathcal{H}_i = -\frac{\hbar^2}{2m_e} \nabla_i^2 + V_i. \tag{3.2.7}$$

Here $\nabla_i$ operates on the coordinates of the $i$-th electron.

Using this Hamiltonian and the eigenfunctions defined in eqns (3.2.3) and (3.2.4) the energy of this system is

$$\mathcal{E} = \int \psi^* \mathcal{H} \psi \, d\tau_1 \, d\tau_2 \cdots d\tau_N = \frac{1}{N!} \int \det[\varphi_{k'}^*] \mathcal{H} \det[\varphi_k] \, d\tau_1 \, d\tau_2 \cdots d\tau_N. \tag{3.2.8}$$

Since the operator is linear, the integral can be written as the sum of integrals over the various terms, namely

$$\mathcal{E} = \sum_{i=1}^{N} \mathcal{E}_i + \frac{1}{2} \sum_{i,j=1}^{N}{}' \mathcal{E}_{ij} + \mathcal{E}_c, \tag{3.2.9}$$

where we have defined

$$\mathcal{E}_i = \frac{1}{N!} \int \det[\varphi_{k'}^*] \, \mathcal{H}_i \det[\varphi_k] \, d\tau_1 \, d\tau_2 \cdots d\tau_N, \tag{3.2.10}$$

$$\mathcal{E}_{ij} = \frac{1}{N!} \int \det[\varphi_{k'}^*] \, \frac{e^2}{r_{ij}} \det[\varphi_k] \, d\tau_1 \, d\tau_2 \cdots d\tau_N, \tag{3.2.11}$$

$$\mathcal{E}_c = \frac{1}{N!} \int \det[\varphi_{k'}^*] \, \mathcal{H}_c \det[\varphi_k] \, d\tau_1 \, d\tau_2 \cdots d\tau_N. \tag{3.2.12}$$

The last term in eqn(3.2.9) involves only the ion cores, and does not interest us here for the study of the electrons. Therefore, we want to evaluate only the first two energy terms.

Let us consider first the $\mathcal{E}_i$ term, and recall some of the properties of a determinant. It has $N!$ terms, each of which is a product of $N$ $\varphi_k$'s. In the latter product there are never two functions which are taken from either the same row or the same column. Therefore, $\mathcal{H}_i$, which contains derivatives only with respect to the particular coordinates $\boldsymbol{\rho}_i$, operates only on one particular $\varphi_k(\boldsymbol{\rho}_i)$ in the above-mentioned product. The other terms may be moved to the left of $\mathcal{H}_i$. Integrating over the coordinates in those terms, the integral vanishes according to the orthogonality assumption in eqn (3.2.5), unless the determinant of $\varphi^*_{k'}$ contains the complex conjugates of *all* the terms of $\varphi_k$ which we moved to the left of $\mathcal{H}_i$. There is one and only one such term in $\det[\varphi^*_{k'}]$ for every term of $\det[\varphi_k]$ which fulfils this condition and all the other terms integrate to zero. Usually, determinant terms may be positive or negative, but here each such term in one determinant is multiplied by the *same* term in the other determinant, so that the product is always positive. The integration over the terms moved to the left is then 1, according to eqn (3.2.5), and we are thus left with the integration over $\boldsymbol{\rho}_i$, which is the only term which remains to the right of $\mathcal{H}_i$. In other words, $\mathcal{E}_i$ is made out of a sum of $N!$ terms, each of which has the *form*

$$\int \varphi^*_k(\boldsymbol{\rho}_i)\, \mathcal{H}_i\, \varphi_k(\boldsymbol{\rho}_i)\, d\tau_i\,.$$

For given $k$ and $i$, the function $\varphi_k(\boldsymbol{\rho}_i)$ appears in $(N-1)!$ terms of the determinant of eqn (3.2.4). Integration of each of them leads to identical results because it does not make any difference which of the *other* functions integrates to 1. Therefore, the whole determinant in eqn (3.2.10) leads to

$$\mathcal{E}_i = (N-1)! \sum_{k=1}^{N} \frac{1}{N!} \int \varphi^*_k(\boldsymbol{\rho}_i)\, \mathcal{H}_i\, \varphi_k(\boldsymbol{\rho}_i)\, d\tau_i. \qquad (3.2.13)$$

For readers who find it rather difficult to follow the above argument, I recommend to work out as an illustration the case of a second-order determinant, for which

$$\mathcal{E}_1 = \frac{1}{2} \int \begin{vmatrix} \varphi^*_1(\boldsymbol{\rho}_1) & \varphi^*_1(\boldsymbol{\rho}_2) \\ \varphi^*_2(\boldsymbol{\rho}_1) & \varphi^*_2(\boldsymbol{\rho}_2) \end{vmatrix} \mathcal{H}_1 \begin{vmatrix} \varphi_1(\boldsymbol{\rho}_1) & \varphi_1(\boldsymbol{\rho}_2) \\ \varphi_2(\boldsymbol{\rho}_1) & \varphi_2(\boldsymbol{\rho}_2) \end{vmatrix} d\tau_1\, d\tau_2\,, \qquad (3.2.14)$$

which leads to

$$\mathcal{E}_1 = \frac{1}{2} \int [\varphi^*_1(\boldsymbol{\rho}_1)\, \mathcal{H}_1\, \varphi_1(\boldsymbol{\rho}_1) + \varphi^*_2(\boldsymbol{\rho}_1)\, \mathcal{H}_1\, \varphi_2(\boldsymbol{\rho}_1)]\, d\tau_1\,, \qquad (3.2.15)$$

and similarly for $\mathcal{H}_2$. The more general case should be clearer then.

The index $i$ in eqn (3.2.13) is that of the argument $\boldsymbol{\rho}_i$ in the integrand. After the integration over this variable, the result cannot depend on this particular $i$. We may as well take any one of these indices, for example the first one, and write

$$\mathcal{E}_i = \frac{1}{N} \sum_{k=1}^{N} \int \varphi_k^*(\boldsymbol{\rho}_1)\, \mathcal{H}_1\, \varphi_k(\boldsymbol{\rho}_1)\, d\tau_1 \,, \tag{3.2.16}$$

which already shows that the normalization factor taken in eqn (3.2.3) is correct, because the result is 1 if $\mathcal{H}_1$ is replaced by 1. The sum over $i$ is thus

$$\sum_{i=1}^{N} \mathcal{E}_i = N\mathcal{E}_i = \sum_{k=1}^{N} \int \varphi_k^*(\boldsymbol{\rho}_1)\, \mathcal{H}_1\, \varphi_k(\boldsymbol{\rho}_1)\, d\tau_1 = \mathcal{E}_e \,, \tag{3.2.17}$$

which is the energy of these electrons when they are separated from each other and do not interact.

When the same kind of algebra is repeated for $\mathcal{E}_{ij}$, it is seen that the coordinates of *two* electrons are involved in $r_{ij} = |\mathbf{r}_i - \mathbf{r}_j|$. Therefore, for each term of $\det[\varphi_k]$ there are two terms in $\det[\varphi^*_{k'}]$ which do not integrate to zero by applying the orthogonality condition of eqn (3.2.5). The non-zero terms will thus be made out of $\varphi_k(\boldsymbol{\rho}_i)$ and $\varphi_{k'}(\boldsymbol{\rho}_j)$ and their complex conjugates. The previous argument about repeating the *same* integrals $(N-1)!$ times applies equally well here, and so does the conclusion that the integrals do not depend on the particular choice of the indices $i$ and $j$, which may as well be replaced by 1 and 2. Hence

$$\frac{1}{2}\sum_{i,j=1}^{N}{}' \mathcal{E}_{ij} = \frac{1}{2}\sum_{k,k'=1}^{N}{}' \int |\varphi_k(\boldsymbol{\rho}_1)|^2 \frac{e^2}{r_{ij}} |\varphi_{k'}(\boldsymbol{\rho}_2)|^2\, d\tau_1\, d\tau_2$$

$$- \frac{1}{2}\sum_{k,k'=1}^{N}{}' \int \varphi_k^*(\boldsymbol{\rho}_1)\varphi_{k'}^*(\boldsymbol{\rho}_2) \frac{e^2}{r_{ij}} \varphi_k(\boldsymbol{\rho}_2)\varphi_{k'}(\boldsymbol{\rho}_1)\, d\tau_1\, d\tau_2. \tag{3.2.18}$$

Now, $|e|\,|\varphi_k(\boldsymbol{\rho}_1)|^2$ is the probability that there is an electron at the coordinates $\boldsymbol{\rho}_1$. Therefore, the first sum is the Coulomb interaction between a pair of electrons, summed over all the pairs. The second summation of integrals cannot be given such a simple classical interpretation. It is clear, however, that it somehow comes out of the Coulomb potential because of the use of a determinant, which means that it is due to the Pauli principle. It may thus be regarded as a kind of 'correction' to the classical Coulomb interaction of the first summation, which does not take into account the Pauli principle. According to this principle, two electrons that have the same spin cannot be in the same position, so that their overlap is smaller than that of classical electrons. The integrals which appear in the second summation of eqn (3.2.18) are called *exchange integrals*. The sum itself is called the *exchange energy* term.

It may be worth noting that the integrals in the energy terms thus obtained here can be evaluated only if all the functions $\varphi_k(\boldsymbol{\rho}_1)$ are known, which is hardly ever the case. It is more common to *evaluate* the functions $\varphi_k(\boldsymbol{\rho}_1)$ by minimizing the total energy obtained when these energy terms are substituted in eqn (3.2.9). There are different techniques for the actual use of this method that mostly differ by certain simplifying assumptions which are introduced before the energy minimization. They are all known by the general name of the *Hartree–Fock* method. These will not be described here.

Consider the second summation in eqn (3.2.18) which has just been described as the exchange energy. The important feature of this energy term is that the integrations in it contain also summation over the spin functions. Since these functions are orthogonal to each other, the integral will vanish if the spins are not parallel. Therefore, this term actually represents the energy *difference* between the state of two parallel spins, and the state when they are antiparallel. When one is interested only in the *magnetic* properties of the material, this term may just as well be replaced by a Hamiltonian which tries to hold the spins parallel (or antiparallel, depending on the *sign* of the appropriate integral) to each other.

In order to state the substitution of a Hamiltonian more precisely, let $\bar{\Psi}$ be the wave functions of the system of electrons of the $M$ atoms, when they are at a very large distance from each other, namely a certain combination of the functions of a single atom (or ion). Let $\Psi$ be the true eigenfunctions of the system when the atoms are put closer together, so that they interact *weakly*, in such a way that the degenerate energy levels are split, but not mixed beyond recognition. It is legitimate to assume that it is possible to have some sort of a one-to-one correspondence between $\Psi$ and $\bar{\Psi}$ for this case. As a substitute for the true Hamiltonian, $\mathcal{H}$, we would like to have an effective Hamiltonian, $\mathcal{H}_{\text{eff}}$, so that its matrix elements with respect to the $\bar{\Psi}$'s will be the same as the matrix elements of the original Hamiltonian with respect to the $\Psi$'s, namely

$$\langle \bar{\Psi}_k | \mathcal{H}_{\text{eff}} | \bar{\Psi}_{k'} \rangle = \langle \Psi_k | \mathcal{H} | \Psi_{k'} \rangle. \tag{3.2.19}$$

Obviously, if the energy difference between parallel and antiparallel spins is going to be the above-mentioned exchange integral, something that contains the sum of terms which are proportional to $\mathbf{s}_i \cdot \mathbf{s}_j$ can do the job, where $\mathbf{s}_i$ are the spin of each electron. However, it is not convenient to deal with each electron separately, and it is better to sum first over all the electrons of each atom (or ion) at a lattice point. Some care in carrying out this summation may be necessary in some cases, the fine details of which can be found in the the treatise of Herring [21], and will not be repeated here. The final result is what is intuitively felt to be the case, namely that

$$\mathcal{H}_{\text{eff}} = -\sum_{i,j=1}^{M} J_{i,j} \mathbf{S}_i \cdot \mathbf{S}_j \,, \tag{3.2.20}$$

where

$$J_{i,j} = 2 \int \varphi_i^*(\mathbf{r}_1)\varphi_j^*(\mathbf{r}_2) \frac{e^2}{|\mathbf{r}_1 - \mathbf{r}_2|} \varphi_i(\mathbf{r}_2)\varphi_j(\mathbf{r}_1)\, d\mathbf{r}_1\, d\mathbf{r}_2 \,. \tag{3.2.21}$$

The convention is to keep the minus sign as in eqn (3.2.18), so that a positive $J_{i,j}$ means a ferromagnetic coupling that tends to align spins parallel to each other, while a negative $J_{i,j}$ means an antiferromagnetic coupling. It should be noted that in the present Hamiltonian $\mathbf{S}_i \cdot \mathbf{S}_j$ can have values between $-S^2$ and $S^2$, whereas the appropriate part in eqn (3.2.18) varies between 0 and $\pm S^2$, which introduces an extra factor of 2 in the definition of eqn (3.2.21). It should also be noted that $\mathbf{S}_i$ is the *total* spin of all the electrons bound to the atom, or ion, at the lattice site $i$. For an insulator, the spin is that of all the electrons. For a metal, only the electrons of the inner shells are counted, which usually means just the d electrons in the metals Ni, Co, and Fe. The conduction electrons of a metal wander around the whole crystal and do not belong to any lattice site, and the innermost electrons are taken as one entity together with the nucleus.

Since the Coulomb interaction is a scalar, the effective Hamiltonian must contain the scalar products of the appropriate spins. However, it does not necessarily mean that eqn (3.2.20) is the only possibility. In some way it may be regarded as only a first-order term in an expansion, the next term of which being

$$\sum_{i,j=1}^{M} J'_{i,j} \left(\mathbf{S}_i \cdot \mathbf{S}_j\right)^2 ,$$

and even higher orders may be added in principle. As has already been mentioned in section 2.6, the higher-order term is indeed encountered in some physical situations. However, nothing beyond the first term can generally be included in a quantum-mechanical calculation.

The *Heisenberg Hamiltonian* of eqn (3.2.20) is, thus, the justification, on a deeper level, for the assumption of a force which tries to align neighbouring spins. When the spin operators are replaced by their eigenvalues, this Hamiltonian leads to, and justifies, the first energy term of eqn (2.2.25). It is thus the basis of all the theory of the Weiss 'molecular field' approximation that has been used throughout chapter 2, and the basis for most of the rest of this book. In chapter 2 it was also assumed that only nearest-neighbour interaction is usually important, and this part can also be readily seen from eqn (3.2.21). Since the integral involves the *overlap* of the wave function, it is quite clear, even without detailed computations, that its value must decrease very rapidly with increasing distance between the ions. In particular, $J$ must be negligible for electrons on farther atoms. Therefore, it is usually sufficient to consider exchange interaction between nearest neighbours only, as has been done in chapter 2.

It may also be added that the treatment here referred specifically to the so-called 'direct' exchange coupling. In many of the ferrites discussed in section 2.5, there is no such direct coupling between the magnetic ions, *e.g.* Fe. Instead, there is an antiferromagnetic coupling between the spin of the iron and that of an oxygen ion, and another antiferromagnetic coupling between that oxygen and the spin of another iron in the same molecule. This coupling, known as a *superexchange*, still tries to align the spins of those two iron ions parallel to each other, and is *effectively* the same as as a direct ferromagnetic interaction. On the level of this book, it is not necessary to distinguish between the two.

The integral in eqn (3.2.21) is symmetric to interchanging $i$ and $j$. Therefore, $J_{i,j} = J_{j,i}$, and it is sufficient to take only half the sum of eqn (3.2.20). This feature allows us to write the Heisenberg Hamiltonian in its more common form, as

$$\mathcal{H}_{\text{eff}} = -2\sum_{i>j} J_{i,j}\mathbf{S}_i \cdot \mathbf{S}_j\,. \tag{3.2.22}$$

## 3.3 Exchange Integrals

If the functions $\varphi_i$ are orthogonal to each other, adding a term with $e^2/r_{12}$ can be expected to contribute a positive value. This is indeed the case for electrons in the *same* atom. In an unfilled shell, electrons tend to have parallel spins as long as that is allowed for the same shell, thus creating a large total spin $S$ for the shell. When the functions $\varphi_i$ are not orthogonal, a rough estimation of the exchange integral $J$ usually leads to a negative value. For a problem like the hydrogen molecule, this negative exchange is easy to understand by a simple physical argument: because of the Coulomb attraction, the two electrons would prefer to be close to both nuclei, which they can do if they share the same orbit that goes around the *two* nuclei. According to the Pauli principle, the orbit sharing is possible only if the spins of the two electrons are antiparallel. Therefore, this antiparallel state has a lower energy than the state in which the two spins are parallel. One can thus expect the exchange integral $J_{ij}$, for interaction between electrons in different atoms, to be generally negative. And indeed computations from eqn (3.2.21), for almost any reasonable assumption on the functions $\varphi_i$, lead to a *negative* exchange integral.

However, it is known from experiment that Fe, Co and Ni (and some rare earths) are ferromagnets, and the exchange integrals for them must be positive, unlike a similar transition metal, *e.g.* Cu, for which the eigenfunctions $\varphi_i$ are nearly the same, but in which that integral must be negative. It used to be stated [21] that nobody has been able to compute a positive exchange integral for Fe, and a negative one for Cu, because rather large positive and negative contributions subtract to a smaller value that is very sensitive to the computational accuracy. More modern computations [22] already have the right sign, but the *magnitude* of the computed exchange integral still differs considerably from the experimental value. Improving the techniques [23, 24] keeps improving the results, but not sufficiently yet. The accuracy is certainly not sufficient to account for the possibility that Cu may become [25] ferromagnetic under certain conditions.

It is, thus, not possible yet to determine the value of the exchange integral in the ferromagnetic metals from basic principles. One can just *assume* the Hamiltonian of eqn (3.2.20), and take $J_{ij}$ as a parameter whose value is obtained by fitting the theory to a certain experimental value (usually the Curie temperature). The theoretical situation is clearer in the case of *ferrimagnets*, discussed in section 2.5. There, $J < 0$, and the basic interaction is antiferromagnetic, but the moments of the two sublattices do not subtract to zero because they are not equal. The net moment is then effectively ferromagnetic, in spite of the negative exchange. The theory is also quite clear for the case of the indirect superexchange, mentioned in the previous section. In those ferrites, one Fe ion is coupled antiferromagnetically to an O ion which in turn is coupled antiferromagnetically to another Fe ion. The net effect is a ferromagnetic coupling between the two Fe ions, but the integrals are both *negative*. However, even in these cases, the values of the exchange integrals have to be taken from experiment, because the theory is not sufficiently well developed to yield reliable values.

## 3.4 Delocalized Electrons

The whole concept of interaction between electrons which are localized on ions at lattice sites is at best very much oversimplified. After all, a strong exchange coupling implies a large spatial overlap of the electron wave functions, which cannot be realized if these electrons are strictly localized. Moreover, at least in the metals Fe, Co and Ni, conduction electrons are moving around, and they must also interact in some way with the electrons at the lattice sites. The picture is clearer when two ferromagnetic layers are separated by a non-ferromagnetic metal, and an exchange interaction is carried [26, 27] by the conduction electrons of the latter. But even in the ferromagnet itself, some interaction is carried by mobile conduction electrons. In these metals, the 3d band is overlapped in energy by a much wider 4s band, and since bands are filled to the Fermi level, the electrons which each atom contributes to the conduction band are not all from the 4s band, and are partly from the 3d band. Therefore, the number of d electrons per atom (or f electrons in the case of rare earths) contributing to the bulk magnetization is *not* an integral number, which is indeed an experimental fact. From the experimental saturation magnetization of these metals, the number of Bohr magnetons per atom is 0.6 for Ni, 1.7 for Co, and 2.2 for Fe. Besides, the measured specific heat at low temperatures in these materials shows a bigger contribution from the electron gas than can be possibly accounted for by valence electrons (4s in Fe, Co and Ni).

The theoretical study which is based on the Heisenberg Hamiltonian, as used throughout the rest of this book, ignores these difficulties, and just puts a non-integral number of Bohr magnetons at the lattice sites. Therefore, another theory has been developed in parallel, which assumes [28] a completely delocalized, free-electron gas, moving in the presence of the fixed background of the positively charged ions at the lattice sites. Calculating the actual energy bands of these electrons can account for the actual specific heat, and can yield theoretical values for the saturation magnetization *vs.* temperature curves, like the one plotted in Fig. 2.1 here, as well as for other transport and magnetic properties in metals. This theory is called *collective electron ferromagnetism*, or *itinerant electron ferromagnetism.*

The itinerant electron ferromagnetism is elegant, and some of its results are easy to follow even without detailed computations. Consider for example Cu, with 11 electrons per atom. These electrons are sufficient to fill the 3d shell, and a filled shell does not have any net magnetic moment, because there is an equal number of electrons with spin up and with spin down. In the 4s shell the exchange interaction is rather low, and the distance between neighbouring levels is too large. Therefore, Cu does not have any magnetic moment. In Ni, there are 10 electrons which have to be subdivided between the 3d and 4s shells. In a gas of free atoms, there are 8 electrons in 3d, and 2 electrons in 4s. In a solid, because of filling up bands up to the same (Fermi) energy level, it can be concluded from the experimental magnetic data that 9.4 electrons per atom are in 3d, and 0.6 electrons per atom in 4s. In the unfilled 3d shell the spins are not balanced, because the exchange interaction within the atom causes more spins to be up than down. The exchange energy gain is more than sufficient to compensate for the energy loss due to the electrons being raised to higher levels in the band when they cannot use the lower ones that can only be occupied by those with an opposite spin to the electrons that are already there. The difference between the moments gives rise to a net magnetic moment of 0.6 Bohr magnetons per atom.

The maximum possible imbalance in 3d is when 5 out of the 10 electrons enter the half band with spin up, and the other 5 split between the other half band with spin down and the 4s. For $n$ electrons per atom, out of which $x$ are in 4s and $n - x$ in 3d, at most 5 can be with spin up, leaving $n - x - 5$ in 3d with spin down. The net magnetic moment is then

$$\mu_H = [5 - (n - x - 5)]\mu_B = [10 - (n - x)]\mu_B. \qquad (3.4.23)$$

In Ni, $n = 10$ and $x = 0.6$, which is concluded from the experimental value of $\mu_H = 0.6\mu_B$, as has been mentioned already. Ignoring the change in the band structure in alloys of Ni with other metals, and assuming that 0.6 electrons per atom still go to the 4s band in these alloys, their magnetic moment should be

$$\mu_H = (10.6 - n)\mu_B, \qquad (3.4.24)$$

which agrees quite well with experiment for Ni alloys. For example, in alloying Ni with Cu, which has 11 electrons per atom, the saturation magnetization decreases more or less linearly with increasing concentration of Cu, reaching zero at about 60% Cu, in accordance with this simple relation.

Similar estimations for the metal itself, and the effect of some alloying, work well enough for Co. In Fe there are deviations of about 20% from the experimental value, which is not surprising because the assumption that the energy band structure does not change between one element and the other is oversimplified. But a linear relation still works for some iron alloys [29], and it is often possible to account for the experimental results in some others [30] by very simple models. For Mn this argument breaks down. Continuing as before, the magnetic moment of Mn (with $n = 7$) should be larger than that of Fe; but actually the moment is 0. Pure Mn is not ferromagnetic, because the exchange in Mn is not strong enough to raise electrons to higher energy levels in the band, leaving the lower energy ones, that become forbidden (by the Pauli exclusion principle) for electrons with the same spin. After all, eqn (3.4.23) gives the *maximum* moment, which can only be achieved for a *strong* exchange coupling. However, in alloys with materials such as Al or Bi, the distance between the Mn atoms decreases sufficiently to increase the exchange integral, and these alloys are ferromagnetic.

It should be quite clear from this outline that the itinerant electron theory, with more detailed and more sophisticated calculations of the energy bands, can be very successful in interpreting many experimental data. For example, already computations which would be considered rather primitive today [31] showed, as later confirmed by more elaborate ones [22, 32], that the energy bands sometimes contain very sharp and narrow peaks. Therefore, a sharp change in some properties may be encountered when a particular composition of an alloy passes a certain peak in the energy band. Also, transport properties, in particular the giant magnetoresistance effect, can be interpreted [33] practically only by the itinerant electron picture, even though localization does play [34] a certain role.

Generally speaking, the itinerant electron theory is quite successful in dealing with the whole crystal. However, it is quite clear that such a theory cannot handle any *spatial* variations in the magnetization, for the simple

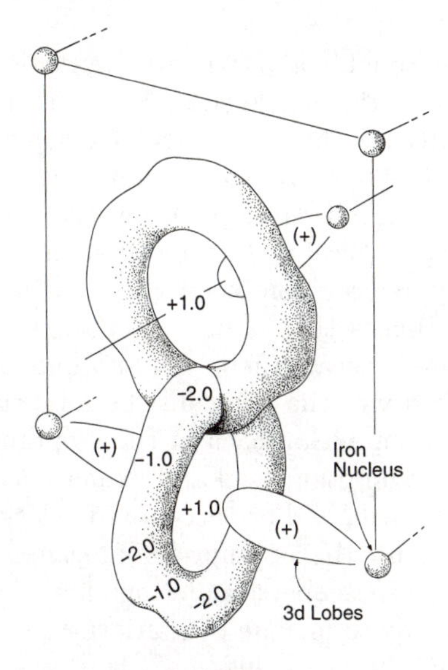


FIG. 3.1. The magnetization distribution in a unit cell of bcc Fe, as 'seen' by neutrons. Reproduced from Fig. 3 of [37] by permission.

reason that the energy band calculation is independent of space. Therefore, an all-itinerant model must assume that the magnetization is the same in the whole space, while there is strong experimental evidence to the contrary, namely that the magnetization in a crystal is a function of space. In particular, bulk ferromagnets have been shown by very many techniques (which will be discussed in section 4.1) to be subdivided into *domains* in which the magnetization points in different directions. These effects, and the whole concept of hysteresis as seen in Fig. 1.1, must be ignored in an all-itinerant theory.

Even besides the subdivision into domains there is strong experimental evidence against the itinerant electron theory, some of which has been listed in a popular review of Stearns [35], along with the experimental evidence against a theory that assumes purely localized electrons. She [35, 36] tried to outline a combined picture, in which *part* of the 3d electrons is localized, the other part being itinerant. An even clearer picture can be seen from Fig. 3 of [37], reproduced as Fig. 3.1 here. It plots the results of shooting neutrons through an iron crystal. Since the neutron has a magnetic moment, it interacts with the magnetic field while passing through the crystal. This figure interprets the experimental neutron data in terms of the magnetic field through which those neutrons pass. It is quite obvious from the figure that *some* of the magnetization is localized at the lattice sites, but it is also clear that this magnetic moment is very much smeared around these sites. The magnetic field distribution is complicated, and the picture cannot be even roughly approximated by the naive assumption of a point magnetic charge at the lattice sites. Some details of this figure were somewhat modified later [38], but they do not change its general properties.

There is thus no doubt in anybody's mind that neither the itinerant electron theory nor the localized electron one can be considered to be a complete picture of the physical reality, and that they should both be combined into one theory. Such a combined approach may come some time, but the present situation is that it has not been seriously tried, on any decent quantitative level, because it is just too difficult. Workers in magnetism stick to one theory or the other just because they are unable to do any better. Actually, even within each of these approaches there are still too many simplifying assumptions and approximations which are as bad as ignoring the other approach, and are accepted for lack of anything better. In this book I choose to use the assumption of localized magnetic moments on lattice sites, and the Heisenberg Hamiltonian as derived in section 3.2, because it is the only way to include the variation of the magnetization in space, which is the main topic of this book. It means using a non-integral number of Bohr magnetons per atom, which is physically strange but can be understood from the foregoing argument. Once such a non-integral value is accepted there is nothing wrong with using it in the calculations, and comparing the results with experiment. It also means ignoring that part of the exchange interaction between the localized 3d electrons which is carried by the conduction electrons. However, such a simplification is inevitable, and in any case expressing this contribution as part of the Heisenberg Hamiltonian, eqn (3.2.20) or (3.2.22), is quite a reasonable approximation. As long as the exchange integral cannot be calculated from basic principles and its value is taken from experiment, the experimental value contains the itinerant electron contribution anyway.

## 3.5 Spin Waves

The foregoing should be an adequate justification for the use of the Hamiltonian

$$\mathcal{H} = -\sum_{\boldsymbol{\ell},\boldsymbol{\ell}'}{}' J_{\boldsymbol{\ell}\boldsymbol{\ell}'} \mathbf{S}_{\boldsymbol{\ell}} \cdot \mathbf{S}_{\boldsymbol{\ell}'} - \sum_{\boldsymbol{\ell}} g\mu_B H S_{\boldsymbol{\ell}}^{(z)} , \tag{3.5.25}$$

where $\boldsymbol{\ell}$ are the lattice vectors, namely the vectors from the origin to the lattice points. This equation is more or less equivalent to eqn (2.2.25) which has already been used in chapter 2, only there $\mathbf{S}$ were classical vectors, whereas here they designate the spin operators. The justification may not be as good as may be expected but it is the best we have at this stage of the theory, which leaves a lot to be desired. In a way, it is not much more than an assumption, to be adopted from now on. It is hoped, however, that the foregoing argument is sufficiently convincing for adopting this assumption, which at least is not just an arbitrary, *ad hoc* assumption as is the Weiss 'molecular field' used in chapter 2.

It is customary to replace the operators $S_{\boldsymbol{\ell}}^{(x)}$ and $S_{\boldsymbol{\ell}}^{(y)}$ by the operators

$$S_{\boldsymbol{\ell}}^{+} = S_{\boldsymbol{\ell}}^{(x)} + iS_{\boldsymbol{\ell}}^{(y)}, \qquad S_{\boldsymbol{\ell}}^{-} = S_{\boldsymbol{\ell}}^{(x)} - iS_{\boldsymbol{\ell}}^{(y)}, \tag{3.5.26}$$

for which the Hamiltonian becomes

$$\mathcal{H} = -\sum_{\boldsymbol{\ell},\boldsymbol{\ell}'}{}' J_{\boldsymbol{\ell}\boldsymbol{\ell}'} \left\{ \frac{1}{2} \left[ S_{\boldsymbol{\ell}}^{+} S_{\boldsymbol{\ell}'}^{-} + S_{\boldsymbol{\ell}}^{-} S_{\boldsymbol{\ell}'}^{+} \right] + S_{\boldsymbol{\ell}}^{(z)} S_{\boldsymbol{\ell}'}^{(z)} \right\} - \sum_{\boldsymbol{\ell}} g\mu_B H S_{\boldsymbol{\ell}}^{(z)}. \tag{3.5.27}$$

As is known from quantum mechanics, for spins at the *same* lattice site, the original components of the spin $\mathbf{S}$ for the *same* $\boldsymbol{\ell}$ fulfil the commutation relations

$$\left[ S_{\boldsymbol{\ell}}^{(x)}, S_{\boldsymbol{\ell}}^{(y)} \right] = i\hbar S_{\boldsymbol{\ell}}^{(z)}; \quad \left[ S_{\boldsymbol{\ell}}^{(y)}, S_{\boldsymbol{\ell}}^{(z)} \right] = i\hbar S_{\boldsymbol{\ell}}^{(x)}; \quad \left[ S_{\boldsymbol{\ell}}^{(z)}, S_{\boldsymbol{\ell}}^{(x)} \right] = i\hbar S_{\boldsymbol{\ell}}^{(y)}, \tag{3.5.28}$$

while these components commute for spins at different lattice sites, and all these commutation relations vanish for $\boldsymbol{\ell} \neq \boldsymbol{\ell}'$. Using these relations, and the definitions in eqn (3.5.26),

$$\begin{aligned} S_{\boldsymbol{\ell}}^{(z)} S_{\boldsymbol{\ell}}^{+} &= S_{\boldsymbol{\ell}}^{(z)} \left[ S_{\boldsymbol{\ell}}^{(x)} + iS_{\boldsymbol{\ell}}^{(y)} \right] \\ &= \left[ S_{\boldsymbol{\ell}}^{(z)}, S_{\boldsymbol{\ell}}^{(x)} \right] + S_{\boldsymbol{\ell}}^{(x)} S_{\boldsymbol{\ell}}^{(z)} + i\, S_{\boldsymbol{\ell}}^{(y)} S_{\boldsymbol{\ell}}^{(z)} - i \left[ S_{\boldsymbol{\ell}}^{(y)}, S_{\boldsymbol{\ell}}^{(z)} \right] \\ &= i\hbar\, S_{\boldsymbol{\ell}}^{(y)} + S_{\boldsymbol{\ell}}^{(x)} S_{\boldsymbol{\ell}}^{(z)} + i\, S_{\boldsymbol{\ell}}^{(y)} S_{\boldsymbol{\ell}}^{(z)} + \hbar\, S_{\boldsymbol{\ell}}^{(x)} \\ &= \hbar\, S_{\boldsymbol{\ell}}^{+} + S_{\boldsymbol{\ell}}^{+} S_{\boldsymbol{\ell}}^{(z)}. \end{aligned} \tag{3.5.29}$$

Hence,

$$\left[ S_{\boldsymbol{\ell}}^{(z)}, S_{\boldsymbol{\ell}'}^{+} \right] = \hbar\delta_{\boldsymbol{\ell}\boldsymbol{\ell}'} S_{\boldsymbol{\ell}}^{+}, \tag{3.5.30}$$

where $\delta_{\boldsymbol{\ell}\boldsymbol{\ell}'}$ is the Kronecker symbol, which is 1 if $\boldsymbol{\ell} = \boldsymbol{\ell}'$, and 0 otherwise. Similarly,

$$\left[ S_{\boldsymbol{\ell}}^{(z)}, S_{\boldsymbol{\ell}'}^{-} \right] = -\hbar\delta_{\boldsymbol{\ell}\boldsymbol{\ell}'} S_{\boldsymbol{\ell}}^{+}. \tag{3.5.31}$$

To complete the transformation it is necessary to consider also

$$\left[ S_{\boldsymbol{\ell}}^{+}, S_{\boldsymbol{\ell}'}^{-} \right] = \left[ S_{\boldsymbol{\ell}}^{(x)}, S_{\boldsymbol{\ell}'}^{(x)} \right] - i \left[ S_{\boldsymbol{\ell}}^{(x)}, S_{\boldsymbol{\ell}'}^{(y)} \right] + i \left[ S_{\boldsymbol{\ell}}^{(y)}, S_{\boldsymbol{\ell}'}^{(x)} \right] + \left[ S_{\boldsymbol{\ell}}^{(y)}, S_{\boldsymbol{\ell}'}^{(y)} \right]. \tag{3.5.32}$$

Here the first and the last term obviously vanish, and the two in the middle can be evaluated by substituting from eqn (3.5.28), leading to

$$\left[ S_{\boldsymbol{\ell}}^{+}, S_{\boldsymbol{\ell}'}^{-} \right] = 2\hbar\delta_{\boldsymbol{\ell}\boldsymbol{\ell}'} S_{\boldsymbol{\ell}}^{(z)}. \tag{3.5.33}$$

Actually, the commutation relations are not usually written in this form in books or papers on this problem. It is customary to write them in the special 'units' in which $\hbar = 1$, and therefore omit $\hbar$. It saves some ink to write equations in this way, but in principle it has a major disadvantage if and when the result is to be expressed in terms of *real* units. If at the end of the calculation the result is a certain measurable quantity that one wants to compare with experiment, it is not always very clear if that result has to be multiplied by $\hbar$, or divided by $\hbar^2$, or what. It is always much easier to substitute the particular value $\hbar = 1$ if it is wanted than to put in another value if $\hbar = 1$ is assumed to begin with. As a matter of fact, theories which use $\hbar = 1$ (or other non-physical units, such as the velocity of light $c = 1$, etc.) can exist only in fields in which theorists compare their results with each other's, and the experiment is far removed. It can never happen in the normal trend of physics, in which theory and experiment are expected to go hand in hand. Therefore, it is always a better policy to keep $\hbar$ in the equations.

Consider now, at each lattice point, another operator defined as

$$N_{\boldsymbol{\ell}} = S\hbar - S_{\boldsymbol{\ell}}^{(z)}, \tag{3.5.34}$$

where $S$ is the spin *number* of the atom (or, rather, the ion) at that lattice site. Since $S\hbar$ is the *largest* eigenvalue of $S_{\boldsymbol{\ell}}^{(z)}$, the eigenvalues $n_{\boldsymbol{\ell}}\hbar$ of the operator $N_{\boldsymbol{\ell}}$ express the *difference* between the maximum possible value, and the actual value, of the $z$-component of the spin at the lattice site $\boldsymbol{\ell}$. Therefore the numbers $n_{\boldsymbol{\ell}}$ are called the *spin deviations* at the lattice point $\boldsymbol{\ell}$. Let $\Psi_{n_{\boldsymbol{\ell}}}$ denote the eigenstate for which the spin deviation is $n_{\boldsymbol{\ell}}$, namely

$$N_{\boldsymbol{\ell}}\Psi_{n_{\boldsymbol{\ell}}} = n_{\boldsymbol{\ell}}\hbar\,\Psi_{n_{\boldsymbol{\ell}}}. \tag{3.5.35}$$

In principle this eigenstate is a function of the spin coordinates at *all* the lattice sites, but such an operator with a *particular* value of $\boldsymbol{\ell}$ operates only on the coordinates which apply to this particular $\boldsymbol{\ell}$.

It is readily verified that the same $\Psi_{n_{\boldsymbol{\ell}}}$ is also an eigenstate of the operator $S_{\boldsymbol{\ell}}^{(z)}$, whose eigenvalue is $\hbar(S - n_{\boldsymbol{\ell}})$. Indeed, by using the definition of eqn (3.5.34), and substituting from eqn (3.5.35),

$$S_{\boldsymbol{\ell}}^{(z)}\Psi_{n_{\boldsymbol{\ell}}} = (S\hbar - N_{\boldsymbol{\ell}})\,\Psi_{n_{\boldsymbol{\ell}}} = \hbar(S - n_{\boldsymbol{\ell}})\Psi_{n_{\boldsymbol{\ell}}}. \tag{3.5.36}$$

For other properties of this function, consider the expression

$$S_{\boldsymbol{\ell}}^{(z)}S_{\boldsymbol{\ell}}^{+}\Psi_{n_{\boldsymbol{\ell}}} = \left\{\left[S_{\boldsymbol{\ell}}^{(z)}, S_{\boldsymbol{\ell}}^{+}\right] + S_{\boldsymbol{\ell}}^{+}S_{\boldsymbol{\ell}}^{(z)}\right\}\Psi_{n_{\boldsymbol{\ell}}}.$$

Substituting for the first term in the curly bracket from eqn (3.5.30), and for the second term from eqn (3.5.36),

$$S_\ell^{(z)}\left(S_\ell^+\Psi_{n_\ell}\right) = \hbar\left(1+S-n_\ell\right)\left(S_\ell^+\Psi_{n_\ell}\right), \tag{3.5.37}$$

which means that $S_\ell^+\Psi_{n_\ell}$ is also an eigenstate of $S_\ell^{(z)}$. In fact, it is the eigenstate which has the eigenvalue $\hbar[S-(n_\ell-1)]$. Since the latter is the eigenvalue of $\Psi_{n_\ell-1}$, the state $S_\ell^+\Psi_{n_\ell}$ must be proportional to $\Psi_{n_\ell-1}$. Similarly, it can be shown that the operator $S_\ell^-$ transforms $\Psi_{n_\ell}$ to something proportional to $\Psi_{n_\ell+1}$.

The behaviour of the operators $S_\ell^-$ and $S_\ell^+$ is, thus, similar to that of the creation and destruction operators. They create or destroy *spin deviations.* However, there is a big difference in that the commutation relation of the conventional creation and destruction operators is $[a,a^*]=1$. If the right hand side of eqn (3.5.33) was a *number*, $S_\ell^+$ and $S_\ell^-$ could be normalized to make the commutation relation equal to 1; but that right hand side is an operator and not a number. The best which can be done is to define the operators

$$a_\ell = \frac{1}{\sqrt{2S}\,\hbar}\,S_\ell^+\,, \qquad a_\ell^* = \frac{1}{\sqrt{2S}\,\hbar}\,S_\ell^-\,, \tag{3.5.38}$$

for which eqn (3.5.33) becomes

$$[a_\ell\,,\,a_{\ell'}^*] = \frac{\delta_{\ell\ell'}}{\hbar S}\,S_\ell^{(z)}\,. \tag{3.5.39}$$

Nevertheless, it has become customary to use the *approximation* in which the *operator* $S_\ell^{(z)}$ in the right hand side of eqn (3.5.39) is replaced by its *eigenvalue* $S\hbar$. The justification is [39] that replacing that operator by its eigenvalue is correct to a first order, and introduces only a second-order error, at low temperatures. The basic assumption is that in the region of interest, almost all the spins are parallel to $z$, and the deviations are small on the average, namely

$$M_z(0) - M_z(T,H) \ll M_z(0).$$

It should be remarked, however, that there is a difference between a proof [39] that the neglected term is small at low temperatures, and a *quantitative* estimate of how small is small. It is easy to be convinced that replacing the operator by its eigenvalue is a good enough approximation if the temperature is not too high. It is less easy to say up to which experimental accuracy, or up to which temperature, such an approximation is justified. Such a *quantitative* estimation has never been done, nor has there ever been any quantum-mechanical treatment of the low-temperature region by any other approximation. Therefore, at the present stage of our knowledge, there is no choice but to accept this *assumption* because there is no other way to continue the calculation. But it must be borne in mind that a certain, unspecified approximation is involved. Therefore, the statement (which is only too often made) that a quantum-mechanical theory is inherently more accurate than something classical (as in chapter 2 here), is at best unproved and unchecked.

Replacing the right hand side of eqn (3.5.39) by just $\delta_{\boldsymbol{\ell}\boldsymbol{\ell}'}$,

$$[a_{\boldsymbol{\ell}}\,,\, a^*_{\boldsymbol{\ell}'}] = \delta_{\boldsymbol{\ell}\boldsymbol{\ell}'}\,, \tag{3.5.40}$$

the operators $a_{\boldsymbol{\ell}}$ and $a^*_{\boldsymbol{\ell}}$ become the same as the conventional destruction and creation operators, so that

$$a^*_{\boldsymbol{\ell}}\Psi_{n_{\boldsymbol{\ell}}} = \sqrt{n_{\boldsymbol{\ell}}+1}\,\Psi_{n_{\boldsymbol{\ell}}+1}\,, \qquad a_{\boldsymbol{\ell}}\Psi_{n_{\boldsymbol{\ell}}} = \sqrt{n_{\boldsymbol{\ell}}}\,\Psi_{n_{\boldsymbol{\ell}}-1}\,. \tag{3.5.41}$$

Moreover, according to the definition in eqn (3.5.35),

$$N_{\boldsymbol{\ell}} = \hbar a^*_{\boldsymbol{\ell}} a_{\boldsymbol{\ell}}, \tag{3.5.42}$$

so that according to eqn (3.5.34),

$$S^{(z)}_{\boldsymbol{\ell}} = \hbar\,(S - a^*_{\boldsymbol{\ell}} a_{\boldsymbol{\ell}})\,. \tag{3.5.43}$$

Substituting from eqns (3.5.38) and (3.5.43) in eqn (3.5.27),

$$\begin{aligned}\mathcal{H} = &-\sum_{\boldsymbol{\ell},\boldsymbol{\ell}'}{}' \hbar^2 J_{\boldsymbol{\ell}\boldsymbol{\ell}'}\left[S\,(a_{\boldsymbol{\ell}} a^*_{\boldsymbol{\ell}'} + a^*_{\boldsymbol{\ell}} a_{\boldsymbol{\ell}'}) + (S - a^*_{\boldsymbol{\ell}} a_{\boldsymbol{\ell}})\,(S - a^*_{\boldsymbol{\ell}'} a_{\boldsymbol{\ell}'})\right] \\ &-\sum_{\boldsymbol{\ell}} g\mu_B \hbar H\,(S - a^*_{\boldsymbol{\ell}} a_{\boldsymbol{\ell}})\,.\end{aligned} \tag{3.5.44}$$

For $\boldsymbol{\ell} \neq \boldsymbol{\ell}'$, the operators commute, and $a^*_{\boldsymbol{\ell}} a_{\boldsymbol{\ell}'}$ may be replaced by $a_{\boldsymbol{\ell}'} a^*_{\boldsymbol{\ell}}$. If the *names* of $\boldsymbol{\ell}$ and $\boldsymbol{\ell}'$ are then interchanged in the summation, the second term in eqn (3.5.44) becomes identical to the first one. A similar argument applies to the terms linear in $S$ when opening the brackets of that equation. The term which contains the product of four $a_{\boldsymbol{\ell}}$ operators is neglected, because according to eqn (3.5.42) it is a product of two spin deviation operators $N_{\boldsymbol{\ell}}$. At low temperatures most of the spins are aligned, the deviations from the fully aligned state are small, and second-order terms are negligible. This argument can easily be made [39] more quantitative. One can even add [40] the neglected second-order term as a perturbation, and find out the range of validity of this approximation. It should be noted, though, that unlike the dropping of the second-order term, the approximation which has already been made in replacing eqn (3.5.39) by eqn (3.5.40) *cannot* be made quantitative, and there is not much point in quantifying one without the other. The result is

$$\mathcal{H} = C - 2\hbar^2 S \sum_{\boldsymbol{\ell},\boldsymbol{\ell}'}{}' J_{\boldsymbol{\ell}\boldsymbol{\ell}'}\,(a^*_{\boldsymbol{\ell}} a_{\boldsymbol{\ell}'} - a^*_{\boldsymbol{\ell}} a_{\boldsymbol{\ell}}) + \sum_{\boldsymbol{\ell}} g\mu_B \hbar H a^*_{\boldsymbol{\ell}} a_{\boldsymbol{\ell}}, \tag{3.5.45}$$

where

$$C = -\hbar^2 S^2 {\sum_{\boldsymbol{\ell},\boldsymbol{\ell}'}}' J_{\boldsymbol{\ell}\boldsymbol{\ell}'} - g\mu_B \hbar H S N, \tag{3.5.46}$$

and $N$ is the total number of ion sites, namely $\sum_{\boldsymbol{\ell}}$.

The basic assumption is that the ground state for a ferromagnet is the state in which all the spins are aligned along $z$. In that state, every $S_{\boldsymbol{\ell}}^{(z)}$ has its maximum eigenvalue, and there are no spin deviations. Therefore, no $a_{\boldsymbol{\ell}}$ operator can destroy any deviation in this state, which is denoted by $\Psi_0$. In other words, if $a_{\boldsymbol{\ell}}$ operates on this state, $\Psi_0$, the result is zero. And since all the terms in eqn (3.5.45), except for $C$, have an $a_{\boldsymbol{\ell}}$ on the right hand side,

$$(\mathcal{H} - C)\,\Psi_0 = 0. \tag{3.5.47}$$

Therefore,

$$\mathcal{H}\Psi_0 = C\Psi_0, \tag{3.5.48}$$

which means that the ground state $\Psi_0$ is an eigenstate of $\mathcal{H}$, and that $C$, as defined in eqn (3.5.46), is the *energy* of this state.

Spin deviations at any particular lattice point are not eigenstates of the Hamiltonian (3.5.45), because a creation at one lattice point, $\boldsymbol{\ell}$, is accompanied by a destruction at *another* lattice point, $\boldsymbol{\ell}'$. Therefore, the excited states are not localized on any one atom. They are made out of spin deviations which are propagated throughout the whole crystal. Its description thus calls for a theory which involves the crystal as a whole, for which one should take advantage of the *periodic structure* of crystalline solids. For that purpose, the $a_{\boldsymbol{\ell}}$ operators are expanded in a Fourier series, as is done [41] in the study of the normal modes of the lattice vibrations, or of any other property of solids. For $N$ unit cells in the lattice, and one atom per unit cell, the Fourier expansion is

$$a_{\boldsymbol{\ell}} = \frac{1}{\sqrt{N}} \sum_{\mathbf{q}} a_{\mathbf{q}} e^{i\mathbf{q}\cdot\boldsymbol{\ell}}, \qquad a_{\boldsymbol{\ell}}^* = \frac{1}{\sqrt{N}} \sum_{\mathbf{q}} a_{\mathbf{q}}^* e^{-i\mathbf{q}\cdot\boldsymbol{\ell}}, \tag{3.5.49}$$

where the summation is over all the allowed vectors $\mathbf{q}$ in the Brillouin zone of the reciprocal lattice, quantized according to periodic boundary conditions. As is the case in any other Fourier expansion, the inverted expansion is

$$a_{\mathbf{q}} = \frac{1}{\sqrt{N}} \sum_{\boldsymbol{\ell}} a_{\boldsymbol{\ell}} e^{-i\mathbf{q}\cdot\boldsymbol{\ell}}, \qquad a_{\mathbf{q}}^* = \frac{1}{\sqrt{N}} \sum_{\boldsymbol{\ell}} a_{\boldsymbol{\ell}}^* e^{i\mathbf{q}\cdot\boldsymbol{\ell}}. \tag{3.5.50}$$

From the commutation relation (3.5.40) it is seen that

$$\left[a_{\mathbf{q}}\,, a_{\mathbf{q}'}^*\right] = \frac{1}{\sqrt{N}} \sum_{\boldsymbol{\ell}} e^{-i\mathbf{q}\cdot\boldsymbol{\ell}}\, \frac{1}{\sqrt{N}} \sum_{\boldsymbol{\ell}'} \left[a_{\boldsymbol{\ell}}\,, a_{\boldsymbol{\ell}'}^*\right] e^{i\mathbf{q}'\cdot\boldsymbol{\ell}'}$$

$$= \frac{1}{N}\sum_{\boldsymbol{\ell}} e^{i(\mathbf{q}-\mathbf{q}')\cdot\boldsymbol{\ell}} = \delta_{\mathbf{q}\mathbf{q}'}, \tag{3.5.51}$$

because all the terms in the last sum vanish unless $\mathbf{q} = \mathbf{q}'$, in which case the sum is $N$. Substituting the Fourier expansion (3.5.49) in the Hamiltonian (3.5.45), using for the summation index $\mathbf{h} = \boldsymbol{\ell} - \boldsymbol{\ell}'$ instead of $\boldsymbol{\ell}'$ and noting that $J_{\boldsymbol{\ell}\boldsymbol{\ell}'}$ is actually a function of this *relative distance* $\mathbf{h}$ between $\boldsymbol{\ell}$ and $\boldsymbol{\ell}'$, and not of $\boldsymbol{\ell}$ and $\boldsymbol{\ell}'$ separately,

$$\mathcal{H} = C - 2\hbar^2 S \sum_{\boldsymbol{\ell},\mathbf{h}} J(\mathbf{h}) \frac{1}{N} \sum_{\mathbf{q}} a^*_{\mathbf{q}} e^{-i\mathbf{q}\cdot\boldsymbol{\ell}} \left[ \sum_{\mathbf{q}'} a_{\mathbf{q}'} e^{i\mathbf{q}'\cdot(\boldsymbol{\ell}-\mathbf{h})} - \sum_{\mathbf{q}'} a_{\mathbf{q}'} e^{i\mathbf{q}'\cdot\boldsymbol{\ell}} \right] + \sum_{\boldsymbol{\ell}} g\mu_B \hbar H \frac{1}{N} \sum_{\mathbf{q}} a^*_{\mathbf{q}} e^{-i\mathbf{q}\cdot\boldsymbol{\ell}} \sum_{\mathbf{q}'} a_{\mathbf{q}'} e^{i\mathbf{q}'\cdot\boldsymbol{\ell}}. \tag{3.5.52}$$

The summation over $\boldsymbol{\ell}$ in both terms is the same as the last sum in eqn (3.5.51), which is just a delta function. Therefore, after rearranging,

$$\mathcal{H} = C + \sum_{\mathbf{q}} \left[ 2\hbar^2 S \sum_{\mathbf{h}} J(\mathbf{h}) \left(1 - e^{-i\mathbf{q}\cdot\mathbf{h}}\right) + g\mu_B \hbar H \right] a^*_{\mathbf{q}} a_{\mathbf{q}}. \tag{3.5.53}$$

Thus $\mathcal{H} - C$ is nearly a set of harmonic oscillators, because $a_{\mathbf{q}}$ and $a^*_{\mathbf{q}}$ are readily seen to act as the destruction and creation operators in reciprocal space. The only difference is that the Hamiltonian of a harmonic oscillator is $a^*_{\mathbf{q}} a_{\mathbf{q}} + \frac{1}{2}$, whereas eqn (3.5.53) does not contain the $\frac{1}{2}$. Each of these oscillators is characterized by a vector $\mathbf{q}$ in reciprocal space, but they are *uncoupled* to each other, and each of them may be considered independently of the others. Therefore, the energy levels of each of the terms in the sum over $\mathbf{q}$ of $\mathcal{H} - C$ are those of a harmonic oscillator without the $\frac{1}{2}$. Adding the $C$ term, the energy levels of $\mathcal{H}$ are

$$\mathcal{E} = C + \sum_{\mathbf{q}} \mathcal{E}_{\mathbf{q}}, \tag{3.5.54}$$

where

$$\mathcal{E}_{\mathbf{q}} = n_{\mathbf{q}} \hbar \left[ 2\hbar S \sum_{\mathbf{h}} J(\mathbf{h}) \left(1 - e^{-i\mathbf{q}\cdot\mathbf{h}}\right) + g\mu_B H \right], \tag{3.5.55}$$

and $n_{\mathbf{q}}$ is a non-negative, integral number, which is the eigenvalue of $a^*_{\mathbf{q}} a_{\mathbf{q}}$. This $n_{\mathbf{q}}$ may be defined as the number of *spin wave quanta*, and the operators $a_{\mathbf{q}}$ and $a^*_{\mathbf{q}}$ destroy and create such spin wave excitations. Each of these elementary excitations is called a *magnon*. The form of eqn (3.5.54) demonstrates again that $C$ is the energy of the ground state, for which the number of magnons, $n_{\mathbf{q}}$, is zero for every $\mathbf{q}$.

Now that the energy levels are known, it is possible to construct the *partition function* from which the physical properties of a system in thermal equilibrium can be derived. Its general, statistical mechanics definition is given by eqn (10-14) of [42],

$$Z = \sum e^{-\beta \mathcal{E}_n}\,, \tag{3.5.56}$$

where $\beta$ is defined in eqn (1.3.12). The summation is over all the allowed quantum states $n$, whose energy is $\mathcal{E}_n$. From this function one can obtain *e.g.* the average internal energy per unit volume,

$$\overline{\mathcal{E}} = k_B T^2 \frac{\partial}{\partial T} \ln Z, \tag{3.5.57}$$

and the specific heat from its derivative, etc. The average component of the magnetic moment in the direction of the magnetic field is

$$M_z = k_B T \frac{\partial}{\partial H} \ln Z. \tag{3.5.58}$$

In the case of eqn (3.5.54) under study here, the partition function is thus

$$Z = e^{-\beta C} \sum_{n_{\mathbf{q}}} \prod_{\mathbf{q}} e^{-\beta \mathcal{E}_{\mathbf{q}}}\,, \tag{3.5.59}$$

where $\mathcal{E}_{\mathbf{q}}$ is defined in eqn (3.5.55). The notation $n_{\mathbf{q}}$ is just carried over here from the foregoing. Actually, the summation over each $n_{\mathbf{q}}$ is a sum over *all* the non-negative integers, and has nothing to do with any particular value of $\mathbf{q}$. Therefore, the order of sum and product in eqn (3.5.59) may be reversed, and the summation may be carried out first. The latter is a sum of a *geometric series*, leading to

$$Z = e^{-\beta C} \prod_{\mathbf{q}} \frac{1}{1 - e^{-\beta\hbar\left[2\hbar S \sum_{\mathbf{h}} J(\mathbf{h})\left(1 - e^{-i\mathbf{q}\cdot\mathbf{h}}\right) + g\mu_B H\right]}}\,. \tag{3.5.60}$$

Hence

$$\ln Z = -\frac{C}{k_B T} - \sum_{\mathbf{q}} \ln\left[1 - e^{-2\beta\hbar^2 S \sum_{\mathbf{h}} J(\mathbf{h})\left(1 - e^{-i\mathbf{q}\cdot\mathbf{h}}\right)}\, e^{-\beta g \mu_B \hbar H}\right]. \tag{3.5.61}$$

According to the definition in eqn (3.5.46),

$$\frac{\partial(-C)}{\partial H} = g\mu_B \hbar SN = M_0, \tag{3.5.62}$$

which is the magnetic moment obtained when all the $N$ spins are aligned along the field direction, $z$, as is the case at zero temperature. Substituting eqn (3.5.62) and eqn (3.5.61) in eqn (3.5.58), and carrying out the differentiation,

$$M_z = M_0 - g\mu_B \hbar \sum_{\mathbf{q}} \frac{1}{e^{2\beta\hbar^2 S \sum_{\mathbf{h}} J(\mathbf{h})\left(1-e^{-i\mathbf{q}\cdot\mathbf{h}}\right)} e^{\beta g\mu_B \hbar H} - 1}. \tag{3.5.63}$$

It is possible to continue this algebra in its general form a little further, but at some stage it will be necessary to specify the particular symmetry of the crystal under study, and it is somewhat clearer to do it at this stage. Other symmetries can be approached in a similar fashion, but the example given here refers specifically to a body-centred cubic, such as Fe, with an interaction between nearest neighbours only. In this case the summation over $\mathbf{h}$ contains only terms for which $|\mathbf{h}| = A\sqrt{3}/2$ (where $A$ is the cube edge) for each of which $J(\mathbf{h})$ is a universal constant, $J$. The atom at (0,0,0) has eight nearest neighbours, at $\frac{1}{2}(\pm A, \pm A, \pm A)$, so that

$$\mathbf{q}\cdot\mathbf{h} = \frac{1}{2}A(\pm q_x \pm q_y \pm q_z). \tag{3.5.64}$$

Also, this theory started by certain approximations which are only justified at low temperatures, and we may as well introduce another one, that the main contribution is from long wavelengths, namely small $\mathbf{q}$. The short wavelengths have a high energy, and it takes high temperatures to excite them. This argument can be easily made quantitative, because what is used is a power series expansion,

$$1 - e^{-i\mathbf{q}\cdot\mathbf{h}} = i\mathbf{q}\cdot\mathbf{h} - \frac{1}{2!}(i\mathbf{q}\cdot\mathbf{h})^2 + \cdots, \tag{3.5.65}$$

and the expansion may be carried out [40] to higher-order terms to check the effect of neglecting them. Here the series is cut off for simplicity at the quadratic term. Since the linear term in eqn (3.5.65) obviously sums to zero in the summation over $\mathbf{h}$, with equal $\pm$ terms, the approximation we use is

$$1 - e^{-i\mathbf{q}\cdot\mathbf{h}} \approx \frac{1}{2!}(\mathbf{q}\cdot\mathbf{h})^2 = \frac{A^2}{8}\left(q_x^2 + q_y^2 + q_z^2\right), \tag{3.5.66}$$

plus terms which sum up to 0. Therefore to this approximation,

$$\sum_{\mathbf{h}} J(\mathbf{h})\left(1 - e^{-i\mathbf{q}\cdot\mathbf{h}}\right) \approx \sum_{\mathbf{h}} JA^2q^2/8 = JA^2q^2, \tag{3.5.67}$$

because there are eight neighbours.

We also set $H = 0$, as is customary in this kind of calculation. All theories of magnetization *vs.* temperature deal only with the case of zero applied field, as has already been mentioned in section 2.6 and will be further discussed in section 4.1. Equation (3.5.63) then becomes

$$M_z = M_0 - g\mu_B\hbar \sum_{\mathbf{q}} \frac{1}{e^{2\beta\hbar^2 SJA^2q^2} - 1} \,. \qquad (3.5.68)$$

As is the case in all solid-state calculations, the summation over $\mathbf{q}$ may be replaced by an integral over the Brillouin zone in $\mathbf{q}$-space, provided the integrand is multiplied [41] by the density of states, $V/(8\pi^3)$, where $V$ is the volume of the crystal. However, since the exponent in the denominator contains $q^2$, the integrand is very small for large values of $|\mathbf{q}|$, and only a small error can be introduced if the integration is extended over the whole $\mathbf{q}$-space, instead of just over the Brillouin zone. In a way this argument also supplies a further justification to the approximation used in terminating the sum of eqn (3.5.65) at the quadratic term, because the contribution of higher orders in $\mathbf{q}$ is rather small. Thus

$$M_z = M_0 - \frac{g\mu_B\hbar V}{8\pi^3} \int_0^{2\pi}\int_0^{\pi}\int_0^{\infty} \frac{q^2 \sin\theta \, dq \, d\theta \, d\phi}{e^{2\beta\hbar^2 SJA^2q^2} - 1} \,. \qquad (3.5.69)$$

Obviously, the latter approximation is justified if the coefficient of $(qA)^2$ in the exponent is sufficiently large. This coefficient is about 3 for iron at room temperature, which is already rather large for an exponent. This value implies that replacing the Brillouin zone by the whole space is a good approximation for iron below, and probably up to, room temperature.

The integration over the angles in eqn (3.5.69) is straightforward. In the integral over $q$ the variable is replaced by $x = 2\beta\hbar^2 SJA^2q^2$. Also, for a bcc the volume can be written as $V = NA^3/2$, and the number of atoms, $N$, can be eliminated by using eqn (3.5.62)

$$M_0 = M_z(0) = g\mu_B\hbar SN, \qquad (3.5.70)$$

leading to

$$\frac{M_z(T)}{M_z(0)} = 1 - \frac{1}{8\pi^2 S}\left(\frac{k_B T}{2\hbar^2 SJ}\right)^{3/2} \int_0^{\infty} \frac{\sqrt{x}\, dx}{e^x - 1} \,. \qquad (3.5.71)$$

Thus, to a first order at low temperatures the deviation of the magnetization from its value at $T = 0$ is proportional to $T^{3/2}$, which is known as the Bloch law. It fits experiment for all known ferromagnets. This Bloch law has been derived here only for the particular case of bcc, but the derivation is essentially the same [40] for fcc or for simple cubic crystals, and the results differ only by a numerical factor. All these three cubic cases involve the *same* integral,

$$\int_0^\infty \frac{\sqrt{x}\,dx}{e^x - 1} = \frac{\sqrt{\pi}}{2}\zeta\left(\frac{3}{2}\right), \tag{3.5.72}$$

where $\zeta$ is the Riemann zeta function. In principle, the exchange integral, $J$, can be evaluated from the experimental value of the coefficient of $T^{3/2}$. However, this method never yields the same value of $J$ which is obtained from the ferromagnetic or the paramagnetic Curie temperature, as mentioned in chapter 2. The discrepancy is not surprising, because these measurements are done at different temperatures, and there is no reason to believe that $J$ is independent of the temperature. Even if there is no other effect, thermal expansion certainly changes the *distance* between the atoms with changing temperature. And it is obvious from the theory that the exchange integral, which depends on the overlap of the wave functions, must be *very* sensitive to this distance. It has also been demonstrated experimentally that both $J$ obtained from $T_c$ and $J$ obtained from the coefficient of the $T^{3/2}$ term change considerably when the distances among atoms are changed by a hydrostatic pressure [43, 44] or by other [45] means. Pressure is also known [46] to affect the hyperfine field of the Mössbauer effect.

This theory of the first-order term at low temperatures can be (and has been) extended to higher-order terms, as has been mentioned during the foregoing derivation. In particular, Dyson [40] continued the power series expansion of eqn (3.5.65) *and* introduced the magnon interaction as a first-order perturbation, to check which power of $T$ it affects. His result is

$$M/M_0 = 1 - a_0 T^{3/2} - a_1 T^{5/2} - a_2 T^{7/2} - a_3 T^4 + O(T^{9/2}), \tag{3.5.73}$$

with specific expressions given [40] for all these coefficients $a_i$ in all three types of cubic crystals. It is even possible [47] to remove some of the approximations of Dyson by the use of Green functions, and obtain what should be in principle a higher accuracy. The difficulty is that the expansion in eqn (3.5.73) does not fit experiment. Accurate data can be fitted better either with an empirical dependence [48] of $J$ in eqn (3.5.71) on $T$, or with a term with $T^2$ before [49] the $T^{5/2}$ term. The detailed empirical expression for the magnetization of iron whiskers that fits the whole range, from low temperatures and up to the Curie point [50],

$$m = \frac{M_z(T)}{M_z(0)} = \frac{(1-t)^\beta}{1 - \beta t + A t^{3/2} - C t^{7/2}}, \tag{3.5.74}$$

where $t = T/T_c$, and $\beta$, $A$ and $C$ are constants, expands to

$$m = 1 - A t^{3/2} + \frac{1}{2}\beta(\beta - 1)t^2 + O(t^{5/2}), \tag{3.5.75}$$

at low temperatures. It has been suggested [49] that the $T^2$ term originates from contributions of the collective electron ferromagnetism, and this idea was made [51] more quantitative later. Therefore, in this case, like in many others, the itinerant and localized electron theories must be combined together before extending either theory to a high accuracy. Also, measuring $M_z(T)$ and *extrapolating it to zero applied field* is not always very accurate, especially at low temperatures. In some cases the accuracy of the experimental data is not even sufficient to go beyond the first $T^{3/2}$ term of the Bloch law, and higher-order terms are mostly of interest to theorists who compare their results with each other's and not with experiments.

For *antiferromagnets* the situation is much more complicated, because even the ground state is not as simple and as clear cut as in the case of a ferromagnet. Approximations must be introduced already for the calculation of the spin waves at the ground level, and the excitations are hopelessly complicated. There are no conclusions which can be compared with a simple experiment, or any obvious improvement on the molecular field approximation presented in chapter 2. Therefore, this whole theoretical field is beyond the scope of this book.

Other theories which use the Heisenberg Hamiltonian of eqn (3.5.25) are not included in this chapter because they either use classical physics, or at least can be *outlined* without specific mention of quantum-mechanical techniques. One is the molecular field approximation, already described in chapter 2. The others will be considered in the next chapter. However, before concluding this discussion of spin waves, there is one important conclusion from the above treatment, which must be emphasized. The integral in eqn (3.5.69) contains the factor $q^2$ in the numerator only in three dimensions. In two dimensions the factor $d\mathbf{q}$ would have been $qdqd\theta$. In this case (or in one dimension) the integrand with $e^x - 1$ in the denominator will *diverge* in the vicinity of $x = 0$, namely near $\mathbf{q} = 0$. Therefore any small perturbation of the ground state will blow up, even at very low temperatures where all the approximations are justified, and the above calculation is rigorous. In other words, *a ferromagnetic ordering is not possible in one or two dimensions.* This proof that ferromagnetism (or antiferromagnetism, for that matter) is possible only in three dimensions was already given by Bloch in 1930. It must be noted that it is a fundamental property, which does not depend on any approximation. The singularity at $x = 0$ will be there even if the integral is over the Brillouin zone, and not over the whole space; and the other approximations only require a sufficiently low temperature. In practice ferromagnetism *has* been observed in some seemingly one- and two-dimensional systems, to be discussed in section 4.5.

# 4

# MAGNETIZATION *VS.* TEMPERATURE

## 4.1 Magnetic Domains

Before continuing with the theories, it seems necessary to pause and explain why all the molecular field approximation of chapter 2, the spin wave series for low temperatures of chapter 3, as well as *all* other theories of $M_z(T)$, are restricted to the case $H = 0$. For a beginner it must seem natural to introduce a magnetic field, at least as a first-order perturbation, and indeed there is no particular *theoretical* difficulty in doing so. The difficulty is that including a magnetic field without any other modification of the Heisenberg Hamiltonian (or of the energy of eqn (2.2.25) for the reader who has skipped chapter 3) does not have any physical significance, because the results of such a calculation cannot be compared with experiment. This fact is sometimes forgotten by theorists, which makes it even more important to keep mentioning it.

The point is that real ferromagnets at zero applied fields are subdivided into *domains* which are magnetized in different directions. In other words, the direction of quantization, $z$, changes between one domain and another. The *reason* for the existence of these domains must obviously be a term of the Hamiltonian which has been neglected so far, but it is too early at this stage to specify what this term is, and it will be further discussed in section 6.2. However, the very existence of these domains is a well-established experimental fact, as has already been mentioned in section 1.2. These domains cannot be ignored, because they are being *observed* by several techniques.

The older observations [52, 53] include the Bitter pattern, in which tiny magnetic particles, immersed in a liquid, are attracted to the high surface field at the *walls* which separate the domains, thus revealing the location of these walls. In metallic films which are thin enough for high-energy electrons to go through, electrons are deflected by the magnetic field in the domains when they pass through them, thus revealing the location of these domains. Similarly, the domain structure on the surface can be seen by electrons that pass near that surface, or that are reflected from the surface. Polarized light reflected from a magnetized surface changes its plane of polarization (Kerr effect) and its detection reveals the different orientations of the magnetization on the surface of the various domains. If the sample is thin enough, the light can pass *through* it and the rotation of its polarization (Faraday effect) shows the direction of magnetization inside

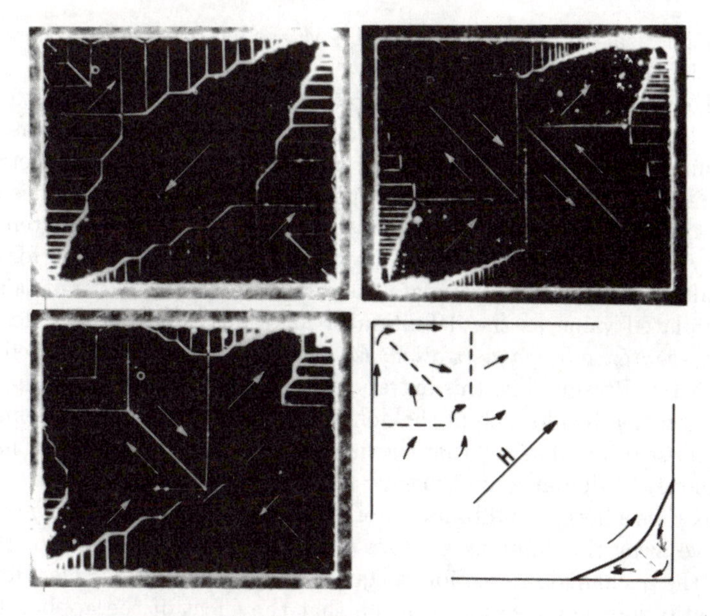


FIG. 4.1. Three different domain configurations in the *same* Ni platelet in zero applied field, after different *histories* of applying and removing a magnetic field. Reproduced from Fig. 9 of [58] by permission.

these domains. Another method was based on scanning the magnetic field near the sample by a very small Hall probe [53]. More recent methods include *e.g.* passing a current through the sample, and measuring the Hall effect at different points [54], where oppositely magnetized domains give an opposite sign of the Hall voltage. They also include scanning electron microscopy with polarized electrons [55] and magnetic force microscopy [56] which allow the study of these domains almost down to atomic size, as well as scanning optical microscopy [57] which increases the resolution of the Kerr effect. These, and other, techniques thus leave no doubt that all bulk ferromagnets are made out of domains, magnetized in different directions, until a sufficiently large field is applied to remove them.

Obviously, the magnetization measured in zero (or small) applied field is an average of its value in the different domains, and has nothing to do with the theoretical value of $M_z(T)$. Moreover, the measured value is not even *unique*, because the domain structure in zero (or small) applied field is not unique. A good example is shown in Fig. 4.1 which is reproduced from Fig. 9 of [58]. It shows three completely different domain configurations observed in the *same* crystal, after subjecting it to a different *history* of the applied field. There are many other possibilities which are not shown, and which can give rise to a different value of the measured remanent magnetization. Actually, in zero field it is possible to measure any value of the magnetization between $-M_r$ and $+M_r$, as has already been mentioned in section 1.1.

Theorists who calculate magnetization *vs.* temperature prefer to ignore these domains, because it is too difficult to take them into account. Their reasoning is that what they calculate in these theories is the magnetization *inside* each of these domains. This quantity is what would have been measured if one domain could be separated from the others, and extended to infinity [59]. The experimentalist's approach is to measure $M_H$ for different values of $H$, then extrapolate the data down to $H = 0$, and take that extrapolated value as the definition of $M_s$. This is the proper definition of the *spontaneous magnetization*, $M_s$, which was somewhat ill-defined in section 1.1. Presumably, this extrapolation should lead to the value of the magnetization inside each of the domains, as in the theoretical definition.

It must be emphasized that the fields applied in the experiment in order to remove the domains are typically a small perturbation, if added as such to the Heisenberg Hamiltonian. For example, in iron the field necessary to drive away the domains *at room temperature* is of the order of $10^3$ Oe, while the exchange interaction is equivalent to a field of $10^6$ Oe. However, this ratio does not necessarily mean that the effect of the applied field is just a third-digit correction, as it seems at first sight. In the first place, the applied field may sometimes be large enough to *create* an appreciable magnetization by distorting the atomic electron orbits. Such an effect is *not* included in the Heisenberg Hamiltonian that assumes fixed spins at the lattice points. But even when this effect is negligible, the mere arrangement of domains has a very large effect on the measured values of the magnetization, and this process is not even linear. Therefore, in many cases a linear extrapolation of the field to zero is not adequate, leading to a relatively large uncertainty in the experimental value at zero field. This uncertainty is very often of the order of the difference between the different theories which are compared with that experiment, and it certainly would not allow any extrapolation to a *non-zero* field. Near the Curie temperature there is a more reliable extrapolation technique, which allows a sufficiently high accuracy by a method known as the *Arrott plots*, which will be described in section 4.6. However, even there the extrapolation works satisfactorily to zero fields only, and there are no experimental data for a small field that can theoretically be added as a perturbation.

It should also be emphasized that all these theories assume an *infinite* crystal. Therefore, they can have a chance of representing the physical reality only if the size of these domains is *much larger* than the *correlation length*, which is the average distance over which fluctuations of the magnetization are correlated. This is not always the case, *especially* near the Curie temperature, where the correlation length diverges.

The magnetization *inside* a domain can in principle be measured, in accordance with the theoretical definition, by using the Mössbauer effect.

In this experiment, a resonant absorption of $\gamma$-rays is obtained when the momentum of the recoil is transferred to the whole crystal instead of the individual nucleus. When either the source or the absorber is in a magnetic field, the energy levels of the nuclei are split by the Zeeman effect, and it is possible to determine from this splitting the value of the effective field at the nucleus. In a ferromagnet, this effective field (called the *hyperfine field*) at the nucleus is essentially proportional to the *magnitude* of the magnetization of the atomic shell around that nucleus. The effect can be observed only in certain isotopes, one of which is $^{57}$Fe, which is particularly convenient for studying ferromagnets.

Since this hyperfine field is proportional to the *magnitude* of the magnetization of the atom to which this nucleus belongs, and is independent of the *direction* of the atomic magnetization, contributions from the $^{57}$Fe nuclei in different domains are the same, and add up. For the same reason, the hyperfine field in an antiferromagnet is the same as that of a ferromagnet. Measurement of the hyperfine field as a function of the temperature thus yields [60] a result proportional to the magnetization in Fig. 2.1 here, and this measurement is indeed carried out in zero applied field. It should be noted that, even though *nuclear* energy levels are studied, the nuclear spin does not enter the calculation of the magnetization: its contribution is negligible compared with that of the spin of the electron, because the Bohr magneton is inversely proportional to the mass. The nuclear spin in this experiment is essentially just a probe used to measure the magnetization due to the *electron* spin.

Similar data can be obtained from measuring [61, 62] the nuclear magnetic resonance. However, both the Mössbauer effect *and* the nuclear magnetic resonance can at best be used for the philosophical definition of the spontaneous magnetization. The accuracy of $M_s(T)$ measured by these techniques is rather poor and inadequate. For good quantitative data one must rely on the measurement of $M_z(H, T)$, and its extrapolation to $H = 0$, as *e.g.* in [48]. Therefore, there is no way to measure the magnetization for $H = 0$, with any reasonable accuracy, and the calculation of such a quantity has no physical meaning, because it cannot possibly be compared with any experiment.

## 4.2 The Landau Theory

The temperature range most popular among theorists is the approach to, and the near vicinity of, the Curie temperature, $T_c$, because in this region it is possible to use first-order approximations in powers of $|T - T_c|$. This approach started with the phenomenological Landau theory [63] which applies to all sorts of phase transitions of the second kind. It will be described here only for the specific case of the magnetization going through the Curie point. The notation

$$m = \frac{M_z(T)}{M_z(0)}, \qquad t = \frac{T}{T_c}, \tag{4.2.1}$$

is used here for brevity.

The basic assumption is that $m$ is small near $T_c$, which makes it possible to expand the thermodynamic potential per unit volume as a power series in $m$. Neglecting higher-order terms, and assuming that there is no magnetic field, this expansion is

$$\Phi(t, m) = \Phi_0(t) + A(t)m^2 + B(t)m^4. \tag{4.2.2}$$

The odd powers, $m$ and $m^3$, are omitted, because $\Phi$ must remain unchanged by a time reversal, which changes the sign of $m$. The coefficients $\Phi_0$, $A$, and $B$ can in principle be functions of other physical properties, which are ignored here. In particular, it is assumed that everything is done at a *constant pressure* so that it is not necessary to specify the dependence on pressure, which is an important part of the Landau theory of *other* phase transitions.

The value of $m$ should be such that the thermodynamic potential is a minimum. A necessary condition is that the first derivative vanishes, namely

$$\frac{\partial \Phi}{\partial m} = 2m\left[A(t) + 2m^2 B(t)\right] = 0. \tag{4.2.3}$$

Also, in order for the solution of eqn (4.2.3) to be a minimum and not a maximum, the second derivative should be positive,

$$\frac{1}{2}\frac{\partial^2 \Phi}{\partial m^2} = A(t) + 6m^2 B(t) > 0. \tag{4.2.4}$$

The solution of eqn (4.2.3) for the region above the Curie point, $t > 1$, is $m = 0$, and in order for this solution to fulfil the requirement of eqn (4.2.4),

$$A(t) > 0, \quad \text{for} \quad t > 1. \tag{4.2.5}$$

The other solution of eqn (4.2.3), which is valid for $t < 1$, is

$$m^2 = -\frac{A(t)}{2B(t)}, \tag{4.2.6}$$

which yields, when substituted in eqn (4.2.4),

$$\frac{\partial^2 \Phi}{\partial m^2} = -4A(t) > 0. \tag{4.2.7}$$

Therefore, noting also that the left hand side of eqn (4.2.6) is positive,

$$A(t) < 0, \quad \text{and} \quad B(t) > 0, \quad \text{for} \quad t < 1. \tag{4.2.8}$$

This result means first of all that $m = 0$ is a solution of eqn (4.2.3) also in the region $t < 1$, but there it is a maximum and not a minimum. The combination of eqns (4.2.8) and (4.2.5) also means that a continuity of $A$ at $t = 1$ requires that $A(1) = 0$. The simplest function (and the first-order approximation for *any other* function) which can fulfil these conditions is obviously

$$A(t) = a(t-1), \quad a > 0, \tag{4.2.9}$$

where $a$ is a constant. From this result it is already possible [63] to draw some conclusions about the entropy, $S = -\partial\Phi/\partial T$, and the specific heat, $C_p = T(\partial S/\partial T)$, in which there turns out to be a *jump* at the Curie point.

Let there now be a magnetic field, $H$, applied to the system. Its energy of interaction with the magnetization is $-\mathbf{M} \cdot \mathbf{H}$ per unit volume, which is $-mHM_z(0)$ in the present notation. With

$$h = HM_z(0), \tag{4.2.10}$$

the thermodynamic potential per unit volume becomes

$$\Phi(t,m) = \Phi_0(t) + a(t-1)m^2 + B(t)m^4 - hm, \qquad a > 0, \tag{4.2.11}$$

where the sign of $a$ is kept from the foregoing, for the limit $h = 0$. By the same token, the sign of $B(t)$ should also be kept as in eqn (4.2.8). The condition that the first derivative vanishes is now

$$\frac{1}{2}h = a(t-1)m + 2B(t)m^3, \tag{4.2.12}$$

which is a cubic algebraic equation for determining $m$.

Above the Curie temperature, the right hand side of eqn (4.2.12) is a monotonic increasing function of $m$. Therefore, for every value of $h$ there is a *single* solution of this equation, and it tends to 0 for $h \to 0$. For $T < T_c$, namely $t < 1$, the first term on the right hand side of eqn (4.2.12) is negative, and the equation has three different real solutions, $m$, for a certain region of not too large $|h|$. One of these solutions is easily seen to be a maximum and not a minimum. Of the other two, one has $m$ antiparallel to $h$, and therefore its energy is higher than that of the solution in which $m$ has the same sign as $h$.

According to the definition in eqn (2.1.22), the initial susceptibility is

$$\chi_{\text{initial}} = \lim_{H\to 0} \frac{\partial\langle M_z\rangle}{\partial H} = [M_z(0)]^2 \lim_{h\to 0} \frac{\partial m}{\partial h}, \tag{4.2.13}$$

in the present notation. Substituting from the derivative of eqn (4.2.12) with respect to $h$,

$$\chi_{\text{initial}} = \lim_{h \to 0} \frac{[M_z(0)]^2}{2\,[a(t-1) + 6B(t)m^2]}\,. \tag{4.2.14}$$

If $h \to 0$, then $m$ is zero for $t > 1$, and is given by eqn (4.2.6) for $t < 1$. Therefore,

$$\chi_{\text{initial}} = \begin{cases} \frac{[M_z(0)]^2}{2a(t-1)} & \text{if } t > 1\,, \\ \frac{[M_z(0)]^2}{4a(1-t)} & \text{if } t < 1\,. \end{cases} \tag{4.2.15}$$

This divergence of the susceptibility at $t \to 0$ is the same as in the Curie–Weiss law, discussed in section 2.4.

The spontaneous magnetization for $t < 1$ at zero field is, according to eqns (4.2.6) and (4.2.9),

$$m_{\text{sp}} = \sqrt{\frac{a(1-t)}{2B(t)}}\,. \tag{4.2.16}$$

The magnetization induced by the field, $M_{\text{ind}} = \chi H$, is in the present notation

$$m_{\text{ind}} = \chi \frac{h}{[M_z(0)]^2} = \frac{h}{4a(1-t)}\,, \tag{4.2.17}$$

according to eqn (4.2.15). These quantities are of the same order at

$$h_t \approx \frac{[2a(1-t)]^{3/2}}{B^{1/2}}\,. \tag{4.2.18}$$

Hence, a field $h \ll h_t$ is a 'weak' field, in the sense that it does not change the thermodynamic properties of the sample. In a field $h \gg h_t$ the thermodynamic properties have values which are determined by the field, and such a field is thus a 'strong' field. Of course, this criterion ignores the effect of the field on the measured $m$ by rearranging the domains, It is obvious from see section 4.1. eqn (4.2.18) that the transition field, $h_t$, vanishes at the point $t = 1$, which is $T = T_c$. Therefore, at the Curie point *any* field is a strong field according to this definition.

## 4.3 Critical Exponents

More modern studies of the *critical region*, namely the near vicinity of the Curie point, are based on two general assumptions, or axioms. The first one is that the *asymptotic* behaviour governing the approach to the critical point (*i.e.* the Curie temperature) of all physical parameters is a power law in $|t-1|$, where $t$ is as defined in eqn (4.2.1). Strictly speaking, this statement does not necessarily mean that any particular physical quantity is *proportional* to a certain power of $|t-1|$. It only means that it *varies as* that power, which is more general than proportionality.

The mathematical definition is that if

$$\lim_{x \to 0} \frac{\ln f(x)}{\ln x} = \lambda, \tag{4.3.19}$$

we say that $f(x)$ varies as $x^\lambda$ when $x$ tends to zero. This statement is written as

$$f(x) \sim x^\lambda \quad \text{as} \quad x \to 0. \tag{4.3.20}$$

The *simplest* possibility for which eqn (4.3.19) is fulfilled is when for small $x$,

$$f(x) = Cx^\lambda(1 + c_1x + c_2x^2 + \cdots), \tag{4.3.21}$$

where $C$, $c_1$, $c_2$, etc., are constants. However, eqn (4.3.19) is also fulfilled in more complicated cases, like *e.g.* if for small $x$,

$$f(x) = C\,|\ln x|^\mu\, x^\lambda(1 + c_1x^\nu + \cdots). \tag{4.3.22}$$

The particular case when the exponent $\lambda$ vanishes may mean that $f(x)$ tends to a *constant* for $x \to 0$, but it may also mean that $f(x) \propto \ln x$.

In accordance with this basic assumption, for the limit $t \to 1$ several *critical exponents* (or *critical indices*) are defined. In particular, for the specific heat,

$$C_p \sim |t - 1|^{-\alpha}, \tag{4.3.23}$$

for the spontaneous magnetization below $T_c$,

$$m \sim (1 - t)^\beta, \tag{4.3.24}$$

and for the initial susceptibility

$$\chi_{\text{initial}} \sim |t - 1|^{-\gamma}. \tag{4.3.25}$$

General thermodynamic considerations can be used to prove [63] that it is the *same* exponent $\alpha$ for the approach of $t$ to 1 from above or from below, and similarly for $\gamma$. Such considerations also impose certain relations between these, and the other, critical exponents. Thus, for example, the induced magnetization is

$$m_{\text{ind}} \sim h\chi \sim h|1 - t|^{-\gamma}, \tag{4.3.26}$$

according to eqn (4.3.25). Using eqn (4.3.24), the field at which this magnetization is of the order of the spontaneous magnetization is

$$h_t \sim |1 - t|^{\beta+\gamma}. \tag{4.3.27}$$

On the other hand, at this transition between a strong and a weak field the *energy* of interaction of the field with the magnetization, $-hm$, should be of the order of the thermal energy, which is of the order of $(1-t)^2 C_p$, because $C_p = -T(\partial^2\Phi/\partial T^2)$. Therefore,

$$h_t \sim |1-t|^{2-\beta-\alpha}. \tag{4.3.28}$$

Combining this equation with eqn (4.3.27) leads to

$$\alpha + 2\beta + \gamma = 2. \tag{4.3.29}$$

In principle, the power laws in eqns (4.3.23)–(4.3.25) are based [64] on experimental observations, and are not just an arbitrary assumption. The experimental values are [64] $\beta \approx 1/3$ and $\gamma \approx 4/3$, whereas the Landau theory, with its particularly oversimplified assumption of eqn (4.2.9), leads to $\beta = 1/2$ according to eqn (4.2.16) and $\gamma = 1$ according to eqn (4.2.15). The molecular field approximation also gives $\beta = 1/2$ and $\gamma = 1$, as will be clarified in section 4.6, or as can also be seen from eqns (2.2.35) and (2.4.47) respectively. This discrepancy illustrates the need for a more sophisticated theoretical approach, and indeed there are more accurate theories of these, and of the other, critical exponents in ferromagnetism as well as in other critical phenomena near their phase transitions. It should be noted, however, that in principle it is not clear if an expansion in powers of $T - T_c$ should be valid *a little* away from $T_c$. The value of $T_c$ itself is determined by the exchange integral, $J$, and the latter may change with temperature, at least because of the thermal expansion which varies the atomic distances. If $J$ varies with temperature, so does the *apparent* value of $T_c$ to which measurements at somewhat lower temperatures seem to lead, thus distorting the apparent value of the critical exponents. Also, in practice it is not always easy to resolve the leading asymptotic term from experimental data, especially when the behaviour is of the type of eqn (4.3.22) here. The next order 'correction' to the leading term, a little away from the Curie point, may be large enough to change the apparent value of the leading term.

Theorists are never concerned with these difficulties, saying that the measurements should be restricted to the very close vicinity of $T_c$, but that is often impractical. It has been noted [65] that the best fit of $\delta = 1 + \gamma/\beta$ varies between 4.2 and 4.7 in the near vicinity of $T_c$. A likely artifact of handling the experimental data [66] can look like a change in the critical exponents when $T_c$ is approached. It is better to use a proper equation of state, over a relatively wide temperature range, as explained in section 4.6. It is possible, of course, to remove most of the experimental data, saying that they are not close enough to $T_c$, but when very few points are left to look at, the data can be fitted to almost any value of the critical exponent. Unfortunately there are indeed some experimentalists who force their data by this technique to fit the current theoretical value, so that in this particular field it is often difficult to say what the experimental result is. On top of that, the theories always consider only the 'bulk' limit, in which the volume of the system is infinite. There is nothing fundamental in this approach, which is only a matter of convenience, but it must be always borne in mind that it may distort the asymptotic behaviour very considerably. It is not only that the sample under study must be very large for such a theory to be a good approximation to reality. It is the size of *each domain* that must be large enough for such a theory to be accurate enough, and such a requirement is hardly ever met. Some of the critical exponents may be more reliable when they are obtained from the analogy with other critical phenomena that do not have the equivalent of a magnetic field and magnetic domains. These are beyond the scope of the present book, which deals only with ferromagnetism, and not with general statistical mechanics.

The second basic assumption of the theories of critical exponents is known as the *scaling hypothesis.* It assumes first of all the existence of a *correlation length*

$$\xi \sim |t-1|^{-\nu}, \tag{4.3.30}$$

which measures the average distance over which fluctuations of the magnetization are correlated. It further assumes that in the critical region, the dominating temperature-dependence of all the physical properties of the system is only through their dependence on this $\xi$. If the length scale is increased by a certain factor, the correlation length shrinks by the same factor. The temperature region, $t-1$, then increases according to eqn (4.3.30), and all the physical properties will also change by fixed power laws. However, $\xi \to \infty$ for $t \to 1$, and the scaled system can be *renormalized*, namely be mapped back on the original one. This procedure is the basis of an important tool for calculating the critical exponents, known as the *renormalization group* theory [67].

Theories of critical exponents use a general space in $d$ dimensions (3 in real space), and a spin vector which has $n$ components (3 in real life). Except for several particularly simple cases, such as the Ising model in one and two dimensions (to be discussed in the next section, 4.4) or the Landau theory etc., the mathematics is complicated. However, it turns out that the problem is very much simplified in the unphysical conditions of a very large $n$ or $d = 4$. Therefore, some power series have been developed which should be a good approximation for a large $n$ and a small

$$\epsilon = 4 - d. \tag{4.3.31}$$

These, as well as more accurate methods, have been reviewed by Fisher [67] and later updated [68] to a certain extent. They are all beyond the scope of this book.

## 4.4 Ising Model

A very popular method for studying the Heisenberg Hamiltonian is an approximation which has already been suggested in the 1925 doctoral thesis of Ising. It is based on leaving out the non-diagonal terms of the spin matrices, and keeping only the components along the field direction, $z$. It means replacing $\mathbf{S}_1 \cdot \mathbf{S}_2 = S_{1x}S_{2x} + S_{1y}S_{2y} + S_{1z}S_{2z}$ by only $S_{1z}S_{2z}$. And since the latter commute, it effectively means dealing with numbers instead of matrices. It also means that instead of the Hamiltonian (3.5.25), or (2.2.25), the *energy* of the system is taken to be

$$\mathcal{E} = -\sum_{\ell,\ell'}{}' J_{\ell\ell'}\sigma_\ell\sigma_{\ell'} - g\mu_B H \sum_\ell \sigma_\ell , \tag{4.4.32}$$

where every lattice point is characterized by a quantum *number* $\sigma_\ell$. I will only mention in passing that there are also theories which do the opposite: leave out $S_z$ and keep only $S_x$ and $S_y$. This assumption, or approximation, is called the *XY-model* and will not be described here.

In principle the Ising model is not a very good approximation for any temperature range. However, it has the advantage of starting directly from the energy levels, and skipping all the steps that lead to them from the Hamiltonian, in other methods. This convenient short-cut makes it possible to concentrate on the details of the statistical mechanics. Therefore, the Ising model is very widely used in a variety of other problems, more than in ferromagnetism for which it was originally developed. For example, in a binary alloy made of atoms $A$ and $B$, one can define for the lattice point $\ell$ the value $\sigma_\ell = +1$ if there is an atom of the type $A$ there, and $\sigma_\ell = -1$ if there is an atom of the type $B$ there. If the interaction is between nearest neighbours only, and if $v_{AA}$ is the potential energy between two neighbours of type $A$, and similarly for $v_{AB}$ and $v_{BB}$, it is readily seen that the energy of any distribution of these atoms is given by eqn (4.4.32), with

$$J = \frac{1}{2}v_{AB} - \frac{1}{4}v_{AA} - \frac{1}{4}v_{BB}, \tag{4.4.33}$$

but without $H$ which has no analogy here. It is thus possible to study theoretically the order–disorder transition (analogous to the Curie or Néel temperature). At high temperatures there is a complete disorder, while below the transition $A$ is regularly a neighbour of $B$ for $J < 0$, and there is a separation to regions of almost pure $A$ and almost pure $B$ for $J > 0$. There is a slight difference from the ferromagnetic (or antiferromagnetic) case, in which the direction of the spin can be reversed at any atomic site, while in the case of an alloy, $A$ cannot be converted into $B$. In this case there is thus the additional constraint that the *total number* of atoms of each type must be conserved. However, the mathematical technique is sufficiently similar to make it the same class of problems. A variation of this case includes the possibility that $A$ is an atom while $B$ is a lattice vacancy, which leads to the thermodynamics of the transition from a solid to a liquid or a gas. There are also other physical problems of cooperative phenomena and phase transitions for which the Ising model is used, which makes it belong more in a book on statistical mechanics than in one on ferromagnetism. However, it is historically a part of ferromagnetism, and cannot be skipped altogether. Besides, the problem has an easy and elegant solution, at least in one dimension, which is well worth noting.

I will restrict the following to interactions between nearest neighbours only, and for the case of spin $\frac{1}{2}$, for which each of the numbers $\sigma_\ell$ can assume either the value $+1$ or the value $-1$. There are some more general studies in the literature, but they are rather complicated. I will also restrict this section to the case of a *one-dimensional* chain, made out of $N$ spins. The spin at the point $\ell$ interacts with the one at $\ell+1$ and the one at $\ell-1$; but since $J$ is assumed to be the same, the *sum* of all the interactions of spins with the one before them is the same as the sum of interactions with the spin after them. Periodic boundary conditions are also assumed here, namely that the spin at point $N$ interacts with the one at point 1. Equation (4.4.32) thus becomes for this case

$$\mathcal{E} = -2J \sum_{\ell=1}^{N} \sigma_\ell \sigma_{\ell+1} - g\mu_B H \sum_{\ell=1}^{N} \sigma_\ell . \tag{4.4.34}$$

This energy is now substituted in the partition function of eqns (3.5.56) and (1.3.12). Only for the sake of the reader who has skipped chapter 3, I will remark that the partition function is a general statistical mechanics function, made out of the energy levels, from which it is possible to derive all the physical properties of a system in thermal equilibrium. For the energy of eqn (4.4.34) this function is

$$Z = \sum_{\sigma_1=\pm 1} \cdots \sum_{\sigma_N=\pm 1} \prod_{\ell=1}^{N} e^{K\sigma_\ell \sigma_{\ell+1} + 2h\sigma_\ell} , \tag{4.4.35}$$

with the notation

$$K = \frac{2J}{k_{\mathrm{B}}T} , \qquad h = \frac{g\mu_B H}{2k_{\mathrm{B}}T} . \tag{4.4.36}$$

Here $k_{\mathrm{B}}$ is the Boltzmann constant, and it is hoped that neither $K$ nor $h$ is confused with these letters used in different meanings in other sections. For the sake of symmetry, $2h\sigma_\ell$ is rewritten as $h\sigma_\ell + h\sigma_{\ell+1}$, because the product over the second term is the same as that over the first one. Equation (4.4.35) is then

$$Z = \sum_{\sigma_1=\pm 1} \cdots \sum_{\sigma_N=\pm 1} \prod_{\ell=1}^{N} e^{K\sigma_\ell \sigma_{\ell+1} + h\sigma_\ell + h\sigma_{\ell+1}} . \tag{4.4.37}$$

The method used here to evaluate $Z$ is more complex than is essential for the one-dimensional problem. There is an easier method, but it works only if $h = 0$ is assumed already at this stage, and it cannot be generalized to two dimensions without complex numerical computations. The method which I am going to describe can be extended to two (but not three) dimensions almost without any change, except that in two dimensions the analytic solution can only be given for $h = 0$. It is actually a one-dimensional formulation of the famous Onsager analytic solution of the two-dimensional Ising model, published in 1944. This solution is considered to be the real breakthrough, and the beginning of all modern statistical mechanics, which makes it worth studying.

This method simplifies $Z$ by defining a $2 \times 2$ matrix whose elements are

$$\langle \sigma_\ell | M | \sigma_{\ell'} \rangle = e^{K\sigma_\ell \sigma_{\ell'} + h\sigma_\ell + h\sigma_{\ell'}} . \tag{4.4.38}$$

When $\sigma_\ell$ and $\sigma_{\ell'}$ pass through the allowed values $\pm 1$, the matrix elements keep going, in a different order, through the elements of the *same* matrix,

$$M = \begin{pmatrix} e^{K+2h} & e^{-K} \\ e^{-K} & e^{K-2h} \end{pmatrix} \tag{4.4.39}$$

where the order chosen for this particular presentation is

$$\begin{pmatrix} ++ & +- \\ -+ & -- \end{pmatrix} .$$

Now, according to the rule for multiplying matrices,

$$\sum_{\sigma_2=\pm 1} \langle \sigma_1 | M | \sigma_2 \rangle \langle \sigma_2 | M | \sigma_3 \rangle = \langle \sigma_1 | M^2 | \sigma_3 \rangle , \tag{4.4.40}$$

$$\sum_{\sigma_2=\pm 1} \sum_{\sigma_3=\pm 1} \langle \sigma_1 | M | \sigma_2 \rangle \langle \sigma_2 | M | \sigma_3 \rangle \langle \sigma_3 | M | \sigma_4 \rangle =$$

$$\sum_{\sigma_3=\pm 1} \langle \sigma_1 | M^2 | \sigma_3 \rangle \langle \sigma_3 | M | \sigma_4 \rangle = \langle \sigma_1 | M^3 | \sigma_4 \rangle , \tag{4.4.41}$$

etc. It is possible to introduce the full formalism of mathematical induction, but even without doing so it should be quite clear by now that when the definition (4.4.38) is substituted in eqn (4.4.37),

$$Z = \sum_{\sigma_1=\pm 1} \cdots \sum_{\sigma_N=\pm 1} \prod_{\ell=1}^{N} \langle \sigma_\ell | M | \sigma_{\ell+1} \rangle \,, \tag{4.4.42}$$

this matrix multiplication leads to

$$Z = \sum_{\sigma_1=\pm 1} \sum_{\sigma_N=\pm 1} \langle \sigma_1 | M^{N-1} | \sigma_N \rangle \langle \sigma_N | M | \sigma_1 \rangle \,. \tag{4.4.43}$$

The second index of the last term is taken here as 1, because for the periodic boundary conditions it is the same as $N+1$ which appears in eqn (4.4.37). Using again the matrix multiplication rule on the two remaining matrices in eqn (4.4.43),

$$Z = \sum_{\sigma_1=\pm 1} \langle \sigma_1 | M^N | \sigma_1 \rangle = \mathrm{trace}\,(M^N). \tag{4.4.44}$$

The matrix $M$ in eqn (4.4.39) is symmetric, and can be diagonalized. However, for the sake of those readers who may not be familiar with the trace of a power of a matrix, let us consider first a transformation $T$ which diagonalizes a general matrix, $M$, namely

$$T^{-1} M T = \begin{pmatrix} \lambda_1 & 0 \\ 0 & \lambda_2 \end{pmatrix}. \tag{4.4.45}$$

Multiplying both sides of the equation on the right by $T^{-1}MT$ leads to

$$T^{-1} M^2 T = \begin{pmatrix} \lambda_1 & 0 \\ 0 & \lambda_2 \end{pmatrix} T^{-1} M T = \begin{pmatrix} \lambda_1 & 0 \\ 0 & \lambda_2 \end{pmatrix}^2 = \begin{pmatrix} \lambda_1^2 & 0 \\ 0 & \lambda_2^2 \end{pmatrix}, \tag{4.4.46}$$

which can obviously be generalized to higher powers. Since the trace does not change by such a transformation, eqn (4.4.44) can be written as

$$Z = \lambda_1^N + \lambda_2^N \,, \tag{4.4.47}$$

where $\lambda_1$ and $\lambda_2$ are the eigenvalues of the matrix defined in eqn (4.4.39).

It can be safely assumed that this theory is only used for very large values of $N$, and even *nearly* equal numbers become very different when raised to a large power $N$. Therefore, if $\lambda_1 > \lambda_2$ the second term in eqn (4.4.47) is negligible compared with the first one, as long as there is no complete degeneracy. It is thus sufficient to take

$$Z = \lambda_1^N \,. \tag{4.4.48}$$

Diagonalizing a $2 \times 2$ matrix can easily be carried out analytically. According to eqn (4.4.39) the equation to be solved is

$$\begin{vmatrix} e^{K+2h} - \lambda & e^{-K} \\ e^{-K} & e^{K-2h} - \lambda \end{vmatrix} = 0, \tag{4.4.49}$$

which is

$$\lambda^2 - 2\lambda e^K \cosh(2h) + 2\sinh(2K) = 0. \tag{4.4.50}$$

This quadratic equation for $\lambda$ has two solutions, the larger of which is the one with a + sign in front of the square root. Substituting in eqn (4.4.48),

$$Z = \left[ e^K \cosh(2h) + \sqrt{e^{2K}\sinh^2(2h) + e^{-2K}} \right]^N . \tag{4.4.51}$$

The magnetization is given by eqn (3.5.58),

$$\begin{aligned} M_z &= k_B T \frac{\partial}{\partial H} \ln Z \\ &= \frac{g\mu_B N}{2} \frac{\partial}{\partial h} \ln \left[ e^K \cosh(2h) + \sqrt{e^{2K}\sinh^2(2h) + e^{-2K}} \right] \\ &= \frac{g\mu_B N e^K \sinh(2h)}{\sqrt{e^{2K}\sinh^2(2h) + e^{-2K}}} . \end{aligned} \tag{4.4.52}$$

Obviously, this magnetization vanishes for $h = 0$, so that the system is not ferromagnetic. However, it is 'almost' ferromagnetic, in the sense that the magnetization (and all other physical properties which may be derived from $Z$) are extremely sensitive to magnetic fields, even when these fields are quite small. This feature is seen in the square root in eqn (4.4.52), or already in eqn (4.4.51). The first term of this square root vanishes for $h = 0$, but already at rather small values of $h$ it becomes bigger than the second term. The reason is that according to eqn (4.4.36), $K$ is small only at high temperatures, while at low temperatures $K \gg 1$, which makes $e^{2K} \gg e^{-2K}$. Therefore, at finite, but small, applied field the second term in the square root is negligible, the hyperbolic sine 'cancels' between the numerator and the denominator of eqn (4.4.52), and the magnetization looks as if it extrapolates to a finite value at zero field. It may be easier to see this effect in the initial susceptibility,

$$\chi_{\text{initial}} = \lim_{H\to 0} \frac{\partial M_z}{\partial H} \propto \lim_{h\to 0} \left(4h^2 e^{2K} + e^{-2K}\right)^{-3/2}, \tag{4.4.53}$$

after dropping higher powers of $h$. It is a constant for $h = 0$, but for non-zero $h$ the second term becomes negligible, and it seems that the susceptibility is diverging as $h^{-3}$.

The average internal energy of the electrons, per unit volume, as given by eqn (3.5.57), is in this case

$$\overline{\mathcal{E}} = k_{\mathrm{B}}T^2 \frac{\partial}{\partial T} \ln Z = -2JN \tanh\left(\frac{2J}{k_{\mathrm{B}}T}\right), \tag{4.4.54}$$

in zero applied field. This energy contributes to the specific heat of a unit volume of the crystal

$$C_v = \frac{\partial \overline{\mathcal{E}}}{\partial T} = k_{\mathrm{B}}N \left(\frac{2J}{k_{\mathrm{B}}T}\right)^2 \operatorname{sech}^2\left(\frac{2J}{k_{\mathrm{B}}T}\right), \tag{4.4.55}$$

which is a *continuous* function of the temperature. Studies which start from the assumption $H = 0$ (as Ising originally did) use this result as an indirect proof that the one-dimensional Ising model predicts no stable ferromagnetism at any temperature. The point is that in a transition from order to disorder the magnetic energy (especially the exchange energy) must be converted into *something*, so that it takes an extra heating at the transition, which must appear as a jump in the specific heat. Indeed, even the simplest theories (such as the Landau theory in section 4.2) predict a discontinuity of the specific heat at $T_c$, and this jump is observed in all experimental evaluations of the specific heat of the electrons. Equation (4.4.55) gives a rounded peak, but not a discontinuity.

Here this conclusion was reached from a more elegant calculation of the actual magnetization and initial susceptibility, in eqns (4.4.52) and (4.4.53). Moreover, this method showed that the study of one dimension is not completely academic, because the system is 'almost' ferromagnetic, which must mean that some small perturbations may make it a real ferromagnet, as will be discussed in the next section, 4.5. This method can also be extended with no particular complication to cover the Ising model in two dimensions. For zero applied field, that problem has an analytic solution not only for a square lattice, but even for a rectangular or a hexagonal one [69]. Details will not be given here, but the result is that the Ising model does give stable ferromagnetism in two dimensions.

In principle this result is wrong, because the true Heisenberg Hamiltonian cannot support ferromagnetism in less than three dimensions, as proved at the end of section 3.5. However, it only takes a rather small modification to have real systems which are *ferromagnetic* because they are 'nearly' one or two dimensional, as will be discussed in the next section. For these cases, the Ising model is a very useful theoretical description, and indeed its results are in reasonably good agreement with experiment. It is not as accurate as the more sophisticated theories, but it applies to the whole temperature range in one analytic solution, which makes it a convenient tool to use.

In three dimensions, or for a spin larger than $\frac{1}{2}$, the Ising model can only be solved by applying further approximations, or by using complicated mathematical techniques and computations, or both [69]. For these cases the Ising model has no particular advantage over other techniques.

## 4.5 Low Dimensionality

Strictly speaking, neither ferromagnetism nor antiferromagnetism can exist in one or two dimensions, at least in as much as the Heisenberg Hamiltonian, with all the study around it, is a good approximation to the physical reality. The proof of this statement was given at the end of section 3.5, and the reader who has skipped chapter 3 should just take my word for it that such a rigorous mathematical proof exists, and that it is undeniable, involving no approximation in it. There is also a different proof [70] which is based on another approach, but leads to the same result, namely that no spontaneous magnetization (or sublattice magnetization) can exist for the Heisenberg Hamiltonian in one or two dimensions. However, it was shown in the previous section that the Ising model for one dimension is 'nearly' ferromagnetic, in the sense explained there, so that even a small perturbation may make it a real ferromagnet. Therefore, a system which is only 'nearly' a one-dimensional Ising system may well be ferromagnetic (or antiferromagnetic), as long as it is not *strictly* one dimensional. One such system can be a set of one-dimensional chains, with a strong exchange interaction within each of the chains, *and* with a weak exchange interaction *among* the chains. This additional interaction can be sufficient, in some cases, to stabilize the ferromagnetism, while being too small to affect the results of the one-dimensional calculations.

Such systems do exist in reality. The one mostly studied is the crystal which is made of molecules of $(CH_3)_4NMnCl_3$, also known as TMMC, for *tetramethyl ammonium manganese chloride.* In this material the chains are separated by about 9 Å, so that the interchain coupling is [71] at least 3 orders of magnitude smaller than the intrachain (antiferromagnetic) exchange interaction. For such materials, the theory of one-dimensional chains is indeed a good approximation [71, 72] to the experimental results, including the measurements [73] of the susceptibility and of the specific heat of the electrons (after subtracting the contribution of the lattice). Of course, the theory is not necessarily just the Ising model and more complex theories [74, 75] have also been developed.

In two dimensions, the Ising model does give a stable ferromagnetism, with a well-defined Curie temperature. A simple *physical* explanation of why ferromagnetism cannot exist in one dimension, but can exist in two, can be seen on pages 309–10 of the book [41] by Ziman, but it cannot change the fact that the more general Heisenberg Hamiltonian does *not* allow ferromagnetism in two dimensions. Obviously, the Ising assumption that the off-diagonal elements of the spins are negligible is a sufficient perturbation that can stabilize the ferromagnetism, in the same way as *other* perturbations can sometimes do it. In some way the Ising model, which allows interaction in the $z$-direction but not along $x$ or $y$, may be regarded as a particular case of an *anisotropic exchange*, which is *not* the same as what is described in section 2.6 under the same name. In the present context it means that the exchange interaction is of the form

$$J_x S_i^{(x)} S_j^{(x)} + J_y S_i^{(y)} S_j^{(y)} + J_z S_i^{(z)} S_j^{(z)},$$

with unequal $J_x$, $J_y$, and $J_z$, and this assumption is sometimes used even in theories about crystals with a high symmetry. It has been noted [70] that such an exchange may be sufficient to stabilize ferromagnetism in two dimensions. However, this kind of an anisotropic exchange exists only in some theoretical studies, and there is no experimental evidence for its possible existence. Among other perturbations which may also do the same, it was shown [76] that with the existence of a dipolar interaction (whose range is *infinite*), a two-dimensional system may become ferromagnetic. A magnetic field may also stabilize ferromagnetism, or at least make the system *look like* a stable ferromagnet. Thus, the initial susceptibility of a two-dimensional system can obey [77] a power law, and diverge above a certain transition temperature, even though it does not really become an 'ordinary' ferromagnet below that temperature. The same was noted by Mermin and Wagner [70] who remarked that their proof rules out only spontaneous magnetization, but it 'does not exclude the possibility of other kinds of phase transitions', such as a diverging initial susceptibility below a certain temperature.

Besides all those cases, the ferromagnetism may be stabilized by a small interaction between layers. As is the case in one dimension, there are also crystals which are 'almost' two-dimensional systems, because they are made of layers with a strong exchange interaction between the ions in them, while the interaction *between* the layers is much weaker. In such crystals [71, 73] the two-dimensional theory fits the experimental results quite well.

Moreover, the experimental study of magnetism in two dimensions is not restricted any more to materials which occur in nature. Since the invention of molecular beam epitaxy (MBE), a whole new class of artificial structures has been made and studied. These are superthin, clean single-crystal films (even down to one atomic layer) separated by all sorts of non-magnetic layers of any desired thickness. These layers are built up [78] into very regular superstructures which allow detailed experimental [79] and theoretical [80] study of both three-dimensional structures, and 'almost' two-dimensional ones. It is a whole new world, which allows the detailed studies of effects that have just been neglected or wrongly evaluated some years ago, such as the properties of the surfaces or the exchange interactions carried by the conduction electrons of a non-magnetic metallic layer between magnetic layers, etc. In particular, it is clear now that a very small interaction between ultrathin layers can make the whole structure ferromagnetic (or antiferromagnetic) but strictly two-dimensional monolayers are paramagnets.

Historically, the problem of magnetization in two dimensions was approached by the study of thin films, evaporated in vacuum, which was not a very high vacuum in those days. Instead of the many layers with weak interaction between them, as in the more recent study mentioned in the foregoing, *one* film was made out of atomic layers in contact, namely with a very strong exchange interaction between them. The question which was very much discussed and argued was how thick such a layer must be before it has the magnetic properties of the bulk material.

Until 1964 the spin wave theory predicted a considerable reduction in the spontaneous magnetization of iron already at 100 Å, or even at a larger thickness. Calculations using the molecular field approximation gave larger $M_s$ down to a much smaller thickness, but nobody took them seriously, because these classical results must be much poorer than the quantum-mechanical ones. Experimentally, the decrease of $M_s$ with decreasing thickness was even faster than the spin wave calculation, and theorists made efforts to modify and correct their calculations in *that* direction. The first exception was the case [81] of films made in a better vacuum than everybody else's, which gave almost the bulk magnetization in Ni films made of only a few atomic layers. This result fitted the molecular field approximation, which also prohibits magnetization at one atomic layer, but predicts $M_s$ which is only a few % below the bulk value at *two* atomic layers. Obviously, this experiment was ignored, as were several which followed.

The real break-through was a zero-field Mössbauer effect experiment [82], which eliminated the problem of reaching saturation for very thin films, and the possibility of magnetization created by the applied magnetic field. A highly enriched (92%) $^{57}$Fe was used, but even that was not sufficient for measuring a single layer. Therefore, many layers were made, separated by SiO, which introduced some uncertainty in defining their average thickness. Still, the results were clear, and at very great discrepancy with a large number of previous experiments: at an iron thickness of 7.5 Å the Curie temperature is 83.5% of its bulk value. The room-temperature hyperfine field is only 4% below its bulk value at 6 Å thickness, and drops to zero only at an average thickness of 4.6 Å. These results are quite close to the prediction of the oversimplified molecular field approximation.

This experiment raised many heated discussions and arguments. Theorists adopted [83] the new results rather quickly, and the theoretical spin wave calculations soon fitted them. Experimentalists took longer to be convinced that all their previous results were wrong, and kept arguing [84] that the thickness of the new [82] films was not measured properly, or that something else was wrong there. They were only convinced after the experiment

[85] in which a *single* Fe layer was used as a *source*, instead of an absorber. This film was measured in the same vacuum chamber at which it had been made, without ever exposing it to the atmosphere, thus avoiding oxidation. Later modifications [86] used a much higher vacuum, and studied the effect of a slow deposition rate, or stopping the deposition for a while, then continuing it, or heating the substrate, etc. The conclusion from all these studies was that there is no ferromagnetism in the limit of one atomic layer, but it takes only a little more thickness than that to stabilize the ferromagnetism. It turned out that the films in the older experiments were not continuous Fe films, because they were heavily oxidized. To avoid oxidation, films must be made rather quickly in a sufficiently high vacuum, then either kept in the vacuum or covered by a protective layer before being exposed to air. Thus the SiO used [82] for separation turned out to be also a protection against oxidation. If the films *are* allowed to oxidize, they become separate islands of Fe *particles* rather than a continuous layer. Thus the magnetization loss already at rather thick films was [83, 87] due to the separation into isolated islands, and *not* the effect of film thickness. The point is that small enough ferromagnetic particles *also* lose their magnetization by an effect known as *superparamagnetism* which will be discussed in section 5.2.

All theories and experiments thus point to the absence of ferromagnetism in two dimensions, unless it is stabilized by one of the ways mentioned in the foregoing. There is, however, one possible exception. The temperature-dependence of the electron spin resonance of $Mn^{2+}$ ions was measured [88] in a 'literally two-dimensional' layer of Mn atoms, made by a certain chemical process. At about 2 K, the resonance field decreased abruptly (by more than $10^3$ Oe within 0.2 K) in a manner which is typical of a phase transition into the *weak ferromagnetism* mentioned in section 2.6. As mentioned in the foregoing, not everything which looks like a magnetic system is one, and the evidence would have been more convincing with a different measurement, instead of the spin resonance which involves a large magnetic field. However, *some* magnetism with a very low Curie or Néel temperature is not really ruled out by the foregoing arguments. If this experiment proves to indicate a real antiferromagnetism, it only means that the Heisenberg Hamiltonian is not the full story, and some additional terms should be added. The possibility of dipole interactions, besides the exchange interaction, has already been mentioned in this section, and a Néel point of 2 K may well be possible for it. After all, most experiments are not carried down to very low temperatures, so that this particular result may not be unique. Besides, the whole concept of a two-dimensional lattice may not be very accurate down to atomic sizes, because the spin is *not* a point charge. Each ion also has a three-dimensional structure, even if it is not necessarily as pronounced as in the case of metallic iron shown in Fig. 3.1 here.

## 4.6 Arrott Plots

The interpolation technique known as the Arrott plots was first suggested orally by Arrott in a conference with no published proceedings, then discussed (together with other methods used at the time) in an unpublished [89] internal report of the General Electric Co. It is based on a power series expansion of the Brillouin function, (2.1.15), whose argument is small in the vicinity of the Curie temperature. The beginning of this expansion has already been given in eqn (2.1.20), but here it is carried to one more term in the expansion of the coth function, yielding

$$B_S(x) = \frac{S+1}{3S}\left[x - \frac{2S^2+2S+1}{30S^2}x^3 + O(x^5)\right]. \tag{4.6.56}$$

Substituting eqn (4.6.56) in the molecular field basic formula (2.2.33) and rearranging the terms,

$$h = \left[\frac{3S}{S+1} - \alpha(T)\right]\mu + \frac{2S^2+2S+1}{30S^2}(h+\alpha\mu)^3. \tag{4.6.57}$$

Near the Curie point, the initial susceptibility diverges, which means that $h/\mu$ is small. Therefore, powers of $h$ higher than the first are neglected. Dividing eqn (4.6.57) by $\mu$, the left hand side should vanish at $T = T_c$, which means that

$$\alpha(T_c) = \frac{3S}{S+1}. \tag{4.6.58}$$

Hence,

$$\frac{H}{M_z} = a(T - T_c) + bTM_z^2, \tag{4.6.59}$$

where $a$ and $b$ are constants. It is not difficult to write these constants explicitly, in terms of the parameters of the molecular field theory, but that is not necessary. The important point to be noted is that they are not functions of $H$, $M_z$ or $T$, and depend only on the type of ferromagnetic material.

The first conclusion from eqn (4.6.59) is that for $H = 0$,

$$M_z^2 \propto (T_c - T)^{-1}, \tag{4.6.60}$$

and that

$$\chi_{\text{initial}} \propto M_z/H \propto (T_c - T)^{-1}. \tag{4.6.61}$$

According to the definitions in section 4.3, this means that the critical exponents for the molecular field approximations are $\beta = 1/2$ and $\gamma = 1$, as stated without proof in that section.

The second conclusion from that equation is that if experimental data for $M_z$ at different fields and temperatures are plotted as $M_z^2$ *vs.* $H/M_z$, at constant temperatures, they should be straight lines in the 'critical region', namely when temperatures are not very far from the Curie point. The intercept of these lines with the $(H/M_z)$-axis is positive if $T > T_c$, and negative if $T < T_c$. The advantage of this kind of plotting for an accurate determination of $T_c$ is very clear and obvious. It should only be noted that the data for too low fields, that do not fit these straight lines, must be discarded, because they represent averaging over domains which are magnetized in different directions. This point was already emphasized in that GE report [89], which warned that in these equations $M_z$ 'represents the measured magnetization of the bulk materials only if domain alignment is complete', which means avoiding too small fields.

Even when these plots are not straight lines (because real materials do not obey the molecular field theory) they are still quite useful [90] for determining the Curie point, because of the clear distinction of the intercept for the temperature to be above or below $T_c$. However, there are difficulties in extrapolating curves, because the human eye can only really deal with straight lines. There are also difficulties [90] in deciding where the limit of the low-field data is. Therefore, it was found better [91] to include the proper *critical exponents* of section 4.3, and try to fit *all* the experimental data in the critical region to the *equation of state*,

$$\left(\frac{H}{M_z}\right)^{1/\gamma} = \frac{T - T_c}{T_1} + \left(\frac{M_z}{M_1}\right)^{1/\beta}, \tag{4.6.62}$$

where the parameters $\gamma$ and $\beta$ are chosen so that a plot of $M_z^{1/\beta}$ *vs.* $(H/M_z)^{1/\gamma}$ at a constant $T$ gives a set of *straight* lines. This kind of a set of plots, which was given the name 'Arrott plots', became the standard technique used by many workers as a routine. However, three points which were emphasized in that paper [91] were later forgotten or ignored, and are worth repeating here:

1. Equation (4.6.62) is only one of many possibilities to keep the same exponents $\beta$ and $\gamma$ at the critical region, and the choice may depend on how wide this region is defined to be.
2. The values of the exponents $\beta$ and $\gamma$ cannot be determined from the fit to the experimental data to any decent accuracy, because the fit looks very much the same over a wide range of the values of these parameters.
3. There is no way to eliminate the curvature of the data for *very low fields*. The best way is to ignore them, see in particular the data points in Fig. 3 of that paper, and also [65].

An equation of state has also been proposed [50] for the whole range of temperatures, not only the critical region, and is given by eqn (3.5.74) here. For $t \approx 1$ it becomes the same as eqn (4.6.62), but only for $H = 0$.

An empirical generalization of eqn (3.5.74) which should apply to any field (from zero up to saturation) has been proposed [92], and compared with some experimental data from the literature. However, these data did not go up to a high field, and the agreement was not really any better than that which could be obtained from eqn (4.6.62). Also, the fit was not very good, mainly because *all* the experimental data were used for the least-square fitting, whereas the low-field data should have been kept out of it, because they represent only the rearrangement of domains. At any rate, this idea did not catch on, and nobody else tried to use that equation for any other experimental data. It should be noted, though, that both Fig. 1.2 and Fig. 2.1 in this book have actually been plotted with the use of *that* formula, and with $\beta = 0.368$ and $\gamma = 1.112$.

It was first noted by Wohlfarth [93] that the Arrott plots should become curved, if the material is not homogeneous. His approach was extended [94] by several others, and later used [95, 96] to explain some features of these plots in *amorphous ferromagnets* which are very heterogeneous indeed. Nevertheless, this theory was forgotten, and for several years the same curvature was attributed to some special properties of amorphous materials, predicted by a theory based on a first-order perturbation of the Heisenberg Hamiltonian, which was supposed to apply at *very low magnetic fields.* In spite of all the warnings against such an approach, as emphasized in the present section and in section 4.1, dozens of experimentalists hurried to produce low-field Arrott plots, to compare with that non-physical theory. Details are beyond the scope of this book, but it should be noted that by introducing the amorphicity as a Gaussian distribution of exchange interactions, an excellent fit to some experimental data was obtained [97] provided *the low-field data were excluded from the fitting.* This theory used 8 adjustable parameters, not all of which were really necessary, and there was actually no difficulty in keeping, for example, the accepted theoretical values of the critical exponents $\beta$ and $\gamma$, which would have given almost as good a fit. It was just a tactical error to insist on showing, in the same paper, that the experimental values of these exponents are unreliable, by fitting the data with very different values of $\beta$ and $\gamma$. In this field of critical exponents theorists got used to telling the experimentalists the 'correct' values to which they should fit their data, which is against the tradition of physics in any other field. Therefore, showing that the data could be fitted very well, for example, with $\gamma = 2.2$ was taken as a heresy, and neither the *Physical Review* nor the *Journal of Applied Physics* would publish it. It was eventually published [97] in *JMMM* and ignored by everybody.

# 5

# ANISOTROPY AND TIME EFFECTS

## 5.1 Anisotropy

The Heisenberg Hamiltonian is completely *isotropic*, and its energy levels do not depend on the direction in space in which the crystal is magnetized. Throughout the previous chapters the measured magnetization was consistently denoted by $M_z$, where the $z$-direction is the direction of the applied field. It does not really have any meaning in the limit of zero applied field, for which most of the calculation has been done. In fact, the conclusion from all the calculations described so far is that a ferromagnetic crystal has a certain magnetic moment $\boldsymbol{\mu}$, whose $z$-component is a certain function of the temperature. We know that at low temperatures most of the spins are parallel to $z$, but this $z$ has not been defined yet, and will be introduced here.

However, before defining this direction, it is illustrative to consider the behaviour of a ferromagnetic particle in the case of complete isotropy, when all directions in space are equivalent, and the choice of $z$ is *arbitrary*. At low temperatures, strong exchange forces hold the spins parallel to each other, and the direction of these spins defines the direction in space of the magnetic moment $\boldsymbol{\mu}$, which is $g\mu_B$ times the vectorial sum of the spins. Let this $\boldsymbol{\mu}$ be at an angle $\theta$ to a fixed magnetic field $\mathbf{H}$. The energy of the interaction between the field and the magnetization of the particle is known to be $-\mu H \cos\theta$. Therefore, at thermal equilibrium the probability of having a particular angle $\theta$ at a temperature $T$ is proportional to $e^{x\cos\theta}$, where

$$x = \frac{\mu H}{k_{\mathrm{B}} T}, \tag{5.1.1}$$

and $k_{\mathrm{B}}$ is the Boltzmann constant. Hence, the average for an ensemble of particles is

$$\langle \cos\theta \rangle = \frac{\int_0^{2\pi}\int_0^{\pi} \cos\theta \, e^{x\cos\theta} \sin\theta \, d\theta \, d\phi}{\int_0^{2\pi}\int_0^{\pi} e^{x\cos\theta} \sin\theta \, d\theta \, d\phi} = \frac{\left[\left(\cos\theta - \frac{1}{x}\right) e^{x\cos\theta}\right]_0^{\pi}}{\left[e^{x\cos\theta}\right]_0^{\pi}} = L(x), \tag{5.1.2}$$

where

$$L(x) = \coth x - \frac{1}{x} \tag{5.1.3}$$

is called the *Langevin function*. It is readily seen that the Langevin function is the limit of the Brillouin function of eqn (2.1.15), for $S \to \infty$.

The left hand side of eqn (5.1.2) is the component parallel to $\mathbf{H}$, of a unit vector in the direction of the magnetization, by the definition of the angle $\theta$, namely

$$\frac{M_H}{|\mathbf{M}|} = \langle\cos\theta\rangle = L\left(\frac{\mu H}{k_B T}\right), \tag{5.1.4}$$

which proves that all ferromagnets are actually just paramagnets. And there is no mistake in this algebra: there are only two differences between this calculation and the study of a gas of paramagnetic atoms in section 2.1. One is that the function $\theta$ is continuous here, while this variable had discrete values in section 2.1, and the other is that the magnetic moment $\mu$ was that of a single atom there, while here it is the moment of a large number of atoms, coupled together. However, the second difference is only quantitative and not qualitative, and the first one should not make any difference, especially since the energy levels of a large spin number $S$ are very close together, and look like a continuous variable. It is thus *true* that if there was no other energy term besides the isotropic Heisenberg Hamiltonian, it would have been impossible to measure any magnetism in zero applied field, and there would be no meaning to a Curie temperature, or critical exponents, or any of the other nice features mentioned in the previous chapters. Theorists who calculate these properties never pay attention to the fact that the possibility of measuring that which they calculate is only due to an extra energy term, which they always leave out.

Of course, a magnetization as in eqn (5.1.4), which is zero in zero applied field, contradicts not only experiments that produce Fig. 1.1. It is also in conflict with everyday experience that, for example, the particles in an audio or video tape stay magnetized, and do not lose the recorded information when the writing field is switched off. It is because real magnetic materials are not isotropic, and not all values of the angle $\theta$ are equally probable. There are several types of anisotropy, the most common of which is the *magnetocrystalline anisotropy*, caused by the spin–orbit interaction. The electron orbits are linked to the crystallographic structure, and by their interaction with the spins they make the latter *prefer* to align along well-defined crystallographic axes. There are therefore directions in space in which it is easier to magnetize a given crystal than in other directions. The difference can be expressed as a direction-dependent energy term.

The magnetocrystalline energy is usually small compared with the exchange energy. The *magnitude* of $M_z(T)$ is determined almost only by the exchange, as in the calculations of the previous chapters, and the contribution of the anisotropy is negligible for almost all the known ferromagnetic materials. But the *direction* of the magnetization is determined only by this anisotropy, because the exchange is indifferent to the direction in space. Therefore, the axis $z$ of the quantization direction is always a direction for which the anisotropy energy is a minimum. It has nothing to do with the direction of the field **H**, even if some of the phrasing in the previous chapters may have led to the conclusion that **H** is always parallel to $z$. In real life the field may be applied at any angle to the internal direction of the anisotropy axis, as has been hinted in Fig. 2.2.

It may be worth noting that theories exist for the case of a *large* anisotropy energy, which is not negligible compared with the exchange. If such materials could be found, their magnetization [98] and even their Curie point [99] would be different when measured in different directions. It should also be noted that adding anisotropy is not sufficient yet for subdividing the crystals into the *domains* mentioned in section 4.1. The exchange tries to align all the spins parallel to each other, and the anisotropy tries to align them along a certain crystallographic direction. Together, they try to align all spins parallel to that direction, and the division into domains must be caused by still another energy term, to be discussed in section 6.2. However, once the domains are there, the anisotropy energy term will try to align the magnetization in each of them along one of the axes of its energy minimum. The domains are thus regularly arranged along well-defined directions, and are not randomly oriented, as Weiss originally assumed.

Quantitative evaluation of the spin–orbit interaction from basic principles is [100] possible, but the accuracy is inadequate, as is the case with the exchange integrals. Therefore, anisotropy energies are always written as phenomenological expressions, which are actually power series expansions that take into account the crystal symmetry, and the coefficients are taken from experiment. Specific expressions can only be written for a specific crystalline symmetry, as is done in the following.

### 5.1.1 *Uniaxial Anisotropy*

The anisotropy of *hexagonal crystals* is a function of only one parameter, the angle $\theta$ between the $c$-axis and the direction of the magnetization. It is known from experiment that the energy is symmetric with respect to the $ab$-plane, so that odd powers of $\cos\theta$ may be eliminated from a power series expansion for the anisotropy energy *density*. Its first two terms are thus

$$w_u = -K_1 \cos^2\theta + K_2 \cos^4\theta = -K_1 m_z^2 + K_2 m_z^4\,, \tag{5.1.5}$$

where $z$ is parallel to the crystallographic $c$-axis, and $\mathbf{m}$ is a unit vector parallel to the magnetization vector,

$$\mathbf{m} = \frac{\mathbf{M}}{|\mathbf{M}|}. \tag{5.1.6}$$

The subscript $u$ is used here, because this kind of anisotropy is usually referred to as a *uniaxial* one. The coefficients $K_1$ and $K_2$ are constants which depend on the temperature. Their values are taken from experiments. In principle the expansion in eqn (5.1.5) may be carried to higher orders, but none of the known ferromagnetic materials seem to require it. In most cases even the term with $K_2$ is negligible, and many experiments may be analysed by using the first term only. And in all known cases $|K_2| \ll |K_1|$, which justifies the power series expansion.

Most workers prefer to rewrite eqn (5.1.5) as

$$w_u = K_1 \sin^2 \theta + K_2 \sin^4 \theta = K_1(1 - m_z^2) + K_2(1 - m_z^2)^2 , \qquad (5.1.7)$$

in which case the coefficient $K_1$ has a *different value* than in the case of eqn (5.1.5), unless $K_2 = 0$, or is negligibly small. Once $K_1$ is properly redefined, the difference between eqn (5.1.5) and eqn (5.1.7) is a *constant*, and a constant energy term does not have any physical meaning: it only means a shift in the definition of the zero energy, which is never important for the problems discussed in this book. Therefore, the choice between eqn (5.1.5) and eqn (5.1.7) is completely arbitrary, as long as the definition is not switched in the middle of a calculation. In either case, both $K_1$ and $K_2$ may be either positive or negative. In most hexagonal crystals, the $c$-axis is an *easy axis*, which means it is an energy minimum and not a maximum. In these cases, $K_1 > 0$ in eqn (5.1.5) or eqn (5.1.7). There are, however, materials for which $K_1 < 0$, and for them the $c$-axis is a hard axis, with an *easy plane* perpendicular to it. Some hexagonal ferrites have also a certain amount of anisotropy *within* [8] the $ab$-plane, but it is always small, and is at most just barely measurable. It will be ignored here.

### 5.1.2 *Cubic Anisotropy*

For *cubic crystals* the expansion should be unchanged if $x$ is replaced by $y$, etc., when the axes $x$, $y$, and $z$ are defined along the crystallographic axes. Again, odd powers are ruled out and the lowest-order combination which fits the cubic symmetry is $m_x^2 + m_y^2 + m_z^2$, but this is just a constant. Therefore, the expansion starts with the fourth power and is actually

$$w_c = K_1(m_x^2 m_y^2 + m_y^2 m_z^2 + m_z^2 m_x^2) + K_2 m_x^2 m_y^2 m_z^2 , \qquad (5.1.8)$$

where here the values of $K_1$ and $K_2$ are also taken from experiments, and they also depend on the temperature. Here again the expansion may be carried to higher orders, but it is not necessary for any known ferromagnet. Some workers prefer to replace the expression with $K_1$ by $-\frac{1}{2}(m_x^4 + m_y^4 + m_z^4)$, but without changing the second term with $K_2$. This substitution does not change the coefficients, because

$$m_x^4 + m_y^4 + m_z^4 + 2(m_x^2 m_y^2 + m_y^2 m_z^2 + m_z^2 m_x^2) = (m_x^2 + m_y^2 + m_z^2)^2 = 1. \qquad (5.1.9)$$

Cubic materials exist with either sign for $K_1$. For example, $K_1 > 0$ in Fe, so that the easy axes are along (100), while for Ni $K_1 < 0$, and the easy axes are along the body diagonals, (111).

If **M** is the same everywhere, the above expressions for the energy density have to be multiplied by the volume of the crystal to obtain the anisotropy *energy*. However, if **M** (or **m**) is a function of space, as is the case in some problems discussed in the following chapters, the energy is

$$\mathcal{E} = \int w \, d\tau, \tag{5.1.10}$$

where $w$ stands for either $w_u$ or $w_c$ (or any other form of anisotropy, as the case may be), and the integration is over the volume of the ferromagnet.

### 5.1.3 *Magnetostriction*

There are other forms of anisotropy besides the magnetocrystalline one. One of them is due to an effect which had already been observed in the 19th century, and given the name *magnetostriction*: when a ferromagnet is magnetized, it shrinks (or expands) in the direction of the magnetization. Strictly speaking, such an effect invalidates even the definition of **M** as the dipole moment per unit volume, because the 'unit volume' itself changes with the magnetization, which changes with the applied field. It is also quite clear that when domains are magnetized (and therefore change their dimensions) in different directions, there can be a misfit of the crystalline lattice at the boundary between such domains, which would lead to an extra strain energy. Such effects have been studied [101, 102] for some simple cases, but the problem of the magnetization in a *deformable* body is outside the scope of this book. Even its mathematical formulation [103] is extremely complicated, and has never been fully developed, even for the case when the sample is [104] magnetically *saturated.* It is therefore assumed here that all bodies are *rigid*, and all these *magnetoelastic effects* will be just ignored. Only three remarks must be made before dropping this subject.

One is that a large part of the *energy* of the internal magnetostriction in a ferromagnetic crystal can be expressed in the same mathematical form as the uniaxial or cubic *magnetocrystalline* anisotropy, given in the foregoing. When the coefficients $K_1$ and $K_2$ are calculated from basic principles, the contribution of this magnetostriction should be added. But when the coefficients are taken from experiment, this contribution is already included, and nothing is really neglected by not mentioning it. The second point is that it is possible to add another dimension, by measuring ferromagnetic crystals under pressure. To a first-order approximation, such experiments can be analysed [105, 106, 107] as an extra anisotropy term. Of course, the easy axes of this term depend on the applied pressure, and do not necessarily coincide with the crystallographic axes of the material. In these cases (and also in some crystals with only an internal strain and no external pressure) it may be possible to have *both* cubic and uniaxial anisotropy terms in the same sample. The third remark is that some first-order theories exist for the effect of internal strains at crystalline imperfections (in particular a certain distribution of dislocations [108, 109] or impurity atoms [110], and others [111]) on the approach to saturation. There are also many experiments which show [112, 113, 114] that the introduction of dislocations (by milling) and their removal (by annealing) has a large effect on the measured *coercivity*. This effect is connected with the large magnetostriction in the vicinity of the dislocations, for which there is a detailed theory [115], but it is outside the scope of this book.

### 5.1.4 *Other Cases*

Other forms of anisotropy include the *shape anisotropy*, originating from magnetostatic properties, which will be discussed in section 6.1. In the case of thin magnetic films there is also another form, known as *induced anisotropy*. It was a very popular subject in the 1950's and 1960's, when there were many experimental investigations of thin films, mostly made out of permalloy, which is an alloy of about 80% Ni and 20% Fe. It was then found that when the film is deposited at an oblique angle to the substrate, or when a large magnetic (or even *electric* [116]) field is applied during the deposition, a uniaxial anisotropy of the form of eqn (5.1.5) or eqn (5.1.7) was developed in the plane of the film. Applying and removing a magnetic field (with or without annealing the sample) could also induce a uniaxial anisotropy that was usually referred to as a *rotatable* anisotropy. The latter was also observed in the bulk [117] and in cobalt[118]. However, in spite of the wide interest at the time, the origin of these phenomena has never been fully established. The conclusion of a 1962 review [119] was that the induced anisotropy 'is very complicated and it is not fully understood', and in 1964 the phrasing was [120] that the problem 'is too complex for a complete quantitative treatment', while a 1969 paper [121] stated that 'the mechanism $\cdots$ is a subject to be investigated'. The vacuum used in those days was not good enough, see section 4.5, and it is quite possible that oxygen played a role [122] in some of these effects. Inhomogeneities of composition [123] and of phase [124] were shown to be part of it, and the possible effect of impurities was demonstrated [125] by the enhancement of this anisotropy when another metal was codeposited with the permalloy. An internal strain may have also [126] played a part. None of these effects was ever fully clarified, nor was there any real advance later, and some features of the induced anisotropy are not easy to explain even in the more modern experiments. It was not so much that the problem was too difficult, but that most people just lost interest in these kinds of experiments, although some are still [127, 128, 129] being reported.

An interesting feature of the permalloy films was that they were polycrystalline, *i.e.* they were made of small crystals whose crystallographic axes were randomly oriented. Therefore, they had a local cubic anisotropy whose easy axes were also randomly oriented, *besides* the overall uniaxial anisotropy. This random anisotropy caused [130, 131] a *ripple structure*: the direction of the magnetization in each of the domains wiggled *slightly* around its average direction. The theory of this ripple (or wiggling) was quite straightforward [132], and was also verified [133] by the ferromagnetic resonance. The exchange and overall anisotropy tend to keep the magnetization in each domain parallel to the uniaxial easy axis, while the random anisotropy tries to tilt it into a different direction for each crystallite. The competition between them results in the former two strong forces keeping the magnetization direction nearly constant, but they yield a little bit to the weaker, random term, allowing a small tilt in each crystallite towards the *local* easy axis of the cubic anisotropy. When amorphous ferromagnets were first made, it was still taken for granted that the *same* argument about random anisotropy in the small permalloy crystallites applies just as well to the random local anisotropy of the ions. Therefore, the effect should be similar [134], namely there should be a ripple structure with a certain smearing [95] of the critical region. The non-physical theory (criticized already towards the end of section 4.6), according to which the occurrence of a random anisotropy, *no matter how small*, must lead to a drastically different *qualitative* behaviour, came only later.

When a certain thickness of the film is deposited with an anisotropy induced along a chosen direction, and the rest of it is deposited with the anisotropy induced along a *different* direction, a special form of a *biaxial* or even *triaxial* [135] anisotropy is obtained. Other special, artificial types of anisotropy have been obtained by deposition [136] on a *scratched* substrate.

### 5.1.5 *Surface Anisotropy*

There are several contributions to this term, the most important of which was suggested back in 1954 by Néel, who pointed out the importance of the reduced symmetry at the surface of a ferromagnet. The spin at the surface has a nearest neighbour on one side, and none on the other side, so that the exchange energy there cannot be the same as in the bulk. A non-magnetic metal deposited on a ferromagnetic one gives [137, 138, 139] an even different environment for the surface spins, and so does [140] the interface between two *different* ferromagnets. The easiest case to consider is that of a thin film, because in this case it is possible to compute the actual wave functions for each atomic layer to a reasonable accuracy. Calculations for a few atomic layers are possible, and show [80, 141, 142] that it is not only the effect of the *last* layer on the surface, but it carries inwards to a few more. The problem is more complicated for other geometries, and it is not even clear to what extent the results on thin films are applicable to them. However, from a phenomenological point of view, any surface energy term should be a tendency of the surface spins to be either parallel or perpendicular to the surface, in the same way as the thin film energy term is an anisotropy whose easy axis is [143, 144] either parallel or perpendicular to the film plane. Therefore, to a first-order approximation any theory should lead to an energy term of the form

$$\mathcal{E}_s = \frac{1}{2} K_s \int (\mathbf{n} \cdot \mathbf{m})^2 \, dS, \qquad (5.1.11)$$

where $\mathbf{m}$ is defined in eqn (5.1.6), the integration is over the surface of the ferromagnet, and $\mathbf{n}$ is a unit vector parallel to the normal pointing out of the surface. The coefficient $K_s$ should be taken from experiment, but there are not many clear-cut experiments which evaluate this parameter, and its value for any given ferromagnetic material is often controversial.

The form of eqn (5.1.11) assumes [145] that the surface anisotropy is a geometrical feature, which depends only on the *shape* of the surface. It is also possible to imagine [146] a surface anisotropy caused by the reduced symmetry of the spin–orbit interaction at the surface. It can lead to an energy that depends on the angle between the magnetization at the surface and the *crystallographic* axes of the material, besides, or instead of, eqn (5.1.11). Computations from basic principles on single-crystal films [142] contain both possibilities together. They could be designed to show the effect of each one separately, and the question could also be clarified by properly designed experiments [79], which has not been done yet.

The energy term in eqn (5.1.11) is the first indication in this book of a possible *space-dependence* of the magnetization. If the surface anisotropy prefers a different direction than that of the anisotropy in the bulk, it is conceivable that the magnetization vector will point along the bulk easy axis in most of the crystal, and will then gradually turn into a different direction when it approaches the surface. Of course, it can happen only if the surface anisotropy energy is large enough to compensate for the work that needs to be done against the exchange energy, that prefers full alignment. It is illuminating to think of this possibility even at this stage because it contains some of the important features of the magnetostatic energy that will be introduced in section 6.1. These different cases also share the common property that they are automatically ignored in a calculation that assumes an *infinite* crystal, which does not have a surface.

### 5.1.6 *Experimental Methods*

There are several methods to *measure* the coefficients $K_1$ and $K_2$ of the magnetocrystalline anisotropy. Usually the *total* anisotropy is measured, and the shape anisotropy of the sample must be known, and subtracted. The most common method is known as the *torque curve*: the crystal is magnetized by a field applied at different angles to the crystallographic axes, and a torsion balance is used to measure the resulting mechanical torque. The applied field must be [147] large enough to remove the magnetic domains [148] but not so large that it affects [149] the measured values. Even an *electric* field may sometimes affect [150] the measured values. The mathematical form of the angular dependence of the torque is usually known as one of the expressions in the foregoing, or a transformation [127] of them, but the analysis of the data is also possible [151] in some cases for which the symmetry is not known in advance. This method is usually applied to single crystals only, but under certain conditions, torque curves can [113, 152] determine the distribution of the magnitudes of anisotropy in a powder with random directions of easy axis. Measuring films with different thicknesses can also yield [153, 154, 155] the *surface* anisotropy.

Other methods have to rely more heavily on theoretical interpretation of what is measured. They include:

1. Measurement of the magnetization in large applied fields, *i.e.* in what is known as the *approach to saturation* region. In this region it is sufficient to use a linear theory, by neglecting higher orders [156] of the magnetization component perpendicular to the applied field, and there are also [157] empirical rules.
2. Ferromagnetic resonance in the geometry of thin films. The theory is well understood, see section 10.1, and the analysis of the data can yield not only the bulk anisotropy constant, but also that of the *surface* [143, 154, 158, 159, 160] anisotropy.
3. The *transverse* initial susceptibility, defined as

$$\chi_t = \lim_{H_x \to 0} \frac{\partial M_x}{\partial H_x}, \qquad (5.1.12)$$

   is plotted versus a *bias field* $H_z$. Old calculations [161] were based on a certain model of Stoner and Wohlfarth, which will be discussed in section 5.4. This model assumes that the sample is made of particles, and that there is no space-dependence of the magnetization within each particle. The old theory predicted *cusps* in this susceptibility when $H_z$ reaches one of the values $-K_1/M_s$ and $\pm 2K_1/M_s$. Such cusps could not be seen [161] in the older experiments. They were later found to exist [162, 163] (although in the form of rounded peaks instead of cusps) in fine-grained ferrites, but not in coarse-grained ones, which must be subdivided into domains. This technique is quite popular nowadays, especially [164] for materials with large values of $K_1$, for which it has been described [165] as 'easy to handle' and 'an interesting alternative' to other methods. Improvements in this method [166] allow the evaluation of $K_2$ as well.
4. Singularities in the *parallel* susceptibility were also predicted [161] but not observed, till a more complete analysis [167] led to the measurement of the *derivative* of the susceptibility, namely $\partial^2 M_z/\partial H_z^2$.

This method works [168] for coarse grains, when each crystallite contains domains. If this derivative is detected by its second harmonic response, the distribution of anisotropies [169] can be measured.

## 5.2 Superparamagnetism

Before introducing another energy term, it should be instructive to study the change which the introduction of the anisotropy made in the calculation at the beginning of the previous section. Consider a group of particles, say spheres for example, having a uniaxial anisotropy as in eqn (5.1.7). Let $K_2$ be neglected for simplicity, although including it does not really complicate the calculation. It only requires to choose a specific value of $K_2/K_1$ for any particular example. If the magnetic moment $\boldsymbol{\mu}$ of a particle is at an angle $\theta$ to the easy axis $z$, and a magnetic field $\mathbf{H}$ is applied along $z$, *i.e.* at $\theta = 0$, the total energy is

$$\mathcal{E} = K_1 V \sin^2\theta - \mu H \cos\theta, \tag{5.2.13}$$

where $V$ is the volume of the particle. This function is plotted in Fig. 5.1 *vs.* $\theta$. Obviously, the Boltzmann distribution cannot be used as in eqn (5.1.2), because not all angles are equally probable *a priori*. There are two minima, one at $\theta = 0$ and one at $\theta = \ pi$, whose energies are

$$\mathcal{E}_1 = -\mu H \quad \text{and} \quad \mathcal{E}_2 = +\mu H \tag{5.2.14}$$

respectively, with an energy barrier between them. At thermal equilibrium, the magnetization will tend to be in the vicinity of these minima.

Actually, for such a configuration the question is not what the thermal equilibrium is, but if that equilibrium is reached at all under normal conditions. As a rough approximation one can assume that the magnetization vectors of the particles spend *all* their time in one of the directions of the minima, and no time at all at any direction in between. In that case, the number of particles jumping over the barrier from minimum 1 to minimum 2 is a function only of the height of the energy barrier, $\mathcal{E}_{\mathrm{m}} - \mathcal{E}_1$, where $\mathcal{E}_{\mathrm{m}}$ is the energy at the maximum, see Fig. 5.1. The latter can be evaluated by equating to 0 the derivative of eqn (5.2.13),

$$\sin\theta\,(2K_1 V\cos\theta + \mu H) = 0. \tag{5.2.15}$$

The solution $\sin\theta = 0$ leads to the two minima, whose energies are given by eqn (5.2.14). The other solution is the maximum, at

$$\cos\theta = -\frac{\mu H}{2K_1 V}. \tag{5.2.16}$$

When it is substituted in eqn (5.2.13), the energy at the maximum is found to be

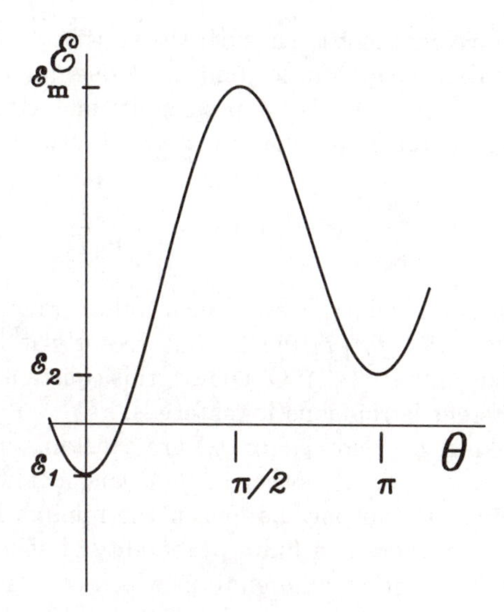


FIG. 5.1. The function of eqn (5.2.13) plotted for $\mu H = 0.15 K_1 V$.

$$\mathcal{E}_\mathrm{m} = K_1 V + \frac{\mu^2 H^2}{4K_1 V} = K_1 V \left[ 1 + \left( \frac{H M_s}{2K_1} \right)^2 \right] . \tag{5.2.17}$$

The second relation is obtained from the definition of $\boldsymbol{\mu}$ as the magnetic moment of each of the particles, and of the magnetization vector **M** as the magnetic moment per unit volume. This definition means that $\boldsymbol{\mu} = V\mathbf{M}$, namely $\mu = M_s V$, since $M_s$ as defined in section 4.1 is the magnitude of **M**, in the absence of magnetic domains. .

Therefore, the number of particles jumping over the barrier from minimum 1 to minimum 2 per unit time can be written as

$$\nu_{12} = c_{12} e^{-\beta(\mathcal{E}_\mathrm{m} - \mathcal{E}_1)} = c_{12} e^{-\beta K_1 V (1 + H/H_K)^2} , \tag{5.2.18}$$

where $c_{12}$ is a constant, $\beta$ is defined in eqn (1.3.12), and

$$H_K = \frac{2K_1 V}{\mu} = \frac{2K_1}{M_s} . \tag{5.2.19}$$

Similarly, the number of particles jumping over the barrier from minimum 2 to minimum 1 per unit time is

$$\nu_{21} = c_{21} e^{-\beta(\mathcal{E}_\mathrm{m} - \mathcal{E}_2)} = c_{21} e^{-\beta K_1 V (1 - H/H_K)^2} , \tag{5.2.20}$$

where $c_{21}$ is another constant. In the particular case $H = 0$ the barrier is the same in either direction, and these two constants must be the same. In this case it is more convenient to consider the *relaxation time* $\tau$ which is the average time it takes the system to jump from one minimum to the other, instead of the probability of this jump per unit time. One is the reciprocal of the other, and the previous equations may be rewritten (for $H = 0$) as

$$\frac{1}{\tau} = f_0 e^{-\alpha} \quad \text{with} \quad \alpha = \frac{K_1 V}{k_B T}, \tag{5.2.21}$$

where $f_0$ is a constant, which has a dimension of frequency. The original estimation of Néel was $f_0 \approx 10^9\,\mathrm{s}^{-1}$, but recently it has become more customary to take $f_0 \approx 10^{10}\,\mathrm{s}^{-1}$. Of course, this constant is not necessarily the same for different ferromagnetic materials.

Strictly speaking, $c_{12}$ and $c_{21}$ (or $f_0$) are constants only if the magnetization cannot ever be at any other angle $\theta$, and is always in one of the two energy minima. It can only happen if the minima have zero widths. In any realistic case, there is a finite probability of spending some of the time in the vicinity of either minimum, in which case the pre-exponential coefficients $c_{12}$ and $c_{21}$ are functions of the temperature, and of the applied field $H$. However, if the minima are rather narrow, and the barrier energy is rather large, it can be expected that $c_{12}$ and $c_{21}$ (or $f_0$) have only a *weak* dependence on $T$ and $H$, which is negligible when compared with the dependence in the exponential, and only a small error is introduced when they are taken as constants. More generally, the same eqn (5.2.21) should also apply to other kinds of anisotropy, when $K_1 V$ is replaced by the *energy barrier* for that particular case. The derivation of this equation assumed a particular form for the barrier $\mathcal{E}_m - \mathcal{E}_1$, and it is obvious that it does not apply *as it is* to other barriers. Strangely enough, this trivial statement had to be emphasized [170] because some workers used eqn (5.2.21) for *other* anisotropies. But if the correct energy barrier is used, eqn (5.2.21) holds, provided the minima are rather narrow, and the barrier is rather high.

This argument about narrow minima was made more quantitative by Brown [171] who considered the magnetization vector in a particle to wiggle around an energy minimum for a while, then jump (from wherever it happens to be then) to somewhere around the other minimum, then wiggle around there before jumping again. It is actually a problem of *random walk* and Brown wrote a differential equation to describe it, and showed that the eigenvalues of that equation should determine more rigorously the above-mentioned $\nu_{12}$ and $\nu_{21}$, or $\tau$.

Brown did not solve his differential equation. Instead he [171] tried some analytic approximations and an asymptotic expansion, which he [172] improved later. From these estimates he concluded that for a uniaxial anisotropy the exact solution would not be drastically different from what is obtained by taking $c_{12}$ and $c_{21}$ as constants, in the range of values of the physical parameters for which this theory is usually applied. Numerical solutions of Brown's differential equation for the case of a uniaxial anisotropy, in zero [173] or non-zero [174] applied field, showed that assuming $c_{12}$ and $c_{21}$ to be one and the same *constant* is a sufficiently good approximation, for all practical purposes. The *same* conclusion may be drawn from a more recent numerical solution [175] which is based on a modification of the method used in the previous [173] computation. However, it does not complicate any analysis of data if a higher accuracy is used, for which case it is better to adopt at least the asymptotic result of Brown, and instead of just a constant $f_0$ take

$$f_0 = \frac{2K_1\gamma_0}{M_s}\sqrt{\frac{\alpha}{\pi}} \quad \text{for} \quad \alpha > 1, \tag{5.2.22}$$

where $\gamma_0$ is the gyromagnetic ratio. For the case when even better accuracy is required, there are several easy-to-use approximations [176] for the exact numerical solution.

The situation is completely different in the case of a *cubic anisotropy.* A slight complication is encountered in a calculation similar to that leading to eqns (5.2.18) and (5.2.20) here, which calls for a solution of a cubic equation to evaluate $\theta$ at the maximum energy. But at least in the particular case $H = 0$ the solution is straightforward, leading to a result which is very similar to eqn (5.2.21) of the uniaxial case, with the only difference that $K_1$ is replaced by $K_1/4$. However, in this case the assumption of a constant factor in front of the exponential turns out to be a bad approximation. There are minima along $x$, $y$ and $z$ (for a positive $K_1$) and very many possibilities of wiggling around each one of them before jumping to one of the others. Evidently, this wealth of possibilities makes a big difference in the random-walk problem. A numerical solution [177, 178] for cubic materials gave results which were considerably different from the simple Néel exponential of eqn (5.2.21). Moreover, this difference is *measurable*, because the relaxation time can be estimated from the line-width of the Mössbauer spectrum. Such measurements for different sizes of some cubic particles at different temperatures were [179] very far from the prediction of eqn (5.2.21), and quite close to that which has been obtained by the numerical solution [177] of the Brown differential equation. For the accuracy used here, this difference will be ignored and eqn (5.2.21) (with $K_1/4$ instead of $K_1$) will be used for cubic symmetry too, because such details are beyond the scope of this book. There are other approximations anyway, *e.g.* the assumption that the particles are spheres with no shape anisotropy is not always [170] justified. Besides, under certain circumstances the assumption that the magnetization in the particle is uniform, and does not depend on space, may not be [180] justified either.

At any rate, the dependence of the relaxation time on the particle size is in the exponent, and an exponential dependence is a very strong one.

**Table 5.1** *Examples of the relaxation time $\tau$ of spherical particles whose radius is $R$, for two materials at room temperature.*

| Material | $R$ (Å) | $\tau$ (s) |
|---|---|---|
| Cobalt | 44 | $6 \times 10^5$ |
| | 36 | 0.1 |
| Iron | 140 | $1.5 \times 10^5$ |
| | 115 | 0.07 |

In order to demonstrate how strong it is, numerical examples are given for two materials, both at room temperature, *i.e.* with $k_B T = 4.14 \times 10^{-14}$ erg, and both are calculated using eqn (5.2.21) with the Néel value of $f_0 = 10^9\,\mathrm{s}^{-1}$. One is hexagonal cobalt, for which $K_1 = 3.9 \times 10^6\,\mathrm{erg/cm}^3$. The other is cubic iron, whose easy axes are along ⟨100⟩, for which $K_1 = 4.7 \times 10^5\,\mathrm{erg/cm}^3$. The values of the relaxation time $\tau$ (in seconds) are listed in Table 5.1, for a certain choice of the radius $R$ of the particle, assumed to be a sphere.

Radii in the table are chosen to demonstrate that within a rather small range of particle size the relaxation time can change from being much larger to much smaller than an arbitrarily chosen time-scale of 100 seconds. A different value of $f_0$ would not change the general form, and would only require slightly different radii to demonstrate the same point. A different magnetic material (namely, a different value of $K_1$) would shift the radii value at which this transition occurs, but it will again show the same feature of quite a sharp change from large to small values of $\tau$ when the particle size is decreased. It may thus be concluded that the behaviour of ferromagnets depends on the particle size, and may be distinctly different for different samples made of the same material. It can also be concluded that measurements may sometimes yield different results for the same sample, if they do not take the same time. It is thus necessary to take into account the *time-scale* of the experiment, or the experimental time, $t_{\mathrm{exp}}$.

If $\tau \gg t_{\mathrm{exp}}$, no change of the magnetization can be observed during the time of the measurement, and for all practical purposes the magnetization does not change with time. This is the region of *stable ferromagnetism.* If a magnetic measurement takes something of the order of seconds, it is seen from Table 5.1 that for iron made of particles whose radius is *at least* 150 Å, no change can be observed during the experiment. In fact, no change will be observed in such a sample of iron even if it is kept for several days. In this size range, almost everything mentioned in this section may be ignored. The only point which may not be ignored is that this stability of the magnetization does not necessarily hold at the *lowest* energy minimum. If it is brought by some means to the *higher* minimum of Fig. 5.1, it will just stay there, practically for ever, or until it is brought down by an appropriate application of a magnetic field. This is the essential part of the *hysteresis* observed in all ferromagnets. It is important to bear in mind that the existence of hysteresis means that it is *not* sufficient to calculate the lowest energy of a ferromagnetic system. It is always possible that a lower-energy state exists, but it is not accessible because the system is stuck in a higher-energy state.

Of course, the scale of 100 s is just an illustration, and for certain experiments, or applications, the scale may be completely different. Thus, for example, if it is required that the information on a magnetic tape is kept for *years*, it is necessary to see to it that the particles in the tape are large enough to make $\tau \gg 10^8$ s. In studying [181, 182] rock magnetism, it is necessary to take into account the decay of the magnetization during geological times, which may be millions of years. On the other hand, in Mössbauer effect measurements the 'experimental time' is the time of the Larmor precession, which is of the order of $10^{-8}$ s. It is thus possible that particles of a certain size may be stable for the Mössbauer effect but unstable for the conventional magnetic measurements; and samples which are stable during a human life-time may change during geological times. The principle is the same, but the time-scale may be different, which may shift the transition, of which only an example is given in Table 5.1.

In the other extreme, when the particles are small enough to make $\tau \ll t_{\mathrm{exp}}$, many flips back and forth of the magnetization occur during the time of the experiment. Therefore, in zero applied field the measured, average value will be zero. In a non-zero field, the thermal fluctuations have their way and ignore the anisotropy altogether. The calculation of the previous section, 5.1, then applies and the average magnetization is given by the Langevin function, as in eqns (5.1.1)–(5.1.4). The behaviour is the same as that of the paramagnetic atoms discussed in section 2.1, with no hysteresis but *with* saturation, which is reached when all the particles are aligned. Each particle in this size range behaves like a huge atom, with the spin number $S$ of the order of $10^3$ or even $10^4$, instead of $S$ of the order of 1 in the conventional paramagnets. Since the argument of the Brillouin or the Langevin function is proportional to $SH$, saturation is reached in such materials in fields which are very easy to obtain, whereas in the more conventional paramagnets saturation requires very high fields, which are often beyond the capability of the most powerful magnets available. For this reason, this phenomenon of the loss of ferromagnetism in small particles became known as *superparamagnetism*, when the 'super' part was taken to mean 'large' as in superconductivity.

A single particle of such a small size cannot be made or handled. Experiments are therefore carried out on an ensemble of particles, which in most cases have a wide distribution of particle sizes. Such particles would give rise to a superposition of Langevin functions with different values of $\mu = M_s V$ in the argument, and the measured curve could not possibly look like the Langevin function. However, since the argument in eqn (5.1.4) contains the field $H$ as $H/T$, when the measured magnetization is plotted as a function of $H/T$, data for different temperatures should superimpose onto one curve. Therefore, the superposition of $M_H$ *vs.* $H/T$ onto one curve, *and* the absence of hysteresis, used to be taken as an indication that the sample is superparamagnetic, even when that curve did not look like a Langevin function. With improved techniques for producing very small particles, their size distribution has become narrow enough for a pure Langevin function [183] to be observed, and this indirect argument is not necessary any more. The calculation of section 5.1 can now be said to have been confirmed by direct experiment. Of course, a Langevin function (or any other similar function) can *always* be fitted to such data [184, 185] for a rather narrow temperature range, but the remarkably narrow distribution of [183] can be fitted to such a function over a *wide* temperature range. In this respect this experiment is still quite unique in the literature.

The argument of eqn (5.2.21) also contains the temperature in the denominator. The dependence is actually not just on the particle size, but on $V/T$. Therefore, the transition from stable ferromagnetism to superparamagnetism, which is demonstrated in Table 5.1 for the case of room temperature, shifts to a smaller particle size when the temperature is decreased. In measurements of the $M_H$ *vs.* $H/T$ curve, some hysteresis appears suddenly at a sufficiently low temperature, when the sample becomes a ferromagnet. Naturally, the data at these low temperatures are excluded [186] from the superposition. The temperature at which such a transition occurs, namely $T$ for which the relaxation time $\tau$ is *equal* to the time of the experiment $t_{\mathrm{exp}}$, is called the *blocking temperature*, $T_B$. If there is a size distribution in the sample, the *same* temperature may sometimes be above $T_B$ for some of the particles, and below $T_B$ for the others. Such a sample may thus look superparamagnetic for some high values of the temperature $T$, ferromagnetic at low values of $T$, and a *mixture* of both at intermediate $T$. A demonstration of this effect can be seen in Fig. 3 of [187] which plots the Mössbauer effect data for the *same* sample at different temperatures. At $T = 5\,\mathrm{K}$ the structure is that of pure six lines of a ferromagnet. At $T = 324\,\mathrm{K}$ there is one central line of a paramagnet, and at the in-between $T$ there is an obvious mixing of both, with the superparamagnetic portion increasing with the increasing temperature.

This pattern is very similar to the changes in the Mössbauer spectrum observed at the *same* temperature when the average particle size is changed. Indeed if the properties depend on $T/V$ the effect of changing $V$ should be the same as that of changing $T$. An illustration of the effect of varying the size at a constant $T$ can be seen for example in Fig. 3 of [188], which is actually taken on a material that is an *antiferromagnet* (and not a ferromagnet) for a large particle size or low temperatures. This experiment (as well as others) shows that the argument used here applies as well to antiferromagnets, which also become superparamagnets when the particle size is small enough for the thermal fluctuations to flip the magnetization back and forth during the time of the experiment. This effect is quite obvious from the derivation in the foregoing, and it is also clear that the same applies [186] to ferrimagnets, but it is always nicer to have an experimental verification for any theoretical conclusion. The same pattern of a transition from one to six lines can also be obtained by the application [189] of various magnetic fields. It has already been mentioned that the field scale of $SH$ makes it possible to reach the alignment of all particles at easily attained fields. Therefore, it is possible[189] to see the whole development from zero to partial to a total alignment, and this change with the applied field is quite similar to the pattern change with changing temperature.

The wide distribution of particle sizes is most probably the main reason for the *gradual* disappearance of hysteresis when the *average* particle size decreases. The sharp change predicted by the theory is *smeared* in measurements [190] of the hysteresis properties (*i.e.* remanence and coercivity) of 'essentially spherical' particles, as a function of their median diameters. Of course, when the sample contains bigger and smaller particles, some of them may be ferromagnets and some paramagnets at a certain temperature, and the measured properties will then show some sort of a partial hysteresis, as in the Mössbauer effect data mentioned in the foregoing. However, it is possible that *part* of this gradual change, at least in the coercivity, may be due to a different effect. When a particle is magnetized along $+z$, it takes a field $H = H_c$ to reverse its magnetization, see Fig. 1.1. If a field $H < H_c$ is applied, there is an energy barrier, which is *also proportional to the volume*, that prevents the reversal. If the particle size is a little above that which allows a spontaneous flip, it may flip anyway at a field which is somewhat below the bulk coercivity.

There are many theories of such a mechanism in uniaxial particles [191, 192, 193, 194] or platelets [195] and some attempts to take it into account [196] in numerical simulations of the magnetization process. There is even some estimate [197] for the thermal fluctuations overcoming a different kind of an energy barrier, for the motion of a domain wall in bigger particles which are subdivided into domains. However, none of these theories has ever been sufficiently developed for even telling if these effects are large or small for any realistic case. And none of them has ever reached the stage of wondering about the random walk of the magnetization which has been mentioned earlier in this section, and which has not been properly solved even for simpler cases. It should be particularly emphasized that most of the experiments mentioned in this section are semi-qualitative, and check only some features of the theory. The theory predicts a loss of the ferromagnetism when the particle is small, or the temperature is high, and *this* prediction is certainly confirmed. But this theory remains quite crude, and is not developed into more accurate estimations, because there are no experiments that call for a higher accuracy. Actually, except for experiments such as [179] and some of those discussed in [180], there is very little comparison of experiments with *quantitative* theoretical predictions of what the relaxation time is and where the transition should occur in real materials. The main reason is that a quantitative experiment is very difficult to carry out, as will be discussed in the next section.

## 5.3 Magnetic Viscosity

Between the size of superparamagnetism and that of stable ferromagnetism there is in principle a particle size for which $\tau$ is of the order of $t_{\mathrm{exp}}$. According to the example in Table 5.1 it is a very narrow size range, and it is usually quite difficult to prepare a sample of the necessary size to see what happens then. For some techniques of making small particles, the size distribution may well be larger than this transition region. Moreover, it is not even always possible to measure the size of these particles, so much so that there have been many suggestions and attempts to use the superparamagnetic transition as a *measure* for the distribution [189, 198] or at least the *average* [182, 199] of the particle size. Such measurements obviously call for a better theoretical interpretation than the oversimplified estimation of the previous section, which assumes that all the spins within each particle are aligned. Besides other challenges [180] to this assumption, the mere fact that a large proportion of the spins is near the surface in such small particles should make one suspicious of any theory that does not take into account the possible effect of the surface anisotropy.

In practice there is very strong evidence [187, 200, 201, 202, 203] that the magnetization near the surface is often quite different from that in the inner part of the particle, see also the last paragraph of section 5.1.5. Iron particles in particular may be oxidized, so that they are actually made of an iron core surrounded by a shell of iron oxides [201, 204, 205], for which the simple theory of the previous section does *not* apply. Surface effects may also be implied from the observation [206] that magnetic properties of small particles are sometimes sensitive to surfactants adsorbed on the surface. It has also been noticed [207] that the shape of fine particles may not be spherical, and that they may tend to stick together, forming long chains [208] or other [209] aggregates, which change [210, 211, 212] the relaxation time considerably. Interactions between particles have been demonstrated [213, 214] to be very important in real measurements, and these interactions may sometimes look like a *size* distribution [215] in analysing Mössbauer effect data. Other effects, *e.g.* magnetostriction, may also be interpreted as if they were [215] a size distribution. There is thus little wonder that the particle size determined from the magnetic measurements can often be very different [216] from their directly measured size, although such two measurements *are* sometimes [217] consistent, for 'quite uniform and well-isolated' particles.

Comparison between theory and experiment in this particular field is further complicated by the unknown physical constants, because both the saturation magnetization [217, 218, 219] and the anisotropy constant [217, 220] of fine particles differ from their bulk values. If these parameters are adjusted for the small particles, there is not really any direct evaluation of what the theory of the previous section predicts. The Curie temperature may also be different for small particles [221, 222] from what it is in the bulk, or there may be some small regions within the particle [223] which flip before the magnetization of the whole particle flips, when the Curie point is approached. And all these unknowns and uncertainties are superimposed on a theory which is extremely sensitive to small mistakes in the particle size [224], or in other physical parameters, as demonstrated in Table 5.1.

In spite of all these difficulties, there is a surprisingly large number of experiments in the literature for particles in this narrow region for which $\tau \approx t_{\text{exp}}$, even though it is not clear in many cases whether it is really the whole sample, or only part of it, for which the particles are in this size range. In this region of $\tau$, the magnetic properties change while being measured, and this change can in principle be observed. Thus, for example, if a magnetic field is applied and then removed, the average magnetization decays, on a time-scale of the order of $\tau$, which should be possible to measure. A decay is usually exponential to a first order, so that the remanent magnetization should behave according to

$$M_r(t) = M_r(0)\, e^{-t/\tau} \tag{5.3.23}$$

where $t$ is the time. Fitting experimental data to this relation can yield the value of the relaxation time, $\tau$, or at least its average when the system has a distribution of the values of $\tau$.

However, nobody ever tries to fit data to eqn (5.3.23), because it is taken for granted that there must be a wide distribution of the particle size, which must cause a wide distribution of $\tau$, and the time decay is actually

$$M_r(t) = M_r(0) \int_0^\infty P(\tau) e^{-t/\tau}\, d\tau, \tag{5.3.24}$$

where $P$ is a distribution function. Old estimations, and more recent numerical computations [225] for *specific* distribution functions $P$, show that under certain conditions, eqn (5.3.24) can be approximated by

$$M_r(t) = C - S \ln(t/\tau_0), \tag{5.3.25}$$

where $C$ and $S$ are constants, and this functional form is used to analyse

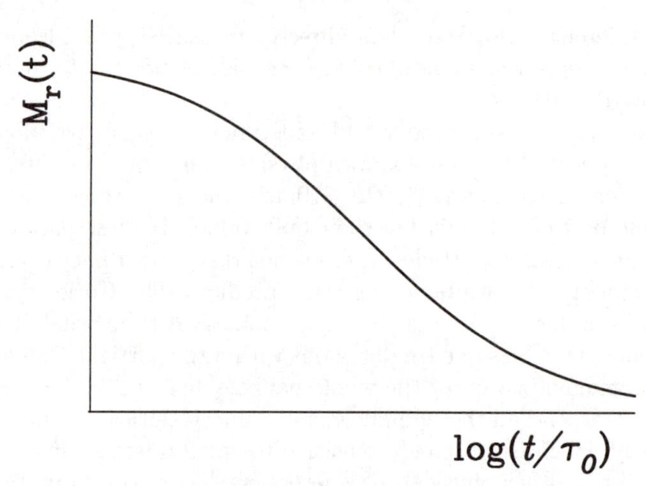


FIG. 5.2. Schematic representation of a magnetization decay on a logarithmic scale.

practically all experimental data. Most workers omit the constant $\tau_0$ and absorb it in $C$, but it is wrong to do so, because a logarithm is only defined for a dimensionless number. It may also be worth noting that most workers do not choose the integrand as in eqn (5.3.24), with a distribution of the values of $\tau$. They prefer a distribution of the particle sizes, or of the energy barriers, and use the dubious assumption that the relation between these parameters and $\tau$ is known and established. It is not, according to the discussion in the previous section.

This choice of the logarithmic function is rather strange, because it is not regular for either small or large values of $t$, and can certainly not represent the beginning or the end of the measurements. The real function may at most be linear in $\ln(t/\tau_0)$ over a limited range, which does not contain the short and the long time. In principle, it can at most look like the schematic plot in Fig. 5.2, and indeed this form is what is observed [226] when data are taken over a wide range of the time. However, many experimentalists just assume that the logarithm is the 'true' form to be used, and they do not report (or do not measure) anything outside the range for which eqn (5.3.25) can be fitted. Cases have been reviewed [227] in which the reported time range was so narrow that it may not even be in the linear region of Fig. 5.2, and in an extreme case $M_r$ was measured at only *two values* of $t$, in order to determine $S$ of eqn (5.3.25).

The logarithm is so inconvenient that even if it were an essential part of the physical problem, there should be some attempts to avoid it as much as possible. Using it as an approximation, *even* if it is a *good* approximation, as is claimed [225] for certain cases, is a completely unnecessary complication. It has been claimed [228] that the criticism of the logarithmic function as breaking down for large and small $t$ is 'incorrect' (*sic*!), because eqn (5.3.25) is only used for a certain time-window $t_{\min} \ll t \ll t_{\max}$, but no reason was ever given for leaving out the time outside this region. Actually, the limits of the region which is linear in $\ln(t/\tau_0)$ are related [229] to the *width* of the distribution. By trying to fit *everything* to eqn (5.3.25), or by avoiding the region outside that time-window, important physical information is thus lost. Moreover, there is reason to believe that in many experiments only *part* of the sample decays with $\tau$ of the order of $t_{\exp}$. The main justification for using eqn (5.3.25) is [225] that it is a good approximation 'for a *wide* distribution of energy barriers'. However, if the distribution is wide, it is easy to imagine that some of the particles are large enough to be stable ferromagnets under the conditions of the experiment, or that some of the particles are small enough to be superparamagnetic, or both. Leaving out the part of the decay curve for short and for long values of the time leaves out all the information about the small and the large particles in the sample. It is a risky procedure, especially since it has been shown that at least one method produces [230] two groups of particles in the same sample: large ones which are ferromagnetic, and small ones which are superparamagnetic. On top of all that, a logarithm may not even be a true representation of a wide distribution, because an *alternative* explanation [229] says that an apparent linear dependence on $\log(t/\tau_0)$ may *also* be caused by magnetostatic interactions among the particles. As long as eqn (5.3.25) is used for the analysis of the data, it is impossible to distinguish between these two effects.

Actually, it is not even necessary to look for an approximation which is easier to use than the logarithm, because it is possible [227] to carry out the integration in eqn (5.3.24) rigorously and analytically, if $P(\tau)$ is taken to be the so-called *gamma distribution function*,

$$P(\tau) = \frac{1}{\tau_0 \Gamma(p)} \left( \frac{\tau}{\tau_0} \right)^{p-1} e^{-\tau/\tau_0} , \tag{5.3.26}$$

where $\Gamma$ is the gamma function, and $p$ and $\tau_0$ are adjustable parameters. This function looks more or less like any other probability function, as can be seen from the three examples plotted in Fig. 5.3 for the particular choice of $p = 2$, 3 and 4. On this reduced scale, the value of $\tau_0$ does not have to be specified, but it will obviously have to be if $\tau$ is given in real units of time. Graphs can readily be plotted for other values of these parameters, and they all look qualitatively the same. It is, therefore, as legitimate to use as any other distribution function, and at least no convincing argument has ever been presented for the use of any *other* distribution function, the choice of which is also quite arbitrary. Any difference between different probability functions is at most a second-order effect, which is better left to be studied only after all the first-order effects have been clarified. The

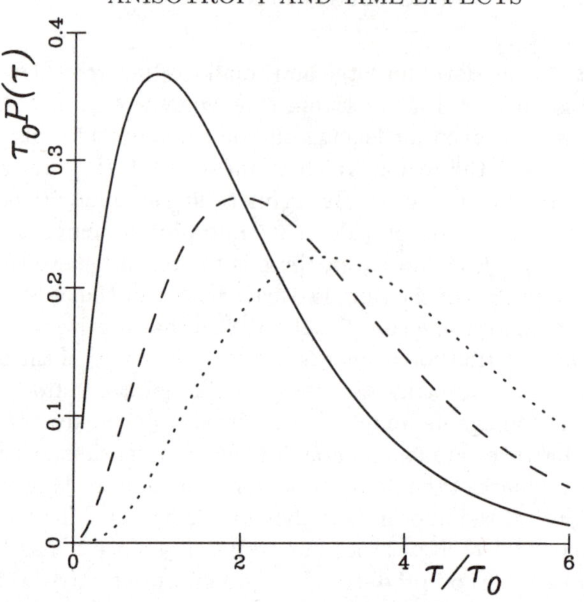


FIG. 5.3. The gamma distribution function of eqn (5.3.26) plotted for $p = 2$ (full curve), $p = 3$ (dashed curve) and $p = 4$ (dotted curve).

*same* function of eqn (5.3.24) was used [231] for a distribution of energy barriers, instead of the distribution of relaxation times used here. For this case the integration cannot be carried out analytically as is done here.

When eqn (5.3.26) is substituted in eqn (5.3.24) and the integration is carried out, the result is

$$\frac{M_r(t)}{M_r(0)} = \frac{2}{\Gamma(p)} \left(\frac{t}{\tau_0}\right)^{p/2} K_p \left(2\sqrt{\frac{t}{\tau_0}}\right) , \tag{5.3.27}$$

where $K_p$ is the modified Bessel function of the third kind. This function is well-defined, its properties have been investigated for any range of the parameters, and $x^p K_p(x)$ has no singularities. Therefore, all sorts of experimental data may be fitted to this function, and no separate treatment is needed for large or small $t$. Such a fitting should determine the two parameters, $p$ and $\tau_0$, of the function $P$ of eqn (5.3.26). Their values then determine the two parameters which are most significant for any physical problem that involves any kind of probabilities. One of them is the *mean value* which in the case of eqn (5.3.26) is given by

$$m = p\tau_0, \tag{5.3.28}$$

and the other is the *variance*, which in this case is

$$\sigma^2 = p\tau_0^2 . \tag{5.3.29}$$

This important physical information about the particular system under study is just lost if eqn (5.3.25) is used.

The possibility of using the gamma distribution function has somehow been ignored, and is not even mentioned in the most recent review [225] of different models. They [225] and others [232] insist on choosing some *other* $P(\tau)$, and carrying out the integration numerically. And even for cases in which $\log(t/\tau_0)$ turns out to be inadequate, they [225] and others [228] suggest using a *power series* in $\log(t/\tau_0)$ thus conserving the inconvenient and non-physical singularity for small and large $t$. Similar suggestions have been reviewed in [227]. There has been an attempt [233] to plot one universal curve for the decay of the magnetization measured for the same sample at different temperatures $T$. However, even for that purpose, it was suggested to plot the data as a *function* of $(T/T_0)\ln(t/\tau_0)$, while it is clear that any function of that parameter is also a *function* of $te^{T/T_0}$. It seems that this field cannot advance before the obsession with logarithms is over.

## 5.4 The Stoner–Wohlfarth Model

When a ferromagnetic particle is large enough, all the time-effects described in section 5.2 are negligibly small. Nevertheless, such particles may still be small enough for the exchange energy to hold all spins tightly parallel to each other, and not allow the space-dependence of the magnetization which enters only at a larger particle size. In this case, as in the case studied in section 5.2, the exchange energy is a constant, and does not enter the energy minimizations. There are then only the anisotropy energy of the particle and the interaction with the applied field to be considered. It is then possible to use the *same* energy relation as in eqn (5.2.13) to solve for the hysteresis curve of these stable but small ferromagnetic particles. Such a calculation is known as the *Stoner–Wohlfarth model.*

Actually, the original study [234] of Stoner and Wohlfarth assumed a shape anisotropy, which will be defined in section 6.1, and not the uniaxial, crystalline anisotropy as in eqn (5.2.13). However, the mathematics is the same, and this model was also used later for the case of this anisotropy. Moreover, a calculation based on this model is being widely used to measure the *crystalline* anisotropy, as mentioned in section 5.1.6.

The main assumption of Stoner and Wohlfarth is that the material is made up of rather small particles, which are sufficiently separated from each other so that interactions between them are negligible. If the magnetic field, $H$, is applied at an angle $\theta$ to the easy axis of the uniaxial anisotropy of the particle, the magnetization vector will rotate to an angle $\phi$ from the field direction, which means that the magnetization will be at an angle $\phi - \theta$ from the easy axis. The energy of this system is the same as in eqn (5.2.13), with the change of the angles into the ones defined here, namely

$$\mathcal{E} = K_1 V \sin^2(\phi - \theta) - \mu H \cos\phi, \qquad (5.4.30)$$

where $V$ is the volume, and where the magnetic moment $\mu$ of the particle may be replaced by $M_s V$, as was later done in section 5.2 as well.

Stoner and Wohlfarth preferred to use a different definition of the energy zero, and replaced $\sin^2$ by the cosine of the double angle. They worked with the reduced energy,

$$\eta = \frac{\mathcal{E}}{2K_1 V} + \text{const} = -\frac{\cos[2(\phi - \theta)]}{4} - h\cos\phi, \tag{5.4.31}$$

where

$$h = \frac{M_s H}{2K_1}. \tag{5.4.32}$$

For given values of $\theta$ and $h$, the magnetization will choose the angle $\phi$ which minimizes this energy, namely the solution of

$$\frac{\partial \eta}{\partial \phi} = \frac{1}{2}\sin[2(\phi - \theta)] + h\sin\phi = 0, \tag{5.4.33}$$

provided the solution represents an energy minimum and not a maximum. This condition can be expressed as

$$\frac{\partial^2 \eta}{\partial \phi^2} = \cos[2(\phi - \theta)] + h\cos\phi > 0. \tag{5.4.34}$$

Because of the multi-valued trigonometric functions, eqn (5.4.33) has always more than one solution for a given $h$ and $\theta$, and it can happen that more than one of these solutions represents an energy minimum. In order to obtain a unique solution, it is necessary to specify, and follow, the *history* of the value of $h$ for each $\theta$. A solution which starts from a particular branch cannot be just allowed to jump into another branch. The jump must be at a field value at which there is no energy barrier between these branches. This important feature is the basis of the hysteresis which is always part of magnetism, and in order to see how it works it helps to look first into the trivial case $\theta = 0$. In this case eqns (5.4.33) and (5.4.34) are

$$(h + \cos\phi)\sin\phi = 0, \quad \text{and} \quad \cos(2\phi) + h\cos\phi > 0. \tag{5.4.35}$$

One solution of the first half is $\cos\phi = -h$, which is a valid solution if $|h| < 1$, but it does not fulfil the second half. This solution represents an energy *maximum* and has no physical significance. The other solution is

$$\sin\phi = 0, \quad \text{and} \quad 1 + h\cos\phi > 0. \tag{5.4.36}$$

The combination means that it is necessary to use $\phi = 0$ for $h > -1$, and $\phi = \pi$ for $h < 1$.

It is thus seen that the solution is unique if $|h| > 1$, but in the region $|h| < 1$ both $\phi = 0$ and $\phi = \pi$ are valid energy minima. At this point it is necessary to introduce the field history. If we start by applying a large positive $h$, then reduce the field to zero, and increase it in the opposite direction, the physical system remains on the branch of the solution $\phi = 0$, till the field $h = -1$ is reached. At this field the solution becomes unstable, and the system must jump to the other branch, $\phi = \pi$. Note in particular that according to eqn (5.4.31), the reduced energy in this case is $\eta = -\frac{1}{4} - h\cos\phi$. Once $h$ passes zero, and becomes even slightly negative, the state $\phi = 0$ has a *higher* energy than that with $\phi = \pi$. However, the magnetization cannot just jump into the lower-energy state, because it is in a minimum energy state, which means that there is an energy barrier that holds it there. The situation is similar to the energy displayed in Fig. 5.1. The system is just stuck in the higher-energy state, till the field reaches the value $h = -1$, at which the barrier is removed and a jump to a lower-energy state becomes possible. A similar, but reversed, argument applies to starting from a large negative $h$, in which case the other branch is held till the field reaches the value $h = 1$. The whole hysteresis curve is then *qualitatively* similar to the limiting curve plotted in Fig. 1.1, and the coercivity as defined there is for the reduced field $h = 1$, which means $H_c = 2K_1/M_s$ according to eqn (5.4.32).

If $\theta \neq 0$, eqn (5.4.33) has to be solved numerically, but the general behaviour is rather similar to the case of $\theta = 0$ which has just been described. Starting from a large positive field, the solution which starts with $\phi = 0$, *i.e.* $\cos\phi = 1$, curves down with decreasing values of $h$ to lower values of $\cos\phi$, namely to smaller values of the component of the magnetization in the field direction,

$$M_H = M_s \cos\phi. \tag{5.4.37}$$

At the point where this branch stops to be a minimum, there is a jump to a second branch, thus displaying something which looks more or less like Fig. 1.1. Obviously, the jump occurs where the left hand side of eqn (5.4.34) passes through zero, making that branch change from a minimum to a maximum. The combination of a zero for this equation together with eqn (5.4.33) gives rise to several relations [234] between the 'critical' values of $h$ and $\phi$ at which the jump occurs for a given $\theta$.

It may be interesting to look also into the other extreme case which does not call for a numerical evaluation. This case is $\theta = \pi/2$, *i.e.* a field perpendicular to the easy axis of the anisotropy, which effectively means no anisotropy at all. In this case eqns (5.4.33) and (5.4.34) become

$$(h - \cos\phi)\sin\phi = 0, \quad \text{and} \quad -\cos(2\phi) + h\cos\phi > 0. \tag{5.4.38}$$

In this case, the solution $\cos\phi = h$, which is a valid solution if $|h| < 1$, also fulfils the second half of eqn (5.4.38), and is an energy minimum. It yields a magnetization proportional to the field, as in a paramagnet, with no hysteresis and with a zero coercivity. At $h = \pm 1$ it changes over to the second solution of $\sin\phi = 0$, which is the saturation of $\phi = 0$ or $\phi = \pi$.

After computing the hysteresis curves for each field angle $\theta$, Stoner and Wohlfarth [234] computed the average for a random distribution of the angles $\theta$, namely a collection of particles with a random distribution of the direction of their easy axes with respect to the direction of the applied field. The resulting curve is very similar to the one shown in Fig. 1.1. Actually, many experimental curves could be analysed in terms of this simple theory, which has been widely used over the years. Even magnetization curves of thin permalloy films obey approximately the Stoner–Wohlfarth theory, although the physical mechanism behind it is not clear.

The main advantage of this theory is that it is sufficiently simple to add some extra features to it. It is just as easy to replace the random distribution of $\theta$ by some other distribution, centred wherever there is an experimental reason to believe that the directions of easy axes are more likely to be, as in the case of an aligned, or a partly aligned, magnetic tape. The case of a cubic, *instead* of a uniaxial, anisotropy has also been worked out [235] in detail. In this case there are more branches than in the uniaxial case, which makes it sometimes more difficult to decide into which branch to jump. But these difficulties can be handled. A random cubic anisotropy *besides* an overall uniaxial one has also been used [236] in the study of the magnetization ripple, mentioned in section 5.1.4. The parallel and perpendicular susceptibilities [161], discussed in section 5.1.6, have also been calculated from this model. Further developments and a look into finer details [237] can even start to offer a physical interpretation for the *difference* between experimental results and the Stoner–Wohlfarth theory. It may sometimes lead to an understanding of the parts neglected in the Stoner–Wohlfarth theory, which are the interactions between the particles and the possibility of *some* space-dependence of the magnetization, within each particle. Interactions of certain groups of ellipsoids have also been computed [238] for this model.

# 6

# ANOTHER ENERGY TERM

## 6.1 Basic Magnetostatics

Besides the energy terms discussed so far, there is another term which has not been mentioned yet, and it is time to introduce it. This term is the *magnetostatic self-energy* which originates from the classical interactions among the dipoles. For a *continuous* material it is described by Maxwell's equations, which the reader is assumed to be familiar with, in the form taught to undergraduates, even if not necessarily familiar with the part which is most relevant for ferromagnets. From a historical point of view it is interesting to note that this energy term was part of the Hamiltonian in the early study of [39] spin waves, which included the anisotropy as well. Dyson [40] objected to some of the approximations used [39] for this term, but did not introduce any other, and since then somehow everybody just got used to leaving out this energy term.

For the meantime it is just assumed for simplicity that the material is continuous, leaving for the next chapter the study of a crystal made out of discrete atoms (or ions). Not all of Maxwell's equations are used in the present discussion of a *ferromagnet*, with no particular reference to its *electric* properties. One of the equations states that

$$\nabla \times \mathbf{H} = 0, \tag{6.1.1}$$

in the absence of any currents, or displacement currents. It should be noted, however, that this assumption of zero currents does not lead to a restrictive, particular case. It is customary in the study of ferromagnetism to separate the magnetic fields into two categories, and treat the field $\mathbf{H}$ in eqn (6.1.1) as separate from the *applied field* produced by currents in coils. As long as these 'different' fields are properly superimposed, there is no loss of generality, and there is nothing wrong with this *convenient* notation.

The most general solution of eqn (6.1.1) is well known. The vector $\mathbf{H}$ is a gradient of a scalar, $U$, called the *potential.* The convention is to define it with a minus sign,

$$\mathbf{H} = -\nabla U. \tag{6.1.2}$$

Another one of Maxwell's equations is

$$\nabla \cdot \mathbf{B} = 0, \tag{6.1.3}$$

where $\mathbf{B}$ is the *magnetic induction*, defined in eqn (1.1.2). See the book by Brown [1] for the derivation of these equations, and for a rigorous definition of the vectors $\mathbf{B}$ and $\mathbf{H}$. It should only be emphasized that it is *wrong* to write $\nabla \cdot \mathbf{H} = 0$, as is done in some books. The latter is equivalent to eqn (6.1.3) only if eqn (1.1.3) holds, which is not the case in ferromagnetism. The factor $\gamma_{\mathrm{B}}$ invented by Brown [1] will be used throughout this chapter, as a way of introduction, in order to make the transition easier for readers who have only used the SI units till now. For the SI units, now used in all undergraduate textbooks, its value is $\gamma_{\mathrm{B}} = 1$, while for the Gaussian, cgs units, used in all the literature on magnetism, $\gamma_{\mathrm{B}} = 4\pi$. More conversion factors are listed in section 6.4, and from there on, for the rest of the book, only the cgs system of units will be used. Rightly or wrongly, this system of units is still used almost exclusively in all the literature on magnetism, even though some usage of SI is starting to creep into some of the more recent papers. For anybody who wants to study this subject there is no alternative to getting used to the cgs units.

Substituting eqns (1.1.2) and (6.1.2) in eqn (6.1.3),

$$\nabla^2 U_{\mathrm{in}} = \gamma_{\mathrm{B}} \nabla \cdot \mathbf{M}, \tag{6.1.4}$$

which should be valid *inside* the ferromagnetic body (or bodies). Outside this body (or these bodies) $\mathbf{M} = 0$, so that $\mathbf{B} = \mathbf{H}$ and the differential equation is

$$\nabla^2 U_{\mathrm{out}} = 0. \tag{6.1.5}$$

It is also known from undergraduate textbooks that Maxwell's equations require that the component of $\mathbf{H}$ parallel to the surface, and the component of $\mathbf{B}$ perpendicular to the surface, are continuous on the boundary of two materials. These requirements lead to the boundary conditions that on the surface of the ferromagnet,

$$U_{\mathrm{in}} = U_{\mathrm{out}}\,, \qquad \frac{\partial U_{\mathrm{in}}}{\partial n} - \frac{\partial U_{\mathrm{out}}}{\partial n} = \gamma_{\mathrm{B}} \mathbf{M} \cdot \mathbf{n}\,, \tag{6.1.6}$$

where $\mathbf{n}$ is the unit normal to the surface of the ferromagnetic body (or bodies), taken to be positive in the outward direction. Besides these boundary conditions, the potential $U$ is required to be *regular* at infinity, which means that both $|rU|$ and $|r^2 \nabla U|$ are bounded as $r \to \infty$. This regularity essentially means that the behaviour of the potential at a large distance from the magnetized bodies is the same as that of the potential of a point charge, which can be expected if the magnetization vanishes outside a certain finite volume.

Instead of the scalar potential, the problem may be formulated equally well by writing $\mathbf{B} = \nabla \times \mathbf{A}$ and deriving a differential equation with boundary conditions for the *vector potential* $\mathbf{A}$. However, this formulation is less convenient for the problems discussed in this book, and will not be used here.

Once the differential equations and boundary conditions have been solved and $U$ is known for the whole space, $\mathbf{H}$ can be calculated from eqn (6.1.2). The *energy* can then be evaluated as

$$\mathcal{E}_{\mathrm{M}} = -\frac{1}{2}\int \mathbf{M}\cdot\mathbf{H}\,d\tau, \tag{6.1.7}$$

where the integration is over the ferromagnetic bodies. This equation will be proved more rigorously in the next chapter. For the meantime it may be taken as the interaction of each dipole with the field $\mathbf{H}$ created by the other dipoles, and a factor $\frac{1}{2}$ introduced in order to avoid counting twice the interaction of $A$ with $B$, and of $B$ with $A$.

### 6.1.1 *Uniqueness*

The most important feature of these differential equations and boundary conditions is that their solution is *unique*. In order to prove this statement, suppose that there are two functions of space, $U_1$ and $U_2$, that fulfil all the equations (6.1.4) to (6.1.6) and are both regular at infinity. Then the function $U_3 = U_1 - U_2$ and its derivative must be *continuous* everywhere, including the surfaces on which the normal derivatives of $U_1$ and $U_2$ are discontinuous. Also, according to eqn (6.1.4), $\nabla^2 U_3 = 0$ everywhere, which means that for an integration over any arbitrary volume,

$$\int (\nabla U_3)^2\,d\tau = \int \left[\nabla\cdot(U_3\nabla U_3) - U_3\nabla^2 U_3\right] d\tau = \int U_3\frac{\partial U_3}{\partial n}dS, \tag{6.1.8}$$

where the second equality is a manifestation of the divergence theorem, and the last integral is over the surface surrounding the chosen, arbitrary volume. It should be noted that such a use of the divergence theorem is not allowed for $U_1$ or $U_2$, because of the discontinuity expressed by eqn (6.1.6), which requires integration over both faces of each discontinuity surface. However, according to the present assumption, both $U_3$ and its normal derivative are continuous everywhere, and the integrations over both faces cancel each other because of the opposite direction of $\mathbf{n}$.

If the volume chosen for the integration in eqn (6.1.8) is now allowed to tend to infinity, $dS$ increases as $r^2$, while the regularity condition requires $\partial U_3/\partial n$ to decrease at least as $r^{-2}$, and $U_3$ to decrease at least as $r^{-1}$, so that the surface integral tends to zero. Hence, the integral of $(\nabla U_3)^2$ over the whole space vanishes. And since the integrand is a square, which cannot be negative anywhere, $\nabla U_3$ must vanish everywhere, which means that $U_3$ =const. But a non-zero constant is not regular at infinity. Therefore, $U_3 = 0$ everywhere, and $U_1 \equiv U_2$.

There is thus only one possible solution to the potential problem of any geometry and any distribution of the magnetization. Therefore, it is never necessary to give the intermediate steps, or to justify in any other way a solution to a potential problem. If a certain function is guessed, or arrived at by any other means, it is sufficient to show that if fulfils the differential equations and the boundary conditions, because if it is a solution of the problem, it is always *the* solution of that problem. It should be noted, however, that while a magnetization distribution determines a unique field outside the ferromagnet, the reverse is not true. A measurement of the field outside a ferromagnetic body is *not* sufficient [239] to determine a unique magnetization distribution that creates this field.

### 6.1.2 *Trivial Examples*

The theorem about the uniqueness of the solution allows quoting without proof the potential for some simple cases. The proof is in substituting each of these functions in eqns (6.1.4) to (6.1.6), and checking that it *is* a solution.

The first case is a sphere, whose radius is $R$, uniformly magnetized along the $z$-direction. In this case, $\nabla \cdot \mathbf{M} = 0$, and in polar coordinates $r$, $\theta$ and $\phi$, the differential equation becomes

$$\left[\frac{1}{r^2}\frac{\partial}{\partial r}r^2\frac{\partial}{\partial r} + \frac{1}{r^2\sin\theta}\frac{\partial}{\partial\theta}\sin\theta\frac{\partial}{\partial\theta} + \frac{1}{r^2\sin^2\theta}\frac{\partial^2}{\partial\phi^2}\right]U = 0, \tag{6.1.9}$$

both inside and outside the sphere. Also, in this case

$$\partial/\partial n = \partial/\partial r \quad \text{and} \quad \mathbf{M}\cdot\mathbf{n} = M_r = M_s\cos\theta, \tag{6.1.10}$$

because $M_s$ is the magnitude of $\mathbf{M}$. It can be verified by substitution that

$$U = \frac{M_s}{3}\gamma_{\mathrm{B}}\cos\theta \times \begin{cases} r & \text{if } r \le R \\ R^3/r^2 & \text{if } r \ge R \end{cases} \tag{6.1.11}$$

satisfies eqn (6.1.9), is continuous on $r = R$, has the appropriate discontinuity of the derivative required by substituting eqn (6.1.11) in eqn (6.1.6), and is regular at infinity. Therefore, it is *the* solution of the potential problem inside and outside a uniformly magnetized sphere.

In particular, the potential inside the sphere is actually

$$U_{\mathrm{in}} = \frac{M_s}{3}\gamma_{\mathrm{B}}\, z. \tag{6.1.12}$$

Substituting in eqn (6.1.2), the field inside the sphere is

$$H_{x_{\mathrm{in}}} = H_{y_{\mathrm{in}}} = 0\,, \qquad H_{z_{\mathrm{in}}} = -\frac{M_s}{3}\gamma_{\mathrm{B}}\,. \tag{6.1.13}$$

It is, thus, a *uniform* field, which is antiparallel to $z$. However, the $z$-direction has no particular meaning for a sphere, and in the present context it only defines the direction of the magnetization. Therefore, the internal field in a homogeneously magnetized sphere is antiparallel to the magnetization. It should be clear now why *spherical* particles were specified in the previous chapter. For any other geometry, the direction of this internal field may not be parallel to the easy anisotropy axis, which complicates the problem studied there.

The magnetostatic *energy* of this uniformly magnetized sphere is obtained by substituting this $\mathbf{H}$ in eqn (6.1.7). Since the integrand is a constant, the integration is only a multiplication by the volume of the sphere, $\frac{4}{3}\pi R^3$. Therefore, the magnetostatic self-energy of a uniformly magnetized sphere is

$$\mathcal{E}_{\mathrm{M}} = \frac{2\pi}{9}\gamma_{\mathrm{B}} R^3 M_s^2. \tag{6.1.14}$$

The second example is an infinite circular cylinder, which is uniformly magnetized along the $x$-axis, where the cylinder axis is defined as $z$. Again, $\nabla \cdot \mathbf{M} = 0$ everywhere, and in the cylindrical coordinates $\rho$, $\phi$ and $z$ the differential equation is

$$\left[\frac{1}{\rho}\frac{\partial}{\partial \rho}\rho\frac{\partial}{\partial \rho} + \frac{1}{\rho^2}\frac{\partial^2}{\partial \phi^2} + \frac{\partial^2}{\partial z^2}\right] U = 0, \tag{6.1.15}$$

while the normal is parallel to $\rho$, and the normal component is

$$\mathbf{M} \cdot \mathbf{n} = M_s \cos\phi. \tag{6.1.16}$$

It can be verified by substitution that the solution for this case is

$$U = \frac{M_s}{2}\gamma_{\mathrm{B}} \cos\phi \times \begin{cases} \rho & \text{if } \rho \le R \\ R^2/\rho & \text{if } \rho \ge R \end{cases} \tag{6.1.17}$$

where this time $R$ is the radius of the cylinder. The internal field inside the cylinder is

$$H_{x_{\mathrm{in}}} = -\frac{M_s}{2}\gamma_{\mathrm{B}}\,, \qquad H_{y_{\mathrm{in}}} = H_{z_{\mathrm{in}}} = 0\,, \tag{6.1.18}$$

which is also a uniform field, antiparallel to the magnetization vector. The energy *per unit length along z* is

$$\mathcal{E}_{\mathrm{M}} = \frac{\pi}{4}\gamma_{\mathrm{B}} R^2 M_s^2. \tag{6.1.19}$$

If the same cylinder is magnetized along the $z$-direction, eqn (6.1.15) is still the differential equation to be solved, but in the boundary condition of eqn (6.1.6) one should take $\mathbf{M} \cdot \mathbf{n} = 0$. The solution is then $U = 0$, which leads to $\mathbf{H} = 0$, and $\mathcal{E}_{\mathrm{M}} = 0$.

### 6.1.3 *Uniformly Magnetized Ellipsoid*

The examples of a sphere and a cylinder are particular cases of a more general theorem about uniformly magnetized ellipsoids, which was already known to Maxwell. It will be stated here without a proof.

Generally, the field inside a uniformly magnetized ferromagnetic body is *not* uniform. However, if and only if the surface of this body is of a second degree, the internal field is uniform. This theorem is often stated as applying only to ellipsoids, rather than to surfaces of a second degree, because all other second-degree surfaces extend to infinity, and cannot be realized in practice. Still, the ellipsoid is usually understood to include the limiting case of an infinite cylinder.

When the Cartesian coordinates are chosen along the *principal axes* of a general ellipsoid, the equation of its surface is

$$\left(\frac{x}{a}\right)^2 + \left(\frac{y}{b}\right)^2 + \left(\frac{z}{c}\right)^2 = 1 \quad \text{with} \quad a \leq b \leq c. \tag{6.1.20}$$

It may sometimes be necessary to define $x$, $y$ and $z$ in other directions, but the reader is supposed to know how to perform the rotation of the axes in this equation, and in the ones that follow. If this ellipsoid is uniformly magnetized, the field inside the ellipsoid can be written as

$$\mathbf{H}_{\mathrm{in}} = -N \cdot \mathbf{M} = -\gamma_{\mathrm{B}} D \cdot \mathbf{M}, \tag{6.1.21}$$

where both $D$ and $N = \gamma_{\mathrm{B}} D$ are *tensors*. In the particular case when $\mathbf{M}$ is parallel to one of the principal axes of the ellipsoid, both $D$ and $N$ are *numbers*, and *both* are known by the name *demagnetizing factors*, or sometimes *demagnetization factors*. Except for the use of the letters $D$ and $N$, which is almost (but not quite) universal, it is sometimes difficult to tell *which* of these demagnetizing factors is being referred to. It should also be noted that this field (also known as the demagnetizing field) is the part created by the magnetization. If there is *also* an applied field, produced by some currents in external coils, these fields superimpose, and have to be summed vectorially. It should also be noted that all this treatment applies only to the case of an ellipsoid which is uniformly magnetized, and not to any other spatial distribution of the magnetization.

The last part of this theorem is that the trace of the tensor $D$ is 1, which also means that the trace of $N$ is $\gamma_{\mathrm{B}}$. Therefore, the results of the foregoing examples for the sphere and the cylinder can be obtained from simple symmetry considerations. For a sphere all the directions are equivalent, and the three demagnetizing factors must be equal. Therefore, $D_x = D_y = D_z = \frac{1}{3}$, because the trace of the tensor (which is the sum of these three numbers) is 1. Substituting in eqn (6.1.21) leads to the same result as in eqn (6.1.13). Similarly, for an infinite cylinder, there is no surface of discontinuity along $z$, so that $D_z = 0$. The other two factors should be equal for a circular cross-section, so that $D_x = D_y = \frac{1}{2}$, which leads to the same result as in eqn (6.1.18).

Substituting eqn (6.1.21) in eqn (6.1.7), the magnetostatic self-energy of a uniformly magnetized ellipsoid, whose volume is $V$, is

$$\mathcal{E}_{\mathrm{M}} = \frac{1}{2}V\left(N_x M_x^2 + N_y M_y^2 + N_z M_z^2\right), \tag{6.1.22}$$

which is some sort of anisotropy energy. It is the *shape anisotropy* term, which was mentioned, but not defined, in section 5.1.4.

Two particular cases are of special interest, and both are ellipsoids for which two axes are equal. One is the case $a = b$ and $c > a$ (which is a kind of egg-shaped particle). It is called a *prolate spheroid.* The other is shaped more or less like a disk, or rather a 'flying saucer', and is the case $a < b$ and $b = c$. It is called an *oblate spheroid.* The sphere, $a = b = c$, is the limit of *both.*

If two axes are equal, the related two demagnetizing factors are the same. Thus, for a prolate spheroid $N_x = N_y$, and for an oblate spheroid $N_y = N_z$. In the case of a prolate spheroid, eqn (6.1.22) can, therefore, be written as

$$\mathcal{E}_{\mathrm{M}} = \frac{1}{2}V\left[N_x(M_x^2 + M_y^2) + N_z M_z^2\right] = \frac{1}{2}V(N_z - N_x)M_z^2 + \text{const}, \tag{6.1.23}$$

because $M_x^2 + M_y^2 + M_z^2 = M_s^2$, which is a constant. This shape anisotropy energy term has the same mathematical form as the first-order uniaxial anisotropy term of section 5.1.1, even though their physical origins are different. A similar expression obviously applies to the case of the oblate spheroid, and in either case this shape anisotropy may exist *besides* a uniaxial or cubic magnetocrystalline anisotropy term, discussed in section 5.1. Moreover, the easy axis of the crystalline anisotropy term is not necessarily parallel, or related in any other way, to the easy axis of the shape anisotropy term, and in principle there can be any angle between them.

The original Stoner–Wohlfarth model, discussed in section 5.4, assumed [234] no crystalline anisotropy, and dealt only with the shape anisotropy of ellipsoids. At first both prolate and oblate spheroids were considered, but later extensions addressed mostly prolate spheroids, which are [240] very common in permanent magnet materials. Comparing eqn (6.1.23) with the first term of eqn (5.4.30) it is seen that for a prolate spheroid all the algebra of section 5.4 is unchanged, but eqn (5.4.32) should be replaced by

$$h = \frac{H}{(N_x - N_z)M_s}. \tag{6.1.24}$$

A Stoner–Wohlfarth theory is also possible for particles which have *both* a crystalline and a shape anisotropy. For example, it has been done [241] for a uniaxial anisotropy with both $K_1$ and $K_2$ superimposed on the shape anisotropy. Such calculations usually assume that the easy axis is the same for both anisotropies. If there is a certain angle between these two axes, the problem becomes a little more complicated, but not prohibitively so.

The demagnetizing factors in a general ellipsoid depend only on the *ratios* $a/b$ and $b/c$, and *not* on the values of $a$, $b$ and $c$. Thus, two ellipsoids which have the same shape, but a different volume, have the same $N$ or $D$. Analytic expressions, albeit in terms of elliptic integrals, are known for the functional form of the dependence of the demagnetizing factors on the axial ratios. Formulae, graphs and tables were published [242] by Osborn and then more tables were computed [243] with a more modern computer program. In the particular case of a prolate spheroid, the expressions for the demagnetizing factors are made out of more elementary functions. Specifically, by using the notation

$$p = c/a \ (> 1), \qquad \xi = \sqrt{p^2 - 1}\,/p, \tag{6.1.25}$$

the demagnetizing factors for a prolate spheroid become

$$D_z = \frac{1}{p^2 - 1}\left[\frac{1}{2\xi}\ln\left(\frac{1+\xi}{1-\xi}\right) - 1\right], \qquad D_x = \frac{1 - D_z}{2}. \tag{6.1.26}$$

For a small $\xi$, a power series expansion of the logarithm in this equation leads to

$$D_z = \frac{1}{p^2}\left[\frac{1}{3} + \sum_{k=1}^{\infty}\frac{1}{2k+3}\left(\frac{p^2-1}{p^2}\right)^k\right]. \tag{6.1.27}$$

In this form it is clear that in the limit $p \to 1$, which is a sphere, $D_z = \frac{1}{3}$, as concluded in the foregoing.

Demagnetizing fields and demagnetizing factors are also defined for non-ellipsoidal bodies. These definitions will be given in section 6.3, after the introduction of the magnetic charge.

## 6.2 Origin of Domains

The stage is now set for a demonstration that it is the magnetostatic self-energy term which is responsible for the existence of the magnetic domains of section 4.1; or at least that this energy term *prefers* the subdivision of a ferromagnetic crystal into domains. I choose a particularly simple example, which also makes a nice example for one of the methods to solve potential problems. It is a somewhat modified version of a calculation of Néel of an infinite circular cylinder which is subdivided into two domains.

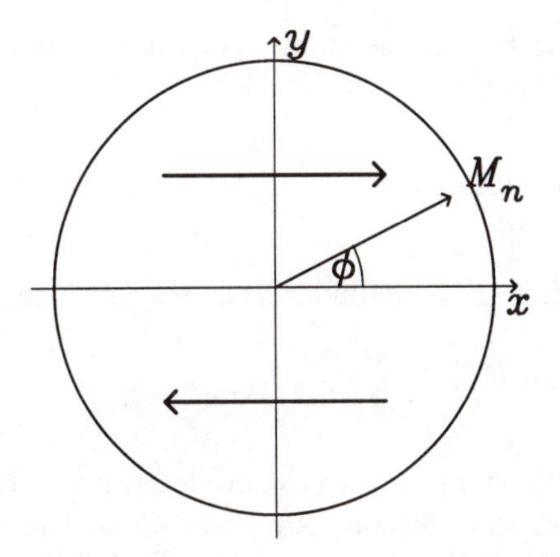


FIG. 6.1. A cross-section of an infinite circular cylinder subdivided into two antiparallel domains.

This cylinder is assumed to be magnetized along the arrows in the cross-section plotted in Fig. 6.1, which means that the magnetization vector is

$$M_y = M_z = 0, \quad M_x = M_s \times \begin{cases} +1 & \text{if } y > 0, \quad i.e. \quad 0 \leq \phi \leq \pi \\ -1 & \text{if } y < 0, \quad i.e. \quad \pi \leq \phi \leq 2\pi \end{cases} \tag{6.2.28}$$

where $y = \rho \sin\phi$ in the cylindrical coordinates, with $0 \leq \phi \leq 2\pi$. The step function can be expressed by its well-known Fourier expansion,

$$M_x = \frac{4}{\pi} M_s \sum_{n=0}^{\infty} \frac{\sin[(2n+1)\phi]}{2n+1}, \tag{6.2.29}$$

which makes the normal component (see Fig. 6.1)

$$M_n = M_\rho = M_x \cos\phi = \frac{2}{\pi} M_s \sum_{n=0}^{\infty} \frac{\sin[(2n+2)\phi] + \sin(2n\phi)}{2n+1} \tag{6.2.30}$$

when the product of the cosine and sine functions is converted into a sum of two sine functions. The sum in this equation can now be broken down into two sums, one with $\sin[(2n+2)\phi]$ and one with $\sin(2n\phi)$. In the first sum the summation index is changed from $n$ to $n-1$, and in the second sum the summation can start from $n = 1$, because the term with $n = 0$ is 0 anyway. Combining again into one sum,

$$M_n = \frac{2}{\pi} M_s \sum_{n=1}^{\infty} \left[ \frac{1}{2n-1} + \frac{1}{2n+1} \right] \sin(2n\phi). \tag{6.2.31}$$

Adding together the two fractions, and substituting in eqn (6.1.6), one of the boundary conditions is

$$\left(\frac{\partial U_{\text{in}}}{\partial \rho} - \frac{\partial U_{\text{out}}}{\partial \rho}\right)_{\rho=R} = \gamma_{\text{B}} M_n = \frac{8}{\pi}\gamma_{\text{B}} M_s \sum_{n=1}^{\infty} \frac{n \sin(2n\phi)}{(2n+1)(2n-1)}, \quad (6.2.32)$$

where $R$ is the radius of the cylinder. It is therefore natural to look for a solution of the form

$$U = \sum_{n=1}^{\infty} u_n(\rho) \sin(2n\phi), \quad (6.2.33)$$

where $u_n$ are functions which have to be determined. Note that because of the uniqueness of the solution, any guess about the functional form is legitimate if it eventually leads to a function which fulfils all the differential equations and boundary conditions.

In the present case $\nabla \cdot \mathbf{M} = 0$ and eqns (6.1.4) and (6.1.5) are both $\nabla^2 U = 0$. Substituting from eqn (6.2.33)

$$\nabla^2 U = \sum_{n=1}^{\infty} \sin(2n\phi)\left(\frac{d^2}{d\rho^2} + \frac{1}{\rho}\frac{d}{d\rho} - \frac{4n^2}{\rho^2}\right) u_n(\rho) = 0. \quad (6.2.34)$$

Obviously, both the differential equations and the regularity at infinity are fulfilled by the functions

$$u_n(\rho) = c_n \times \begin{cases} (\rho/R)^{2n} & \text{if } \rho \leq R \\ (R/\rho)^{2n} & \text{if } \rho \geq R \end{cases} \quad (6.2.35)$$

where $c_n$ are constants. Moreover, all these functions are continuous at $\rho = R$. Substituting these $u_n$ in eqn (6.2.33), and substituting the latter in eqn (6.2.32), it is seen that *all* the requirements are satisfied for the choice

$$c_n = \frac{2\gamma_{\text{B}} M_s R}{\pi(2n+1)(2n-1)}. \quad (6.2.36)$$

It has thus been shown that the potential inside the cylinder is

$$U_{\text{in}} = \frac{2\gamma_{\text{B}}}{\pi} R M_s \sum_{n=1}^{\infty} \frac{\sin(2n\phi)}{(2n+1)(2n-1)} \left(\frac{\rho}{R}\right)^{2n}. \quad (6.2.37)$$

The (demagnetizing) field inside this two-domain cylinder is given by eqn (6.1.2), and in particular its $x$-component is

$$H_{x_{\text{in}}} = -\frac{\partial U_{\text{in}}}{\partial x} = -\left(\cos\phi \frac{\partial}{\partial \rho} - \frac{\sin\phi}{\rho}\frac{\partial}{\partial \phi}\right) U_{\text{in}}. \quad (6.2.38)$$

Substituting $U$ from eqn (6.2.37) and performing the differentiations,

$$H_{x_{\text{in}}} = -\frac{4\gamma_{\text{B}}}{\pi} M_s \sum_{n=1}^{\infty} \frac{n \sin[(2n-1)\phi]}{(2n+1)(2n-1)} \left(\frac{\rho}{R}\right)^{2n-1} . \qquad (6.2.39)$$

In this case there is also a $y$-component, but it does not enter the calculation of the energy, because it is multiplied by $M_y$ which is zero. According to eqn (6.1.7) the magnetostatic energy per unit length along $z$ is in this case

$$\mathcal{E}_{\text{M}} = -\frac{1}{2} \int_{\rho \le R} M_x H_x \, dS = -\frac{1}{2} \int_0^{2\pi} \int_0^R M_x H_x \rho d\rho d\phi. \qquad (6.2.40)$$

Substituting for $M_x$ from eqn (6.2.28) and for $H_x$ from eqn (6.2.39), and carrying out the integration over $\rho$,

$$\mathcal{E}_{\text{M}} = \frac{2\gamma_{\text{B}}}{\pi} R^2 M_s^2 \sum_{n=1}^{\infty} \frac{n}{(2n+1)^2(2n-1)} \int_0^{2\pi} \sin[(2n-1)\phi]\, \Phi(\phi)\, d\phi, \qquad (6.2.41)$$

with

$$\Phi(\phi) = \begin{cases} +1 & \text{if } 0 \le \phi \le \pi \\ -1 & \text{if } \pi \le \phi \le 2\pi. \end{cases} \qquad (6.2.42)$$

If this integral is separated into integration over the regions $\phi \le \pi$ and $\phi \ge \pi$, the integration is elementary, leading to

$$\begin{aligned} \mathcal{E}_{\text{M}} &= \frac{\gamma_{\text{B}}}{\pi} R^2 M_s^2 \sum_{n=1}^{\infty} \frac{8n}{(2n+1)^2(2n-1)^2} \\ &= \frac{\gamma_{\text{B}}}{\pi} R^2 M_s^2 \sum_{n=1}^{\infty} \left( \frac{1}{(2n-1)^2} - \frac{1}{(2n+1)^2} \right) = \frac{\gamma_{\text{B}}}{\pi} R^2 M_s^2. \end{aligned} \qquad (6.2.43)$$

The reader should not be so naive as to jump to the conclusion that all potential problems have such a nice, analytic solution. Obviously, only such cases are chosen for demonstration here, but there are others. Actually, this case of an infinite cylinder is used here only because it has this simple (or relatively simple) solution. The problem of a *sphere* subdivided in a way similar to Fig. 6.1 has also been solved [244], but only by a more complex technique which is beyond the scope of this book. At any rate, the important conclusion is obtained by comparing this result with eqn (6.1.19), namely

$$\frac{\mathcal{E}_{\text{M}}^{\text{one domain}}}{\mathcal{E}_{\text{M}}^{\text{two domains}}} = \frac{\pi^2}{4} > 1. \qquad (6.2.44)$$

Note that this result does not depend on $R$ or $M_s$. Therefore, for any ferromagnetic material, of any size, the magnetostatic energy term is *reduced* by subdividing the crystal into at least two domains.

It is not difficult to extend this calculation to more than two domains, and see that further subdivision can reduce further the magnetostatic self-energy. And in case this example of a cylinder may seem to some readers to be a unique case, a *qualitative* but convincing argument will be given in the next section, showing that this case is quite general, and that the magnetostatic energy prefers a subdivision into domains in *any* geometry. However, just because one energy term prefers this configuration does not necessarily mean that it can always have its way. There are other energy terms which must be considered.

As far as the anisotropy energy is concerned, there is no difference between a uniform magnetization and the two domains shown in Fig. 6.1, because if $x$ is an easy axis, so is $-x$. The anisotropy will only dictate that $x$ is parallel to a particular crystallographic direction, and is not just any direction within the cylinder for either the uniform magnetization or the magnetization in each domain. However, the exchange energy in a ferromagnet prefers neighbours to have parallel spins, and in Fig. 6.1 there is a whole surface for which the neighbouring spins on each side of it are antiparallel to each other. Therefore, in order to create this configuration work has to be done against exchange, and even a very rough estimation shows that this loss of exchange energy is much larger than the gain in the magnetostatic energy. The *total* energy of the configuration of Fig. 6.1, if taken *exactly* as in the foregoing calculation, is larger than that of the uniform magnetization, and the physical system will prefer the latter case.

### 6.2.1 *Domain Wall*

Still, it takes only a slight modification of the foregoing picture to change the argument. The main point is that the magnetostatic forces are very long ranged. They control the behaviour over large distances, and do not change considerably if a distance of several hundred unit cells is inserted between the two domains of Fig. 6.1. It is very different from the exchange, which is a very *short-ranged* force. It should be quite clear from chapters 2 and 3 that it affects nearest, or maybe next-nearest, neighbours only. It is a very strong force between such neighbours, but it does not extend to spins which are much farther away. With small angles between neighbouring spins, large changes of the angle over a distance of many atoms do not involve a large exchange energy. Therefore, the loss in the exchange energy can be very much reduced, if the picture of Fig. 6.1 is approximately maintained, but a *wall* is introduced, in which the direction of the magnetization vector changes *gradually*, instead of an abrupt jump of the magnetization from $\phi = 0$ to $\phi = \pi$. A more complete theoretical treatment will be given in the next chapter, but for the meantime the main features can be understood from a simple, semi-quantitative estimation.

When the spin operators are approximated by classical vectors, as in chapter 2, the exchange energy is as in eqn (2.2.25). And if $J$ is non-zero between nearest neighbours only,

$$\mathcal{E}_{\text{ex}} = -\sum_{ij}{}' J_{ij}\mathbf{S}_i \cdot \mathbf{S}_j = -JS^2 \sum_{\text{neighbours}} \cos\phi_{i,j}, \qquad (6.2.45)$$

where $\phi_{i,j}$ is the angle between $\mathbf{S}_i$ and $\mathbf{S}_j$. A one-dimensional structure is considered, in which *planes* with $n$ spins in each interact with neighbouring planes. The interaction of plane $i$ is taken only with that at $i+1$ and not with the other neighbour at $i-1$, and a factor 2 is introduced instead, as in the transition to eqn (2.2.26). Then the energy loss from the state at which all spins are aligned is

$$\delta\mathcal{E}_{\text{ex}} = 2JS^2 n \sum_i [1 - \cos\phi_{i,j}] = 4JS^2 n \sum_i \sin^2\left(\frac{1}{2}\phi_{i,j}\right) \approx JS^2 n \sum_i \phi_{i,j}^2, \qquad (6.2.46)$$

for small angles. Let this calculation be applied now to the case where the direction of the spins changes from $\phi = 0$ to $\pi$ over $N$ such planes. The angle change between planes need not be the same for all planes, and a better scheme will be given in the next chapter, but for simplicity this angle *is* taken here to be the same. It means that in order to obtain a total change of $\pi$ after $N$ such angles, $\phi_{i,j} = \pi/N$ and the energy loss is

$$\delta\mathcal{E}_{\text{ex}} = JS^2 n \sum_i \left(\frac{\pi}{N}\right)^2 = \frac{JS^2 n\pi^2}{N}. \qquad (6.2.47)$$

The exchange energy loss over this wall is, thus, $N$ times smaller than that of one jump from $\phi = 0$ to $\phi = \pi$. Obviously, if this $N$ is sufficiently large, the loss in exchange energy can be small enough to be compensated by the gain in the magnetostatic energy, which makes the subdivision into domains energetically favourable.

In principle, if the jump from one domain to the other is not abrupt, the calculation of the magnetostatic energy in the foregoing should also be modified, but that correction is rather small if the wall is not too thick. However, there is another complication in that the anisotropy energy now enters as well. Two antiparallel domains *can* be along an easy axis of the anisotropy energy, namely along one of the directions for which this energy term is a minimum. But the spins in the wall must turn out of an easy direction, so that an anisotropy energy has to be spent in creating such a wall. Qualitatively, the anisotropy energy tries to enforce a thin wall, while the exchange energy tries to enforce a thick wall. The above-mentioned number of planes, $N$, is therefore determined by minimizing the sum of exchange and anisotropy energies.

The inevitable conclusion from all these semi-qualitative arguments is that none of these three energy terms (exchange, anisotropy and magnetostatic) can be neglected. Since domains are an experimental fact in sufficiently large crystals (see section 4.1), any realistic calculation of a bulk ferromagnet must contain all these three energy terms. The foundations of such a theory will be given in the next chapter, 7. Only before going into mathematical details, it should be very instructive to continue a little more with the semi-qualitative discussion, and establish more clearly the *physical* picture of a ferromagnet, and the nature of the forces governing its behaviour.

### 6.2.2 *Long and Short Range*

Every undergraduate nowadays studies Maxwell's equations, and most of them can quote the fact that the electric and magnetic forces are *long ranged*, because the potential decreases with distance as $1/r$, which is *a slow* decrease. However, there are relatively few, even among professionals working on the theory of magnetism, who actually try to understand what this statement means.

Consider the simple case of a uniformly magnetized ellipsoid. The field which is measured at a point inside this ellipsoid is given by eqn (6.1.21), where $D$ is determined by the *ratios* of its axes. The absolute size does not enter. Suppose that this ellipsoid is inflated in such a way that its size increases, but its shape is held the same, *i.e.* its axial ratios are kept constant. Then the field is still the same as it was for the small ellipsoid, which is a function of the axial ratios only. If this inflation continues, even in the limit of the ellipsoid extending to infinity, the demagnetizing field in it still depends on the axial ratios of the *surface* which is now an infinite distance away. It is this much of a long range.

The obvious conclusion is that in ferromagnetism there is no physical meaning to the limit of an infinite crystal *without a surface*. It is not just the technical problem that infinite crystals cannot be made in reality. This technicality does not cause any difficulty in *other* fields of physics, where the assumption of infinity can be taken as the limit of a crystal which is large compared with some sort of a measure for the properties under discussion. In this case, even in the theoretical limit of the crystal actually tending to infinity, the *shape* of its surface still determines at least part of the magnetostatic energy term, and surface effects cannot be avoided. Therefore, all calculations of the types described in chapters 3 and 4, which ignore the surface by saying that the crystal is infinite, introduce an error.

This error would not be important if the whole magnetostatic energy term was rather small. But it is not. It is only too often pointed out that the exchange energy *density* is orders of magnitude larger than the magnetostatic energy density. However, the physical system goes by the *total* energy, and not by its density. As stated several times in the foregoing, the exchange force has a very short range. It acts essentially only between neighbouring atoms, so that its effective range is of the order of the unit cell of the crystal. Therefore, the total exchange energy is of the order of its density integrated over the volume of a unit cell. The magnetostatic energy density is small, but having a long range, it is integrated over the whole volume of the crystal. For a sufficiently large crystal, which contains very many unit cells, the total magnetostatic energy is much larger than the total exchange energy. It is not negligible. On the contrary, the exchange energy controls only the microscopic properties, as in the inside of the domain wall, but it is the magnetostatic energy term which mostly determines the structure of the magnetization distribution over most of the crystal.

It must be emphasized again that a large error is introduced not only when the magnetostatic energy is neglected altogether, as it is in most of the calculations described in chapters 3 and 4. Sometimes a certain approximation for the dipole interaction [76, 245] is included in the spin wave theory of $M_s$ *vs.* temperature. And the review [67] of the renormalization group calculations cites several cases into which such interactions have been introduced. However, *all* these cases assume an infinite crystal without a surface, which is inadequate. In ferromagnetism there is always a surface, even for an infinite crystal, and it is the surface which is responsible for the subdivision into domains.

All the calculations in this section were for cases in which the only contribution to the magnetostatic energy term is from the discontinuity of the derivative (which depends *only* on the shape of the surface) because all the examples were *chosen* to be such that $\nabla \cdot \mathbf{M} = 0$. If this term is not 0, the solution of eqn (6.1.4) may be very different from the cases of this section, as will be further discussed in the next section. However, it can already be stated here that although these other cases exist, they are much less common in real life than those for which $\nabla \cdot \mathbf{M} = 0$. The reason is that the main tendency of the magnetostatic energy term is to subdivide large crystals into domains, in each of which $\nabla \cdot \mathbf{M} = 0$. And even when it is not so in the *walls*, only a small part of the spins are in the walls, so that they have a small effect on the overall properties of the crystal. Therefore, a theory which introduces the dipolar interaction but leaves out the surface treats a less important term while neglecting the more important one.

However, omitting this largest energy term is not always as bad as it may sound. Paradoxically, the magnetostatic energy term may often be neglected *because* it is the largest energy term. The point is that being the largest term domains are arranged to minimize the magnetostatic energy, with very little effect of the exchange, and only a minor modification by the anisotropy. For large crystals one may even do quite well [246, 247] by neglecting the exchange altogether. But at the minimum $\mathcal{E}_{\mathrm{M}}$ usually reaches a small value, which may often turn out to be much smaller than the other energy terms, so much so that it is sometimes possible [246, 248, 249, 250] to approximate the energy minimization by a configuration for which $\mathcal{E}_{\mathrm{M}} = 0$. But this energy is only small at the minimum and a deviation from that configuration can cost a large amount of magnetostatic energy. Therefore, when calculating the energy of the correct magnetization distribution, $\mathcal{E}_{\mathrm{M}}$ may often be neglected, but if it is neglected *a priori* a wrong magnetization distribution is reached, which has a very large $\mathcal{E}_{\mathrm{M}}$ term.

Because of this property, and because each domain is homogeneously magnetized, it is often possible to get away without the $\mathcal{E}_{\mathrm{M}}$ term, and with the wrong assumption that the whole crystal is homogeneously magnetized. As explained in section 4.1, it is possible to calculate $M_s(T)$ as if the domains did not exist, and it works. But one must always bear in mind that it is wrong in principle, and that it works only with some tricks, and only for limited applications. It is dangerous ground to step on, and it is necessary to remember this fact, and check each case for compatibility with the assumption of no domains. This assumption can never be taken for granted, and one should certainly not try to extend it beyond its natural limits of validity, where a different theory is required. For example, adding a non-zero magnetic field blows up this theory, see section 4.6. This approach has never been used for the calculation of critical exponents, and it is not clear at all if neglecting the domain structure does or does not have a large effect on these calculations for any specific case.

This distinction between a long-range and a short-range force is already sufficient to resolve (at least qualitatively) the difficulty which I have left open in chapter 1. The exchange force in iron is of the order of $10^6$ Oe, but it takes an application of an extra field of about of $10^3$ Oe to wipe out the domains, and even a really negligible field of 1 Oe can make a large difference to the domain structure. Why cannot the $10^6$ Oe field accomplish what a much smaller field can? The answer is that the very large exchange field has a very short range and only enforces small angles between neighbouring spins. It is not capable of preventing subdivisions into domains over a long range. When a magnetic field is applied, it does *not* do work against exchange forces. It works against magnetostatic forces (which are of this order of $10^3$ Oe in Fe) in removing or rearranging domains. And it can accomplish it because it is applied all over the crystal, and not only among neighbours.

It should be especially noted that the argument about a long and a short range applies only to large crystals, which contain a sufficiently large number of unit cells. In small particles the long range of $\mathcal{E}_{\mathrm{M}}$ does not make a difference, because the integration is only carried over the limited size of the crystal. In spite of the claims of some theorists that they omit the magnetostatic energy term because the crystal is very large, and its surface is far away, it actually works for the opposite extreme. It is in *small* particles that the exchange is sufficiently strong to enforce a uniform magnetization all over the crystal, although anisotropy also plays a role, as in chapter 5. If such a particle is a sphere, the magnetostatic energy does not enter at all. If it is an elongated ellipsoid, the magnetostatic energy plays only the role of a shape anisotropy, which adds to the other anisotropy terms. In either case, the exchange in these small particles is too strong to allow subdivision into domains, or other features of the large particles. A table of typical numerical values of these energy terms can be found, for example, in [251]. It is, thus, in the small, not the large, crystals where one should look for a possible validity of the spin wave theory, and the critical exponents. But then, superimposing the assumption of an infinite sample cannot be a very good approximation for very small particles.

Sufficiently small particles are, therefore, homogeneously magnetized in zero applied field. They are then called in the literature *single-domain* particles. Calculating the exact size at which a multi-domain particle turns into being a single-domain one is not a simple problem, and will be further discussed in a later chapter. At this stage it will only be remarked that semi-qualitative estimations of the energy of the domains and the wall between them, as done in this section, are all right for rather large particles, for which the accuracy is less important. Near the transition, the energy balance is rather delicate, and a higher accuracy is needed, even though this point was ignored in estimations published in the 1940's and 1950's. The first rigorous calculation [252] considered a sphere sliced into planes, as roughly done in section 6.2.1 here, but calculated exactly the exchange, anisotropy and magnetostatic energies for these slices. It reached the value of 37 nm for the radius below which a cobalt sphere is a single domain, and above which it should divide into two domains. Even this calculation turned out to be inaccurate, because it was later found [253] that the total energy in a sphere can be further reduced by making the domains curved, with a cylindrical symmetry. This modification reduced the 'critical radius' for being a single domain to 34 nm in cobalt.

## 6.3 Magnetic Charge

Undergraduate textbooks give a formal solution to the differential equations and boundary conditions in eqns (6.1.4)–(6.1.6), which can be written in the form

$$U(\mathbf{r}) = \frac{\gamma_{\mathrm{B}}}{4\pi}\left(-\int \frac{\nabla'\cdot\mathbf{M}(\mathbf{r}')}{|\mathbf{r}-\mathbf{r}'|}\,d\tau' + \int \frac{\mathbf{n}\cdot\mathbf{M}(\mathbf{r}')}{|\mathbf{r}-\mathbf{r}'|}\,dS'\right), \tag{6.3.48}$$

where $\nabla'$ contains derivatives with respect to the components of $\mathbf{r}'$, the first integral is over the ferromagnetic bodies, the second integral is over their surfaces, and $\mathbf{n}$ is the outward normal.

This solution does not solve the problem in the sense that we can forget about the differential equations. It is often easier to solve the differential equations, as in the examples given in the previous section, than to carry out the integrations in this equation. As a nice illustration, the reader may try to obtain the solution of eqn (6.1.11) for a homogeneously magnetized sphere, by carrying out the integrations in eqn (6.3.48). It is certainly possible to do it, because the two expressions are mathematically identical. But the integration is definitely not trivial. Actually, more often than not, analytic integration from this solution is rather cumbersome and not easy to perform, unless some transformation is first applied to fit the particular case. It is not very useful for numerical integration either, except for certain special cases, because most of the contribution to the integrand is usually from the vicinity of the singularity at $\mathbf{r}' = \mathbf{r}$ where it is not easy to attain an adequate accuracy. Also, the first term in eqn (6.3.48) is an integration over the volume, which is a three-fold integration. In order to calculate the energy, the result has to be substituted in eqn (6.1.2), and then substituted in eqn (6.1.7), which involves another three-fold integration. It may change in the near future, but right now a six-fold numerical integration to any decent accuracy is beyond the capability of existing computers, even though *some* six-fold integrations of this sort [254] *have* been carried out.

This formal solution is more useful when the first integral vanishes, and there is only the second one with a two-fold integration. The energy calculation then involves only a four-fold integral, and if one or two of these integrations can be carried out analytically, the numerical problem becomes quite manageable. But the most important application of eqn (6.3.48) is based on its *qualitative* properties, which allow an insight into what the magnetostatic energy prefers without actually doing any calculations. This possibility of using a physical intuition is due to the formal *form* of the integrals in eqn (6.3.48), which contain $1/r$. This factor also appears in the electrostatic potential of a *point charge*, which allows the first integral to be interpreted as the potential due to a spatial distribution of a *volume charge*, with a charge density $-\nabla \cdot \mathbf{M}$. Similarly, the second integral can be considered as if it was expressing the potential due to a *surface charge* whose surface density is $\mathbf{M} \cdot \mathbf{n}$. Of course, these charges do *not* exist. Many books explain that the difference between electrostatics and magnetostatics is that there is no magnetic charge, and that these integrals have only a mathematical meaning, and do not express any physical reality. However, it is never necessary for any *useful* mathematical tool to have a physical meaning. There is no real physical charge, but the mathematical identity between these integrals and those which involve a charge makes it possible to use the knowledge about a real charge to guess the qualitative properties of the magnetostatic potential.

In particular, we know that similar charges *repel* each other. Therefore, a volume distribution of such a charge can be sustained only if it is held by *other* forces. Left to itself, the charges anywhere in the volume will repel each other as far as they can, which is all the way to the surface.

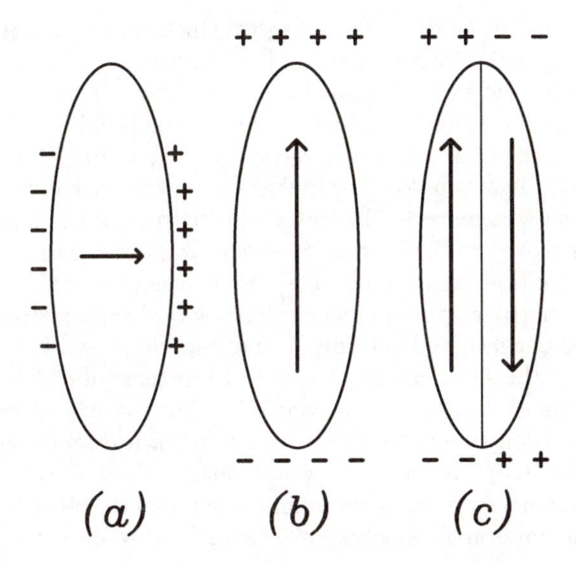


FIG. 6.2. Schematic representation of the surface charge in a particle magnetized along the long and the short axis, and the same particle subdivided into two antiparallel domains.

Therefore, the magnetostatic energy term by itself will prefer to avoid the volume charge altogether, and create only a domain structure with a charge on the outer surface. If there is any volume charge at all, it can only come out of a compromise with another energy term, *e.g.* within certain types of the wall between domains. Because the magnetostatic term is usually the largest force in sufficiently large crystals, and most of the magnetization structure is arranged to fit *this* term, a volume charge will hardly ever be encountered. However, it must always be borne in mind that if a structure which involves a volume charge is introduced into a certain calculation, the magnetostatic energy due to this volume charge is *not* negligible. It is often convenient to introduce such a structure in certain problems, and it is very tempting then to forget about the volume charge, and assume that it probably does not have a large effect. It is, therefore, necessary to warn again that if the volume charge does not enter *other* calculations, it is not because it is negligibly small, but because it is extremely large. It is only by avoiding it that the magnetostatic energy can be minimized to a small value, and if this charge is allowed to creep in, $\mathcal{E}_{\mathrm{M}}$ can increase enormously.

A similar argument applies to the surface charge as well. Consider, for example, the cases shown in Fig. 6.2. The single-domain structure in (a) has the same charge density as the one in (b). But in (a) this charge is spread over a larger area than in (b). Therefore, the energy of the case (b) is *smaller* than that of case (a). For ellipsoids, this result can also be obtained from the known analytic solution, according to which the demagnetizing factor is smallest along the longest axis, and tending to zero in the limit of an infinite cylinder. However, the conclusion from Fig. 6.2 is easier to see, and it also applies to *other* bodies, and not only to ellipsoids.

In case (c) the total surface charge is the same as in case (b). However, the subdivision into two domains makes some of the negative charge from the bottom surface move to the top, replacing part of the positive charge there that is moved to the bottom. Since unlike charges attract each other, the structure in (c) is more favourable, and its energy is smaller than that of the structure in (b). Again, this conclusion obviously applies to any shape of the magnetic particle, and not only to the ellipsoid shown schematically in Fig. 6.2. Also, the same argument applies to further subdivision into more than two domains. It may thus be concluded that the magnetostatic energy term prefers a domain configuration over a uniform magnetization for any ferromagnetic body, and that it would rather continue this subdivision indefinitely, unless stopped by the competition with the other energy terms. Therefore, a uniform magnetization can only exist either in sufficiently small particles, or in a crystal to which a sufficiently large magnetic field is applied. A large magnetic field can wipe out the domains and rotate the magnetization to its own direction.

Before concluding this discussion of eqn (6.3.48) it will be remarked for the sake of completeness that this formal solution is also useful for two other purposes, even though these are usually listed in undergraduate textbooks. One is that eqn (6.3.48) may be taken as an existence theorem. It proves that there is at least one solution to the set of eqns (6.1.4) to (6.1.6), thus completing the proof given in section 6.1 that this set of equations cannot have more than one solution. The second remark is that eqn (6.3.48) proves the principle of *superposition*. Since everything is linear in those integrals, it is always possible to calculate separately, and even by different methods, the potential created by different parts of the charge, then add them together.

### 6.3.1 *General Demagnetization*

A case of special practical interest is a homogeneously magnetized body. The domain configuration in zero, or small, applied field is very complicated, and very difficult to reproduce. It varies from one sample to another, and even for the same sample it depends on the *history* of the applied field, see Fig. 4.1. In order to calibrate measurements, one must start with something which can be related to the material, and not to any specific sample. The best case is a sample in a sufficiently large field, for which one can at least hope that the magnetization is held parallel (or almost parallel) to the direction of the field, throughout the whole sample. However, the field inside a ferromagnet is not the same as the field outside, and the difference (called the demagnetizing field) is a function of the *shape* of the sample. A reasonable estimation of this demagnetization must be subtracted in order to remove the effects of the particular sample, and reach the intrinsic properties of the material. Therefore, the definition of this field, given only for an ellipsoid in the previous section, is extended here to other bodies.

In a uniformly magnetized body, $\nabla \cdot \mathbf{M} = 0$, and the first integral in eqn (6.3.48) vanishes. Substituting the second term in eqn (6.1.2), the (demagnetizing) field inside the ferromagnetic material is

$$\mathbf{H} = -\frac{\gamma_{\mathrm{B}}}{4\pi}\nabla\left(\mathbf{M}\cdot\int\frac{\mathbf{n}}{|\mathbf{r}-\mathbf{r}'|}\,dS'\right), \tag{6.3.49}$$

where $\mathbf{M}$ was taken out of the integral, because it is assumed to be a *constant.* For the same reason, $\mathbf{M}$ can be moved to the left of the differentiations, so that eqn (6.3.49) essentially means that each component of $\mathbf{H}$ is a *linear function* of the components, $M_x$, $M_y$ and $M_z$. Also, eqn (6.1.7) is in this case,

$$\mathcal{E}_{\mathrm{M}} = -\frac{1}{2}\mathbf{M}\cdot\int\mathbf{H}\,d\tau, \tag{6.3.50}$$

where again the constant $\mathbf{M}$ is moved in front of the integral. These two relations mean that the magnetostatic energy, in this case of a homogeneously magnetized body, is a *quadratic form* in the components of $\mathbf{M}$. It can, therefore, be written in the form

$$\mathcal{E}_{\mathrm{M}} = \frac{1}{2}V\left(N_{11}M_x^2 + N_{12}M_xM_y + \cdots\right), \tag{6.3.51}$$

where $N_{11}$ etc. are constants that depend only on the shape of the particle. It is always possible to rotate the axes so that this quadratic form becomes the same as eqn (6.1.22). The latter is, thus, the most general form of the magnetostatic energy of a *uniformly magnetized* ferromagnetic body, which applies to any shape, and not only to ellipsoids. Moreover, by using the properties of the function $1/r$, it can readily be shown [1, 255] that in the diagonalized form of eqn (6.1.22) all three components $N_x$, $N_y$ and $N_z$ are non-negative numbers, whose sum (which is the trace of the tensor $N$) is $\gamma_{\mathrm{B}}$. Therefore, as far as the energy is concerned, any uniformly magnetized ferromagnetic body behaves the same as an ellipsoid which has the same volume. This statement is known as the Brown–Morrish theorem.

It should be particularly noted that this theorem does not even require a simply connected body, and applies even to a body that contains *cavities.* Of course, symmetry considerations may be used just as in the case of an ellipsoid. For example, a cube must have three *equal* demagnetizing factors. Therefore, the demagnetizing factor of a cube is the same as that of a sphere which has the same volume, if that cube is uniformly magnetized. However, such a statement has nothing to do with the question of whether a cube *can* be brought to this state of being uniformly magnetized, and *how* to do it. It is generally assumed that a sufficiently large, uniform applied field can bring the magnetization in the cube to be nearly uniform, but it takes a special, non-uniform field to make the cube completely uniformly magnetized.

The main difference between an ellipsoid and any other body is that the demagnetizing field inside an ellipsoid is *uniform*, namely it is the same at every point inside the ellipsoid, which is not true for any non-ellipsoidal shape. Although the energy of the latter is the same as that of a certain ellipsoid, this energy is an average over a certain field distribution. For non-ellipsoidal bodies in a large applied field, $\mathbf{H}_{\text{appl}}$, it is still customary to define a demagnetizing factor, $N$, and take the internal field as

$$\mathbf{H}_{\text{in}} = \mathbf{H}_{\text{appl}} - N\mathbf{M}, \tag{6.3.52}$$

because it is the only way to eliminate the effect of the shape of the sample and reach the intrinsic properties of the material. However, in non-ellipsoidal bodies it is only an approximation, and it gives only an average of the internal field. Only in an ellipsoid the average is the same as the field at every point.

In principle, the demagnetizing factors (namely, the components of the tensor $N$) can be calculated by evaluating the potential of the surface charge in eqn (6.3.48) for the particular geometry, substituting in eqn (6.1.2) to find the field, and then taking the appropriate average of that field. Two different definitions of averages are used in practice. One uses a field average over the whole volume of the sample, leading to a demagnetizing factor which is called the *magnetometric* demagnetizing factor. The other definition is an average over the middle cross-section of the crystal perpendicular to the direction of the applied field, leading to what is known as the *ballistic* demagnetizing factor. Some of these calculations can be carried out analytically, and some call for a numerical evaluation, with or without certain approximations.

Details of such evaluations, and tables of both kinds of demagnetizing factors, can be found in the literature, and are outside the scope of this book. No specific example will be given here, and only several leading references will be mentioned. Tables of both demagnetizing factors in a rectangular prism exist for the case of [1] one dimension extending to infinity, and for the case of [256] one square cross-section. For a finite circular cylinder there are tables [256, 257], a long review with formulae, tables and graphs [258] and a sophisticated computational scheme [259]. There is a special study of single or double thin films [260, 261] and there are also some theorems [262, 263] of a more general nature, and an attempt [264] at a first-order correction for the case of a slightly non-uniform magnetization distribution. And there is also a detailed discussion [265] of certain drawbacks in practical applications.

## 6.4 Units

Older textbooks used the cgs units system, in which the basic units are the centimetre, gram and second. Modern textbooks for undergraduates have switched completely to the system called SI, for *Système International d'Unités*, and it can safely be assumed that the reader is more familiar with it than with the cgs. It may thus seem simpler to adopt the SI units for this book as well. However, rightly or wrongly, practically all the modern literature on magnetism still uses the so-called Gaussian cgs system of units. And the reader must become familiar with it, if only in order to be able to read all this published literature.

For some people, converting into the SI has become an obsession, bordering on a religious conviction to abolish heresy and make everybody use the 'true' units. However, there is no way of ignoring the fact that there are many researchers who have not been converted, and it seems that they will not be for many years to come. In any case, the use of units is only a matter of convenience, or as Brown [266] phrased it: 'dimensions are the invention of man, and man is at liberty to assign them in any way he pleases, as long as he is consistent throughout any one interrelated set of calculations'. This tutorial [266] also defines, and describes the history of, unit systems, and its reading is highly recommended.

The best source for the definitions of the Gaussian system, and their conversion to SI, is the appendix to the I.U.P.A.P. report [267] on units. I will only state briefly the important conversion factors, in words and not as a table, according to the good advice of Brown [266]: 'At all costs, avoid conversion tables; with them, you never know whether to multiply or divide.' And then, for the rest of the book, only the Gaussian cgs system will be used. Even the factor $\gamma_B$ of Brown which has been used in this chapter and in section 1.1 will not be carried any further. It will be replaced from the next chapter by the cgs value of $4\pi$.

The cgs unit for a magnetic field, $H$, is the oersted (Oe). The SI unit is $1\ \mathrm{A/m} = 4\pi \times 10^{-3}\,\mathrm{Oe}$. Or $1\,\mathrm{Oe} \approx 79.6\ \mathrm{A/m}$. The Oe is the same as Gb/cm, where the gilbert (Gb) is the cgs unit for the magnetic potential, $U$. The latter is measured by ampere (A) in SI, and the number of A has to be multiplied by $0.4\pi$ to obtain the number of Gb.

The magnetic induction, also known as the magnetic flux density, $B$, is measured in gauss (G) in the cgs system. In this system, $H$ and $B$ have the same dimensions, see eqn (1.1.2), and some years ago $H$ was also measured in gauss. However, now the units have different names. The SI unit is $\mathrm{Wb/m^2}$, also called tesla (T), and $1\,\mathrm{T} = 10^4\,\mathrm{G}$.

The permeability $\mu$ is a dimensionless number in the cgs system, and $\mu_0$ of eqn (1.1.2) should be just replaced by the number 1. In SI, for which eqn (1.1.2) is written, the permeability of free space is $\mu_0 = 4\pi \times 10^{-7}\,\mathrm{H/m}$. In this system the *relative* permeability, $\mu_r = \mu/\mu_0$, is also used, as defined in eqn (1.1.4). The value of $\mu_r$ is equal to that of the cgs $\mu$.

The magnetization, $M$, sometimes called the *volume* magnetization, is the dipole moment per unit volume. In cgs it is measured in emu, or emu/cm$^3$, even though emu is not really a unit in any sense of the word. The number in emu/cm$^3$ has to be multiplied by $10^3$ to convert it to A/m. Often $4\pi M$ is specified instead of $M$, and then it is measured in G as $B$ is, see eqn (1.1.2). If $M$ is divided by the density of the material, it is known as the *mass* magnetization, and measured by emu/g in the Gaussian system. It has the *same* numerical value as A.m$^2$/kg, which is the SI unit. The susceptibility and permeability are dimensionless numbers in the cgs system, and the permeability of the vacuum is numerically 1.

The demagnetization factors $D$ and $N$ are dimensionless both in cgs and SI, but there is the factor $4\pi$ in $N$ as defined in this chapter by the two values of $\gamma_{\mathrm{B}}$. The anisotropy constant $K$, defined in chapter 5, has the dimension of an energy density, namely energy per unit volume. In cgs it is measured by erg/cm$^3$, and in SI the unit is J/m$^3$ which equals 10 erg/cm$^3$. All the other conversion factors should be obvious now, and it is hoped that the reader can figure them out.

# 7

# BASIC MICROMAGNETICS

It can be concluded from the last chapter that there is no way to neglect any one of the three energy terms, exchange, anisotropy and magnetostatic, and all three must be taken into account in any realistic theory of the magnetization processes. It would have been nice if the other terms could be added to the Heisenberg Hamiltonian, at least as a perturbation. But this Hamiltonian cannot even be solved quantum mechanically *without* these terms unless quite rough approximations are introduced. Therefore, until a better theory can be developed, the only way is to 'neglect' quantum mechanics, ignore the atomic nature of matter, and use classical physics in a continuous medium.

Such a classical theory has been developed in parallel with the quantum-mechanical studies of $M_s(T)$ which just ignore magnetostatics. Its history is told in [268], from the start with a 1935 paper of Landau and Lifshitz on the structure of the wall between two antiparallel domains, and several works of Brown in 1940–1. Brown gave this theory the name *micromagnetics*, because what he had in mind at first was the study of the details of the walls which separate domains, as distinguished from the *domain theory* which considered the domains, but took the walls to be a negligible part of space. The name is somewhat misleading, because the microscopic details of the atomic structure are ignored, and the material is considered from a macroscopic point of view by taking it to be continuous.

Part of the classical approach is to replace the spins by classical vectors, which has already been done in chapter 2. But on top of that, a classical theory which can be used together with Maxwell's equations must have a classical energy term that can replace the quantum-mechanical exchange interaction, in the limit of a *continuous* material.

## 7.1 'Classical' Exchange

As seen in section 6.2.1, the exchange energy among spins can be written in terms of the angles $\phi_{i,j}$ between spin $i$ and spin $j$, as in eqn (6.2.45). As has been explained there, the angles between neighbours are expected to be *always* small, because the exchange forces are very strong over a short range, and will not allow any large angle to develop. For small $|\phi_{i,j}|$ it is possible to use the same approximation as in the particular case of parallel planes with $n$ spins in each, leading to eqn (6.2.46), and write

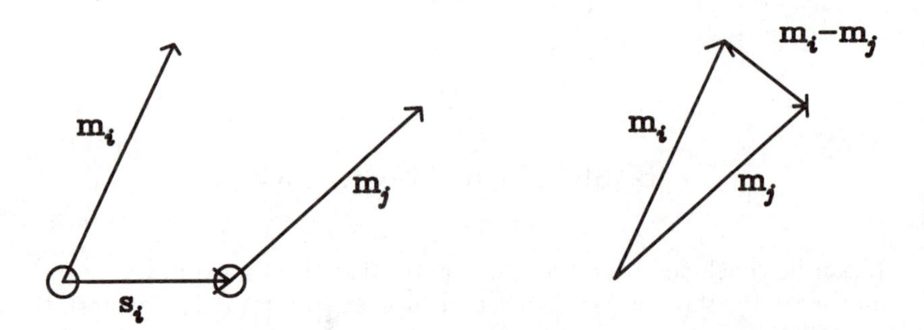


FIG. 7.1. Schematic representation of the change in the angle between neighbouring spins $i$ and $j$, and the position vector $\mathbf{s}_i$ between them.

$$\mathcal{E}_{\text{ex}} = JS^2 \sum_{\text{neighbours}} \phi_{i,j}^2, \tag{7.1.1}$$

after subtracting the energy of the state in which all spins are aligned, and which is used as a reference state in this calculation. It means redefining the zero of the exchange energy, which is always legitimate, provided it is done consistently.

For small angles, $|\phi_{i,j}| \approx |\mathbf{m}_i - \mathbf{m}_j|$, where $\mathbf{m}$ is a *unit* vector which is parallel to the local spin direction, see Fig. 7.1. Note that this definition also means that $\mathbf{m}$ is parallel to the local direction of the magnetization vector, $\mathbf{M}$, and it is actually the same $\mathbf{m}$ as in eqn (5.1.6) whenever $\mathbf{M}$ is a continuous variable, which is defined not only in the lattice points. For such a variable, the first-order expansion in a Taylor series is

$$|\mathbf{m}_i - \mathbf{m}_j| = |(\mathbf{s}_i \cdot \nabla)\mathbf{m}|, \tag{7.1.2}$$

where $\mathbf{s}_i$ is the *position vector* pointing from lattice point $i$ to $j$, see Fig. 7.1. Substituting in eqn (7.1.1),

$$\mathcal{E}_{\text{ex}} = JS^2 \sum_i \sum_{\mathbf{s}_i} \left[(\mathbf{s}_i \cdot \nabla)\,\mathbf{m}\right]^2 \tag{7.1.3}$$

where the second summation is over the position vectors from lattice point $i$ to all its neighbours. For example, for a simple cubic lattice with a lattice constant $a$, this sum is over the *six* vectors $\mathbf{s}_i = a(\pm 1, \pm 1, \pm 1)$. This summation is readily carried out for all three types of a cubic lattice, and it is seen that they all lead to the same expression, and differ only in a multiplicative factor.

Changing the summation over $i$ to an integral over the ferromagnetic body, the result is that for cubic crystals the exchange energy is

$$\mathcal{E}_{\text{ex}} = \int w_e d\tau, \qquad w_e = \frac{1}{2}C\left[(\nabla m_x)^2 + (\nabla m_y)^2 + (\nabla m_z)^2\right], \tag{7.1.4}$$

where

$$C = \frac{2JS^2}{a}c, \tag{7.1.5}$$

$a$ is the edge of the unit cell, and $c = 1$, 2 and 4 for a simple cubic, bcc and fcc respectively. For a hexagonal close-packed crystal, such as cobalt, summation over the $\mathbf{s}_i$ vectors leads to the same result as in eqn (7.1.4), only with

$$C = \frac{4\sqrt{2}JS^2}{a}, \tag{7.1.6}$$

where $a$ is the distance between nearest neighbours.

For lower symmetries, eqn (7.1.4) has to be somewhat modified. But for most cases of any practical interest this equation can be taken as a good approximation for the exchange energy, in as much as the assumption of a continuous material is a good approximation to the physical reality. The constant $C$ is then taken as one of the physical parameters of the material, whose value is obtained by fitting the results of the theory to one of the measurements. Of course, it *can* be obtained from the theoretical expressions in eqn (7.1.5) or eqn (7.1.6), whenever the exchange integral $J$ is known. However, $J$ depends on the temperature, as explained in section 3.5, and the value of $J$ near $T_c$ is not useful for micromagnetics calculations which are usually applied at, or near, room temperature. The best values for this *exchange constant* $C$ are usually obtained from ferromagnetic resonance. The order of magnitude for both Fe and Ni is $C \approx 2 \times 10^{-6}$ erg/cm.

The factor $\frac{1}{2}$ in the definition of $C$ in eqn (7.1.4) is quite arbitrary, and was introduced by Brown [145] in order to avoid a factor of 2 in the differential equations which minimize the energy, and which will be introduced in section 8.3. Many workers prefer to write the energy in eqn (7.1.4) without the factor $\frac{1}{2}$ and define a different constant of the material, $A$, where $C = 2A$. It often causes confusion because *both* $A$ and $C$ are referred to as the 'exchange constant of the material' and it is not always clear which of the two is used in any particular calculation.

The exchange energy of eqn (7.1.4) is a very powerful and useful tool for solving problems in which the direction of the magnetization vector varies from one point to another in the crystal. Its *size* is assumed to be $M_s(T)$ everywhere, as discussed in section 4.1. This energy term is zero for the case of aligned magnetization, when all the derivatives vanish, which is the way its zero has been defined here. It is large for large spatial variations, with large derivatives, which is what one expects the exchange energy to try to avoid. However, there are certain limitations for the application of this energy expression, which must be emphasized. As is the case with *any* theory, one should never be carried away and try to apply this theory beyond the natural validity of its approximations. It is therefore important to specify what these limits are.

The most obvious restriction is connected with the basic assumption of a continuous material, which can only be valid as long as any characteristic length it deals with is very large compared with the size of a unit cell. It is not something which can be guaranteed in advance. It is just necessary to bear in mind that if any micromagnetics calculation comes up with a parameter that has a dimension of length, the result is reliable only if this quantity is much larger than the unit cells.

The second, and less obvious, limitation is that the temperature is not too high. In changing over from the spins at the lattice points to a continuous variable, $\mathbf{M}$, the magnitude of this vector $\mathbf{M}$ comes out automatically as a constant over the whole crystal. It is also an experimental fact that the magnitude of $\mathbf{M}$ within the domains is a constant of the material, $M_s(T)$, which depends only on the temperature, as discussed in section 4.1. However, the picture of fixed spins at the lattice points is not a good approximation to real materials, as discussed in chapter 3, and the experimental fact that

$$|\mathbf{M}| = M_s(T) \tag{7.1.7}$$

is only true as an average, over a rather large volume. It cannot be strictly so at every point when there is enough thermal fluctuation to make a difference between one point and another. For lack of a better model, the theory of micromagnetics assumes that eqn (7.1.7) holds *everywhere*. Therefore, this theory, as it is, cannot be carried all the way to the vicinity of $T_c$, where even small *local* fields may change the magnitude of $\mathbf{M}$.

The necessary modifications of the theory, before it can be applied to high temperatures, are not very clear, even though there have been some attempts to generalize it. The biggest step in this direction was that of Minnaja [269] who showed that in the presence of thermal fluctuations, the exchange energy density in eqn (7.1.4) should be replaced by

$$w_e = \frac{C}{2M^2}\left[(\nabla M_x)^2 + (\nabla M_y)^2 + (\nabla M_z)^2\right], \tag{7.1.8}$$

where $M$ is the magnitude of the vector $\mathbf{M}$, and is a function of space. However, Minnaja did not do the next necessary step, which is to replace eqn (7.1.7) by another relation, which should be used to determine this $M$. Minnaja [269] just ignored eqn (7.1.7) and it left him with too much choice of possible solutions for the differential equations, which cannot do for a general theory. A true generalization of micromagnetics should [270] replace eqn (7.1.7) by something which tends to it in the limit of low temperatures, and has a better physical meaning at high temperatures. This part has not been done yet, and an attempt to solve a special case [271] was not very successful, and was never extended to other problems. It was later suggested

[272] to replace eqn (7.1.7) at high temperatures by adding an extra energy term whose density is proportional to $(|\mathbf{M}| - M_s)^2$, and this form was used [272] to solve a certain problem under some approximations. It was noted [273] that these approximations were not really needed for that solution, but there was no further development of this idea.

In the case of *nucleation* which will be discussed in chapter 9, eqn (7.1.7) can actually be ignored, for reasons which will be explained there. Minnaja [269] solved his high-temperature equations for *this* case of nucleation (in an infinite plate), which is legitimate. Similar nucleation at high temperatures was then calculated [274] for the case of an infinite cylinder.

It should also be noted that the approximation used here is valid only for *small angles* between neighbouring spins. Since the exchange is the largest force over a short range, it can be expected that these angles are *generally* very small indeed. However, this general rule does not exclude some exceptions in unusual cases, such as a corner where the magnetization must turn around due to some constraints on other energy terms. Formally, a discontinuous jump of an angle calls for an infinite exchange energy, if eqn (7.1.4) is taken to be literally correct. But the point is that it should *not* be taken to be literally correct. This equation is, after all, only an approximation to eqn (6.2.45), and the latter has no infinities. Even eqn (7.1.1) is always finite, and approximating it by something that becomes infinite only means that the approximation is not applicable for that particular case, which must be studied by other methods.

It can be argued that an occasional angular jump in some place means that a particular pair of spins has a much higher energy than any other pair in the crystal, which does not seem like an energy *minimum*. However, this argument cannot *rule out* the possibility that this arrangement will be an energy minimum under some special conditions, and it certainly does not justify taking the apparent infinity of the exchange too seriously. This point has often been overlooked [270] and led, for example, to special solutions [275] for a certain, 'singular' point in a particular type of a domain wall. That solution minimizes only the exchange energy, because this term 'goes to infinity for $r \to 0$ proportional to $1/r^2$, and exceeds all other energy terms'. Even Brown, who was always very careful with his definitions, made this mistake, and in a footnote on p. 67 of [145] he ruled out a certain configuration, because it 'would entail infinite exchange energy'.

This problem is a real one, for certain special cases, but it has no general solution, and a certain attempt [276] to solve it has essentially failed. There is no alternative to the use of eqn (7.1.4) for most problems, and some special techniques for special problems. It should only be borne in mind that there *are* cases for which the general rule fails, and should not be used.

In the summation over the position vectors $\mathbf{s}_i$ that led to eqn (7.1.4), it was implicitly assumed that all of them are inside the crystal. When the lattice point $i$ is on the surface, some of these neighbours may be missing, and the sum may come up different than at internal lattice points. It is not a serious problem, and for all practical purposes it is sufficient to keep eqn (7.1.4) as it is, assuming it applies everywhere, and add another energy term which affects the surface only. Actually, this modification of the exchange near the surface is only one of several contributions [145] to the *surface anisotropy energy* term, already mentioned in section 5.1.5.

The form of eqn (7.1.4) is particularly suited only for Cartesian coordinates. It calls for certain transformations in cases for which other coordinate systems are preferable for any reason. It is not very difficult to carry out these transformations, but it is easier if they can be avoided altogether. For this purpose it was suggested [277] to replace the exchange energy density in eqn (7.1.4) by

$$w_e = \frac{C}{2M_s^2}\left[(\nabla \cdot \mathbf{M})^2 + (\nabla \times \mathbf{M})^2\right], \tag{7.1.9}$$

because for a vector of fixed magnitude the difference between this expression and the one in eqn (7.1.4) is [277] a divergence of a certain vector. The volume integral over the latter can be transformed to a surface integral, by using the divergence theorem. Therefore, this difference redefines only the surface anisotropy term, and does not change the exchange in the bulk, and in the form of eqn (7.1.9) it is easier to change to a different coordinate system. However, this suggestion has never been used by anybody else, and will not be used here either.

## 7.2 The Landau and Lifshitz Wall

As a first illustration for the use of this classical exchange energy, a better solution will be given here for the best structure of the wall between antiparallel domains. This wall has already been discussed in section 6.2.1, but very rough approximations were used there, which can at best demonstrate the feasibility of its existence. A much better approach is to minimize the energy of the problem, using the same approximations which were first introduced [268] in 1935.

For this purpose, consider an *infinite* crystal, which has a uniaxial anisotropy of the type of eqn (5.1.7). The domains will arrange themselves with their magnetization parallel and antiparallel to the easy anisotropy axis, which is defined here as the $z$-axis, see section 5.1. We define the $x$-axis along the direction in which the magnetization changes from the $-z$- to the $+z$-direction, namely from $m_z = -1$ to $m_z = +1$, where $\mathbf{m}$ is defined by eqn (5.1.6). In the wall between the domains, $\mathbf{m}$ tilts out of the $z$-direction, which can be either towards $x$ or towards $y$. However, an $m_x$ which is a function of $x$ means a non-zero $\nabla \cdot \mathbf{M}$, and gives rise to a large magnetostatic energy contribution. Obviously, the energy is lower if $m_x = 0$, and only $m_y$ and $m_z$ are left to be functions of $x$. Combining the anisotropy energy density from eqn (5.1.7) with the exchange energy density from eqn (7.1.4), the total energy density for this case is

$$w = K_1 m_y^2 + K_2 m_y^4 + \frac{1}{2} C \left[ \left( \frac{dm_y}{dx} \right)^2 + \left( \frac{dm_z}{dx} \right)^2 \right]. \tag{7.2.10}$$

The magnetostatic energy is left out, because $\nabla \cdot \mathbf{M} = 0$, so that there is no volume charge, and the surface charge is neglected by the assumption of an 'infinite' crystal. The reader has already been warned in section 6.2.2 that such an assumption has no physical meaning, and that leaving out the surface charge by such an argument is never justified. However, this knowledge came much later, and for many years everybody was convinced that this approximation was fully justified, at least for bulk materials. Actually, there are still many who believe, against strong evidence, that at least the *energy* calculated here is a good approximation to the walls in very large crystals. Now it is known that the approximation is not really justified, and that the magnetization structure in a wall does not look at all like the one calculated in this section. Still, this structure is very important from a historical point of view, being the first study in micromagnetics. It is also a nice and easy problem to solve, and as such it makes a good introduction to the more difficult problems of micromagnetics.

The vector $\mathbf{m}$ is a *unit* vector, which means that its magnitude is 1, and $m_y^2 + m_z^2 = 1$. The easiest way to enforce such a constraint is to define an angle, $\theta$, by the relation

$$m_z = \cos\theta \quad \text{and} \quad m_y = \sin\theta. \tag{7.2.11}$$

Substituting in eqn (7.2.10), and integrating the energy *density* over $x$, the total energy per unit area in the $yz$-plane is

$$\mathcal{E} = \int_{-\infty}^{\infty} \left[ K_1 \sin^2\theta + K_2 \sin^4\theta + \frac{1}{2} C \left( \frac{d\theta}{dx} \right)^2 \right] dx. \tag{7.2.12}$$

The Euler differential equation for minimizing this integral is

$$C \frac{d^2\theta}{dx^2} - 2K_1 \sin\theta\cos\theta - 4K_2 \sin^3\theta\cos\theta = 0, \tag{7.2.13}$$

with the boundary condition

$$\left( \frac{d\theta}{dx} \right)_{x=\pm\infty} = 0. \tag{7.2.14}$$

It is easy to integrate such an equation once, and obtain what is known as a *first integral.* It can be achieved for example by multiplying the differential equation by $d\theta/dx$ and integrating over $x$. The result is

$$\frac{1}{2}C\left(\frac{d\theta}{dx}\right)^2 - K_1\sin^2\theta - K_2\sin^4\theta = \text{const}, \qquad (7.2.15)$$

where the right hand side is an integration constant. This integration is actually a particular case of a general theorem [270], which will be proved in section 10.2, according to which *all* one-dimensional problems in static micromagnetics have at least one first integral.

The integration constant can be determined from the condition that the structure must be a *wall* separating two domains magnetized along $\pm z$. It implies that $\sin\theta = 0$ at $z = \pm\infty$, and when this condition is substituted in eqn (7.2.15), together with eqn (7.2.14), this constant is seen to be 0. Hence

$$\frac{d\theta}{dx} = \pm\sqrt{\frac{2K_1}{C}}\sqrt{1+\frac{K_2}{K_1}\sin^2\theta}\,\sin\theta. \qquad (7.2.16)$$

Integration of this equation is obvious for either choice of the sign in front of the square root, and one of the branches is

$$\cos\theta = \frac{\sqrt{1+\kappa}\tanh(x/\delta)}{\sqrt{1+\kappa\tanh^2(x/\delta)}} = m_z, \qquad \delta = \sqrt{\frac{C}{2K_1}}, \qquad \kappa = \frac{K_2}{K_1}. \qquad (7.2.17)$$

Actually, instead of $x$ the argument should contain $x - x_0$, where $x_0$ is the second integration constant of the original second-order differential equation. However, the origin does not have any meaning in an *infinite* crystal, and $x_0$ may be omitted.

The *wall energy* can also be calculated analytically, by substituting eqn (7.2.17) in eqn (7.2.12), and carrying out the integration. The wall energy per unit wall area is thus found to be

$$\mathcal{E} = \sqrt{2K_1C}\left[1+\frac{1+\kappa}{\sqrt{\kappa}}\arctan(\sqrt{\kappa})\right], \qquad (7.2.18)$$

which depends only on the anisotropy and exchange constants of the material. The spontaneous magnetization, $M_s$, does not enter, because it is only connected with the magnetostatic energy term which has been eliminated from the present calculation.

Theoretically, the magnetization in eqn (7.2.17) becomes parallel to $\pm z$ only at infinity, and the wall has an infinite width, but of course this infinity need not be taken too seriously. The scale of $x$ in this equation is $\delta = \sqrt{C/2K_1}$, at least when $\kappa$ is small, and this expression is usually defined as the *wall width.* Any reasonable definition of the width as the distance over which most of the rotation from $m_z = -1$ to $m_z = +1$ takes place, will lead to something of the order of this quantity. A more accurate definition is given in [278].

If the anisotropy is cubic, as is the case *e.g.* in iron or nickel, there are three easy axes along the three cubic axes. The magnetization in some of the domains is at 90° and in some at 180° to the one in the neighbouring domain, see Fig. 4.1. The structure and energy of both 90° and 180° walls have been calculated [279] in a similar way to the calculation in this section, at least for a negligible $K_2$. The results are also similar to the foregoing. Magnetostriction, which has been added [279] as a uniaxial anisotropy superimposed on the cubic one, has some effect on the wall structure, but its effect on its energy is negligible.

The calculation of the energies of the different walls is the basis of what became known [145] as the *domain theory*. In calculating the energy of different configurations of domains, the walls between them are taken to have a zero width, as in the calculation of the magnetostatic energy of the two domains in section 6.2. But then the energy of the walls is added, using expressions such as the one for the uniaxial anisotropy in eqn (7.2.18) here, and multiplying by the wall area according to the assumed geometry. This technique allows the comparison of the total energy of all sorts of configurations, in an attempt to find [52, 279] the one whose energy is lower than that of the others. It is even possible to add the interaction of each configuration with an applied magnetic field and try to follow theoretically the whole hysteresis curve. For large and complex systems it is the only theory, and these studies still continue [280, 281] today. For small particles there are better and more reliable methods, which will be described in chapter 9, but this technique is being used [282, 283] for them as well. More about walls will be given in chapter 8, but most details of the domain theory are outside the scope of this book. Only before concluding this section, the reader must be warned not to be misled by the elegance of the solution into believing that the calculation presented here is the final result on the wall structure or its energy. Even a large crystal ends *somewhere*, and the structure presented here creates much too much charge on the surface. This charge 'demagnetizes' the wall, and distorts its shape to reduce its magnetostatic energy, and this distortion propagates into the internal parts of the wall. The resulting structure becomes quite complex, and cannot be expressed by a one-dimensional function of space. The whole problem then becomes much more complicated than the one presented here, but then complication is inevitable in ferromagnetism.

## 7.3 Magnetostatic Energy

The magnetostatic energy term has been introduced in section 6.1, as eqn (6.1.7), but it has not been proved there. It will be put here on a sounder basis than the argument given there. In order to satisfy those who may feel uneasy about a mere acceptance of Maxwell's equations as they are, it is necessary to start from the atomic nature of real materials, which was not even known at the time of Maxwell. The approximation of a continuous material is inevitable at the end, but it is important to notice that it is not introduced as an arbitrary assumption. It comes as a well-justified approximation for the limit of gradual variation over a size which is large compared with the lattice constant of the material. It is nearly the same justification and the same limit as in the classical approximation for the exchange energy in section 7.1. If anything, the approximation for the magnetostatic term is even more justified than that for the exchange term, as will be seen in the following.

### 7.3.1 *Physically Small Sphere*

Consider a lattice made of magnetic dipoles, with the magnetic moment $\boldsymbol{\mu}_i$ at the lattice point $i$. Let $\mathbf{h}_i$ be the field intensity at the lattice point $i$ due to all the *other* dipoles. In the absence of thermal fluctuations, the potential energy of this system is

$$\mathcal{E}_{\mathrm{M}} = -\frac{1}{2}\sum_i \boldsymbol{\mu}_i \cdot \mathbf{h}_i, \tag{7.3.19}$$

where the factor $\frac{1}{2}$ is introduced because the summation contains each of the interactions twice: once as the interaction of the dipole $i$ with the field due to $j$, and once as that of the dipole $j$ with the field due to $i$.

Let a sphere be drawn around the lattice point $i$. If its radius $R$ is large compared with the unit cell of the material, all the dipoles *outside* this sphere may be taken as a continuum for calculating the field which they create at this particular point, $i$. Therefore, the field $\mathbf{h}_i$ at this point may be evaluated by taking the field due to a continuous material everywhere, *subtracting* from it the field due to a continuous material inside this sphere, and adding the field due to the discrete dipoles within the same sphere. The first of these terms is the field calculated in section 6.1 from Maxwell's equations. It will be denoted from now on by $\mathbf{H}'$, in order to keep the notation $\mathbf{H}$ for the applied field, due to currents in some external coils. As has already been explained in section 6.1, these two fields may be taken as separate entities, and then superimposed. It is necessary to subtract from this field $\mathbf{H}'$ the contribution of a continuous magnetization inside this sphere. If this magnetization does not vary very much inside the sphere, the latter field is approximately the demagnetizing field of a homogeneously magnetized sphere, given by eqn (6.1.13), namely $-(4\pi/3)\mathbf{M}$. Hence,

$$\mathbf{h}_i = \mathbf{H}' + \frac{4\pi}{3}\mathbf{M} + \mathbf{h}'_i, \tag{7.3.20}$$

where $\mathbf{h}'$ is the contribution of the dipoles inside the sphere.

The use of the demagnetizing field is justified if the radius $R$ of the sphere is small compared with the scale over which the direction of the magnetization can be taken as a constant, or at most as a linear function of space. It is necessary to make sure that this assumption is not in conflict with the first assumption, that $R$ is much laıger than the lattice constant of the material. The second requirement is that $R$ is small compared with the so-called *exchange length*, which is the length over which $\mathbf{M}$ changes, namely something of the order of the Landau and Lifshitz wall width, $\sqrt{C/2K_1}$. For a typical case of permalloy, with $C \approx 2 \times 10^{-6}$ erg/cm and $K_1 \approx 10^4$ erg/cm$^3$, this wall width is about 100 nm, namely about 300 unit cells, and about the same number applies to iron. In these cases it is indeed possible to define an intermediate value of $R$, fulfilling both requirements for being sufficiently large *and* sufficiently small, which is usually referred to as a *physically small* sphere. It should be emphasized again that this possibility of defining such a physically small sphere is due to the exchange being very strong over a short range, keeping the spins almost aligned over distances of the order of a unit cell. There are cases of certain rare earths, or their alloys, for which $K_1$ is much larger, and the exchange length is only a few lattice constants. In these cases the continuum approach is not justified, and it is necessary [284] to consider a finite change in the direction of $\mathbf{M}$ from one lattice point to the next.

The last term in eqn (7.3.20) is a sum over fields due to dipoles,

$$\mathbf{h}_i' = \sum_{|\mathbf{r}_{ij}|<R} \left[ -\frac{\boldsymbol{\mu}_j}{|\mathbf{r}_{ij}|^3} + \frac{3(\boldsymbol{\mu}_j \cdot \mathbf{r}_{ij})\mathbf{r}_{ij}}{|\mathbf{r}_{ij}|^5} \right], \tag{7.3.21}$$

where $\mathbf{r}_{ij}$ is the vector pointing from lattice point $i$ to lattice point $j$. In a physically small sphere, $\boldsymbol{\mu}_j$ is actually a constant, which does not depend on $j$. In this case, it is possible to write, for example, the $x$-component of the field in Cartesian coordinates as

$$h_{i_x}' = \sum \left[ -\frac{\mu_x}{r_{ij}^3} + \frac{3x_{ij}(\mu_x x_{ij} + \mu_y y_{ij} + \mu_z z_{ij})}{r_{ij}^5} \right]. \tag{7.3.22}$$

If the crystal has a cubic symmetry, a sum over a *sphere* of the term with $x_{ij}y_{ij}$ or with $x_{ij}z_{ij}$ vanishes because there is an equal contribution from the positive and negative term, and actually this statement is true for almost any other symmetry. Also, for a cubic symmetry, $x$, $y$ and $z$ are interchangeable, and therefore

$$\sum \frac{x_{ij}^2}{r_{ij}^5} = \sum \frac{y_{ij}^2}{r_{ij}^5} = \sum \frac{z_{ij}^2}{r_{ij}^5} = \frac{1}{3} \sum \frac{x_{ij}^2 + y_{ij}^2 + z_{ij}^2}{r_{ij}^5} = \frac{1}{3} \sum \frac{1}{r_{ij}^3}. \tag{7.3.23}$$

Thus the total sum in eqn (7.3.22) is zero, and so is any other component of eqn (7.3.21). In a non-cubic symmetry the sum is not zero, but it is obvious from the form of eqn (7.3.22) that $\mu_x$ can be taken in front of a sum which is just a *number*, and the same is true for the other components. On the whole, under the assumption that $\mathbf{M}$ may be approximated by a constant inside the physically small sphere, $\mathbf{h}'_i$ is a linear function of the components of this $\mathbf{M}$, with coefficients which depend only on the crystalline symmetry. In other words,

$$\mathbf{h}'_i = \Lambda \cdot \mathbf{M}, \tag{7.3.24}$$

where $\Lambda$ is a tensor which depends on the crystalline symmetry, and which vanishes for a cubic symmetry. Substituting eqns (7.3.24) and (7.3.20) in eqn (7.3.19), and changing the sum to an integral, the magnetostatic energy is

$$\mathcal{E}_{\mathrm{M}} = -\frac{1}{2}\int \mathbf{M} \cdot \left(\mathbf{H}' + \frac{4\pi}{3}\mathbf{M} + \Lambda \cdot \mathbf{M}\right) d\tau, \tag{7.3.25}$$

where the integration is over the ferromagnetic body.

It must be emphasized that the approximation of the physically small sphere does not really require that $\mathbf{M}$ is a constant inside this sphere. A *linear* change over the dimensions of the sphere will not make any difference to the foregoing, because it is easy to see by symmetry considerations that its contribution is zero. A generalization [285] of the foregoing derivation considered the case of a quadratic change over the sphere, and showed that its contribution is also zero for a sufficiently high crystalline symmetry. For a lower symmetry, the contribution of a quadratic term is not zero, but it has been shown [285] to be negligibly small for all cases of interest. The proof of this theorem can be summarized qualitatively by the following argument: if the change of the magnetization is rather slow, the foregoing is correct. If it is not slow, the magnetostatic energy may be assumed to be small compared with the exchange energy, and a certain mistake in the smaller term does not affect the total energy. The physically small sphere may also be generalized [285] to be an ellipsoid, but this generalization does not have any real effect on the present calculation.

The middle term in eqn (7.3.25) contains $\mathbf{M} \cdot \mathbf{M}$, which is the constant $M_s^2$ that depends only on the temperature, and does not depend on the spatial distribution of $\mathbf{M}$. Therefore, it is omitted, which only means re-defining the *zero* of the magnetostatic energy and has no effect on energy minimizations. The last term is an energy density $\mathbf{M} \cdot \Lambda \cdot \mathbf{M}$ which has the same formal form of the anisotropy energy density discussed in section 5.1. Therefore, it may be included in the anisotropy energy instead of here. It is particularly convenient to do so because the anisotropy constants in most cases are taken from the experimental values, which *already include* this term. It should only be noted that when the spin–orbit interaction is calculated from basic principles, this term should be added to the resulting anisotropy. The magnetostatic energy has thus been shown to be

$$\mathcal{E}_{\text{M}} = -\frac{1}{2}\int \mathbf{M}\cdot\mathbf{H}'\,d\tau, \tag{7.3.26}$$

which is the same expression already used in chapter 6 without a proof.

As was the case with the exchange energy, the physically small sphere is assumed here to be entirely inside the ferromagnetic body, and this assumption fails for lattice points near the surface. Here, part of the necessary correction is already included in the surface charge, given in section 6.3 as part of the classical energy calculation. The rest of this error affects only spins which are quite close to the surface, and can be expressed [145] as a term with the same functional form as the surface anisotropy energy term. Therefore it can be taken into account as another contribution to the surface anisotropy.

The magnetostatic energy as expressed by eqn (7.3.26) is *non-local.* The volume integral in this equation contains $\mathbf{H}'$, which in turn has to be evaluated by another volume integral, see section 6.3. It effectively means integrating *twice* over the same volume. In this respect this energy term is very different from the exchange and anisotropy energy terms, which are local, namely involve one volume integration of an energy density. This property is another aspect of the long-range nature of the magnetostatic forces which require to take into account the interaction of each dipole with every other dipole in the ferromagnet. There have been some attempts to approximate this double integration by a single one, and these attempts have failed, as reviewed in [270]. The point is that in principle, a long-range force cannot be replaced by a short-range one, without losing some of its important properties. Therefore, the complication of a six-fold integration is part of the physical problem, and as such it is inevitable.

### 7.3.2 *Pole Avoidance Principle*

There are other forms to express the magnetostatic energy, which are mathematically equivalent, but one may be more useful than the other in some problems. In order to establish them, it is necessary first to prove a theorem concerning the magnetic induction,

$$\mathbf{B}' = \mathbf{H}' + 4\pi\mathbf{M}, \tag{7.3.27}$$

which has already been defined in eqn (1.1.2). It is used here as $\mathbf{B}'$ in order to emphasize that it is only the part of $\mathbf{B}$ which is related to $\mathbf{H}'$, and not to the applied field $\mathbf{H}$.

According to eqn (6.1.2),

$$\int \mathbf{H}'\cdot\mathbf{B}'d\tau = -\int \mathbf{B}'\cdot\nabla U d\tau = -\int [\nabla\cdot(U\mathbf{B}') - U\nabla\cdot\mathbf{B}']\,d\tau, \tag{7.3.28}$$

where the second equality is an identity, and the integration is assumed to be over a large enough volume to contain all the ferromagnetic bodies.

The second term in the last expression vanishes according to eqn (6.1.3), and the first term can be transformed according to the divergence theorem. Hence,

$$\int \mathbf{H}' \cdot \mathbf{B}' d\tau = -\int \mathbf{n} \cdot U\mathbf{B}' dS, \qquad (7.3.29)$$

where $\mathbf{n}$ is the normal. Now, the boundary conditions of Maxwell's equations assure a continuity of both $U$ and $B_n$ everywhere. Therefore, the surface integrals cancel when done on both sides of any surface of a ferromagnetic body, and the right hand side of eqn (7.3.29) is an integral over the outside surface of the volume which has been assumed to contain all these bodies. If this surface is allowed to tend to infinity, outside all ferromagnets $\mathbf{B}' = \mathbf{H}' = -\nabla U$ which tends to zero at least as fast as $1/r^2$, see the boundary conditions in section 6.1. Therefore, $UB_n$ tends to zero at least as $1/r^3$, while $dS$ increases as $r^2$, and the whole integral on the right hand side of eqn (7.3.29) tends to 0 at infinity, which means that

$$\int_{\text{all space}} \mathbf{H}' \cdot \mathbf{B}' d\tau = 0. \qquad (7.3.30)$$

This theorem is of some interest in its own right. But it is also useful for a transformation of the expression for the magnetostatic energy. It is attained by substituting eqn (7.3.27) in eqn (7.3.30), which yields

$$\int_{\text{all space}} \mathbf{H}' \cdot (\mathbf{H}' + 4\pi\mathbf{M}) d\tau = 0. \qquad (7.3.31)$$

Breaking this integral into a sum of two integrals, it is seen that the one which contains $\mathbf{M}$ is proportional to the integral for the magnetostatic energy in eqn (7.3.26). To begin with, the integral in eqn (7.3.31) is over all space, which includes parts in which there are no ferromagnetic bodies. However, $\mathbf{M} = 0$ in those parts of space, and they do not contribute to the second part of the integral. Therefore, this second integral is also over the ferromagnetic bodies, as is the integral in eqn (7.3.26). Rearranging,

$$\mathcal{E}_{\text{M}} = \frac{1}{8\pi} \int_{\text{all space}} H'^2 \, d\tau. \qquad (7.3.32)$$

This form of writing the magnetostatic energy demonstrates the *pole avoidance principle.* The integrand is positive everywhere, which makes the magnetostatic energy always positive. The smallest possible value for this energy term is 0, and this value can only be achieved when $H'$ is identically zero everywhere. Therefore, the magnetostatic energy term always tries to avoid any sort of volume or surface charge. A complete avoidance is not usually possible, unless the geometry is that of a toroid, in which a magnetization with no divergence can be parallel to the surface. However, the principle is that this energy term tries to achieve configurations with as little charge as possible. This principle has already been used in the qualitative arguments in section 6.3, which showed for example that an ellipsoid would rather be magnetized along its longest axis, etc. However, in that section this argument was a little premature, because a clever reader may have wondered why avoid the charge rather than think of some sophisticated arrangements with a combination of a positive and negative charge, whose energy may be lower than that of no charge at all. Only now, after the proof of eqn (7.3.32), with an integrand which is always positive, it should be clear that such a 'sophisticated' arrangement cannot exist.

There is still another form to express the magnetostatic energy term, which is also derived from eqn (7.3.30), which can be written as

$$\int_{\text{all space}} (\mathbf{B}' - 4\pi\mathbf{M}) \cdot \mathbf{B}' d\tau = 0, \tag{7.3.33}$$

in accordance with eqn (7.3.27). Rearranging, and using eqn (7.3.27) again,

$$\frac{1}{8\pi} \int_{\text{all space}} B'^2 \, d\tau = \frac{1}{2} \int \mathbf{M} \cdot \mathbf{B}' d\tau = \frac{1}{2} \int \mathbf{M} \cdot (\mathbf{H}' + 4\pi\mathbf{M}) d\tau, \tag{7.3.34}$$

where the integrations on the right hand side are over the volumes in which $\mathbf{M} \neq 0$. Substituting from eqn (7.3.26),

$$\frac{1}{8\pi} \int_{\text{all space}} B'^2 \, d\tau = -\mathcal{E}_{\text{M}} + 2\pi \int M^2 d\tau. \tag{7.3.35}$$

And since $M^2$ is the constant $M_s^2$,

$$\mathcal{E}_{\text{M}} = 2\pi M_s^2 V - \frac{1}{8\pi} \int_{\text{all space}} B'^2 \, d\tau, \tag{7.3.36}$$

where $V$ is the volume of the ferromagnetic body, or bodies.

It must be emphasized again that an energy calculated from eqn (7.3.36) for any particular case is going to yield exactly the same numerical value as the energy calculated from eqn (7.3.32), because these two equations are mathematically identical. However, the minus sign in eqn (7.3.36) does not allow any *physical* interpretation of what configurations of the magnetization this energy term prefers, which can even come close to the simple picture of pole avoidance implied by eqn (7.3.32). It can be said that the magnetostatic energy term prefers the average $\mathbf{B}'^2$ to be as large as possible, but this statement does not help at all to see the actual preferable distribution of the field $\mathbf{B}'$, or that of $\mathbf{M}$. It has been claimed [286] that eqn (7.3.36) has to be preferred over eqn (7.3.32), because $\mathbf{B}'$ has a more direct physical meaning than $\mathbf{H}'$, which is essentially the same as saying that surface or volume charge should not be used because there is no physical meaning to this charge. Sometimes a pure mathematical concept can be more convenient, and allow a better physical intuition into the problem, than a true physical approach. The same disadvantage of a lack of physical intuition applies also to other forms [287] of the magnetostatic energy term.

Beginners may wonder why the first term in eqn (7.3.36), which is just a constant, is not omitted by redefining the energy zero, as has already been done several times in this book. Of course, it is quite legitimate to do so, as long as this new definition is used *consistently*. However, it is not done because it is not useful, and will only mislead people to believe that the magnetostatic energy prefers $\mathbf{B}'$ to be as large as possible everywhere. Redefining the zero will not change the mathematical fact that whatever $\mathbf{B}'$ is, the newly defined energy cannot possibly be more negative than $-2\pi M_s^2 V$, which is the energy of a configuration with no volume or surface charge in the new system. Definitions are chosen to be helpful, and confusing definitions are better avoided, even if they are quite legitimate in principle.

### 7.3.3 *Reciprocity*

A very powerful tool for calculating the magnetostatic energy of certain configurations can be obtained from a generalization of eqn (7.3.30).

Consider two distributions of magnetization in space, $\mathbf{M}_1$ and $\mathbf{M}_2$. Let $\mathbf{H}'_i$ be the magnetic field produced by $\mathbf{M}_i$, for $i = 1, 2$ respectively, and let $\mathbf{B}'_i = \mathbf{H}'_i + 4\pi\mathbf{M}_i$. Using the same proof used to prove eqn (7.3.30), it is readily seen that

$$\int_{\text{all space}} \mathbf{H}'_1 \cdot \mathbf{B}'_2 d\tau = \int_{\text{all space}} \mathbf{H}'_2 \cdot \mathbf{B}'_1 d\tau = 0. \tag{7.3.37}$$

The properties of the functions used for proving eqn (7.3.30) were that $\mathbf{H}'$ is a gradient of a potential which is continuous everywhere and is regular at infinity, and that $\mathbf{B}'_n$ is continuous everywhere, and all these properties are also fulfilled by $\mathbf{H}'_i$ and $\mathbf{B}'_i$ separately. Therefore, the proof is the same, and in the same way as writing eqn (7.3.31), it is possible to conclude that both

$$\int_{\text{all space}} \mathbf{H}'_1 \cdot (\mathbf{H}'_2 + 4\pi\mathbf{M}_2) d\tau = 0 \tag{7.3.38}$$

and

$$\int_{\text{all space}} \mathbf{H}'_2 \cdot (\mathbf{H}'_1 + 4\pi\mathbf{M}_1) d\tau = 0. \tag{7.3.39}$$

Subtracting these two equations, the part with $\mathbf{H}'_1 \cdot \mathbf{H}'_2$ is common to both expressions and cancels, leaving

$$\int_{\text{all space}} \mathbf{H}_1' \cdot \mathbf{M}_2 d\tau = \int_{\text{all space}} \mathbf{H}_2' \cdot \mathbf{M}_1 d\tau. \tag{7.3.40}$$

Of course, it is never actually necessary to integrate over the whole space, and each integral is over the volume in which its integrand is not zero.

This equation is known as the *reciprocity theorem*, and is very useful in solving magnetostatic problems. The integrals in eqn (7.3.40) are parts of the integral in eqn (7.3.26), if $\mathbf{M}_1$ and $\mathbf{M}_2$ are parts of the total magnetization distribution, $\mathbf{M}$, and the theorem is true for any arbitrary subdivision of the magnetization into these two entities. Therefore, an appropriate choice of the way in which $\mathbf{M}$ is subdivided into $\mathbf{M}_1$ and $\mathbf{M}_2$ may often simplify the evaluation of the magnetostatic energy. Of course, this choice has to be fitted to the particular case under study, and there are no guidelines to facilitate the decision. An example of this use of the reciprocity theorem will be given in the derivation of the Brown differential equations in section 8.3.

In principle, the reciprocity theorem is not limited to $\mathbf{M}_1$ and $\mathbf{M}_2$ which add up to the total magnetization distribution, $\mathbf{M}$. The proof of this theorem is quite general, and applies also to cases in which $\mathbf{M}_1$ and $\mathbf{M}_2$ overlap in some part of the space. However, nobody has ever used this theorem for such an overlap. The most direct application of the theorem is for the case in which $\mathbf{M}_1$ and $\mathbf{M}_2$ are the magnetizations in two separate bodies, and eqn (7.3.40) is interpreted to mean that the interaction of the magnetization in one body with the field created by that of the other body is the same as the interaction of the magnetization of the second body with the field created by the first one. The most common example [288] is in the calculation of the energy of interaction between a recording head and the bits it records on a disc or tape, or in calculating the signal on the reading head, which involves the same integral as in eqn (7.3.40). It is rather easy to know the field due to the magnetization in the head, and the magnetization distribution in the recorded tape. It is much more difficult to estimate the field due to the recorded tape and the magnetization distribution in the head. Equation (7.3.40) makes it possible to know the interaction without evaluating the more difficult part.

Brown [1] has listed this reciprocity theorem as only one of several theorems to which he gave the general name of reciprocity theorems. All the others are less common, and less applied, in the literature, and are outside the scope of this book.

### 7.3.4 *Upper and Lower Bounds*

When it is difficult to evaluate the magnetostatic energy exactly, it may often be sufficient to have a reliable estimate for its value. An approximation may do, but an approximation is reliable only with a good estimation of the error involved. In principle, the best estimation is obtained when the exact value can be put between two bounds, especially when these two bounds do not differ very much from each other. Sometimes such bounds are found by a certain trick which is applicable only to a particular problem. However, for magnetostatic energy calculations, Brown [289] has devised a rather general method for finding both an upper bound and a lower bound. If properly used, these bounds may be sufficiently close together, so that the exact value may not be needed. These bounds will be specified here, but the *proof* that they are indeed a lower and an upper bound is different from the one originally presented by Brown. The latter was not very easy to follow or to understand.

Let $\mathbf{M}$ be the actual distribution for which the magnetostatic energy is to be calculated, and let $\mathbf{H}'$ be the true field due to this magnetization. Let $\mathbf{H}_1' = -\nabla \cdot \Phi$ be the field due to some *other* distribution, which will be specified later. Obviously,

$$\mathcal{E}_{\mathrm{M}} \geq \mathcal{E}_{\mathrm{H}} = \mathcal{E}_{\mathrm{M}} - \frac{1}{8\pi} \int_{\text{all space}} \left(\mathbf{H}_1' - \mathbf{H}'\right)^2 d\tau, \tag{7.3.41}$$

where the inequality results from subtracting an integral which cannot be negative, because of the square. Opening the brackets in the integrand, using first eqn (7.3.32) then eqn (7.3.27), and omitting the 'all space' which is implied from now on for all the integrals in this section,

$$\mathcal{E}_{\mathrm{H}} = \frac{1}{8\pi} \int \left(2\mathbf{H}_1' \cdot \mathbf{H}' - {\mathbf{H}_1'}^2\right) d\tau = \frac{1}{8\pi} \int \left[2\mathbf{H}_1' \cdot (\mathbf{B}' - 4\pi\mathbf{M}) - {\mathbf{H}_1'}^2\right] d\tau. \tag{7.3.42}$$

The part of the integral which contains $\mathbf{H}_1' \cdot \mathbf{B}'$ is zero according to eqn (7.3.37). Note that the proof of that equation required only that $\mathbf{H}_1'$ is a gradient of a potential which is continuous and regular at infinity. It is not even necessary that this potential is due to any real distribution of a magnetization. Therefore, by writing this potential explicitly and substituting eqn (7.3.42) in eqn (7.3.41)

$$\mathcal{E}_{\mathrm{M}} \geq \mathcal{E}_{\mathrm{H}} = \int \mathbf{M} \cdot \nabla\Phi \, d\tau - \frac{1}{8\pi} \int (\nabla\Phi)^2 \, d\tau, \tag{7.3.43}$$

where the first integral is over the ferromagnetic body (or bodies), and the second integral is over the whole space.

This result provides a lower bound to the correct magnetostatic energy $\mathcal{E}_{\mathrm{M}}$ of a given magnetization distribution $\mathbf{M}$, in terms of an *arbitrary* function of space, $\Phi$. The only limitation on the arbitrary choice of $\Phi$ is that it is continuous everywhere, and that it is regular at infinity, because only these properties were used in the proof of eqn (7.3.43). A discontinuity of the derivative is allowed, and may be introduced anywhere. However, it is not usually sufficient to have just a lower bound, which may be correct but not useful. After all, a zero is also a lower bound to the magnetostatic energy, which is always positive according to eqn (7.3.32), but this lower bound does not help to solve many problems. A useful lower bound is one for which $\mathcal{E}_{\mathrm{H}}$ is not very different from $\mathcal{E}_{\mathrm{M}}$, which is intuitively understood to be more likely when $\Phi$ is chosen to have at least some of the features expected from the real potential of the problem, $U$. It should be noted that if $\Phi = U$, the inequality in eqn (7.3.43) becomes an equality, according to eqns (7.3.26) and (7.3.32). Therefore, the best choice should always be a $\Phi$ which approximates, or at least imitates, $U$. Thus, for example, it somehow does not seem right to choose a function $\Phi$ which has a discontinuous derivative inside the ferromagnetic body, even if such a choice is allowed in principle, and even though it has never been proved to be a wrong choice. At any rate, such a choice has never been tried in any of the applications of this theorem in the literature, as cited in [288].

In practical applications, $\mathcal{E}_{\mathrm{M}}$ is not known, and it is impossible to determine how good the choice of $\Phi$ is by checking if $\mathcal{E}_{\mathrm{H}}$ is close to $\mathcal{E}_{\mathrm{M}}$. Therefore, a lower bound by itself does not help at all, and the only criterion for the usefulness of $\Phi$ is when an upper bound can also be found, which is not very different from $\mathcal{E}_{\mathrm{H}}$. Only in such a case one can claim that the exact energy value $\mathcal{E}_{\mathrm{M}}$ is sufficiently well determined, because it must be between these two values. The importance of Brown's bounds are thus in the combination of *both* of them, and not in each of them by itself.

To obtain an upper bound, a positive integral is *added* to the true energy, $\mathcal{E}_{\mathrm{M}}$, in the form

$$\mathcal{E}_{\mathrm{M}} \le \mathcal{E}_{\mathrm{B}} = \mathcal{E}_{\mathrm{M}} + \frac{1}{8\pi} \int (\mathbf{B}_1 - \mathbf{B}')^2 \, d\tau, \tag{7.3.44}$$

where $\mathbf{B}_1$ is an arbitrary vectorial function of space. It can be seen from the proof of the relations used in this derivation that it is sufficient to require that $\mathbf{B}_1$ is continuous everywhere, and that $\nabla \cdot \mathbf{B}_1 = 0$. As is the case with $\Phi$, this $\mathbf{B}_1$ does not have to be connected with the real $\mathbf{B}'$ of the problem, but it helps if they are not too different. Substituting from eqn (7.3.36), opening the brackets, and using eqn (7.3.27),

$$\mathcal{E}_{\mathrm{B}} = 2\pi M_s^2 V + \frac{1}{8\pi} \int \left[ \mathbf{B}_1^2 - 2\mathbf{B}_1 \cdot (\mathbf{H}' + 4\pi \mathbf{M}) \right] d\tau. \tag{7.3.45}$$

According to eqn (7.3.37), for which all it takes to assume (as mentioned above) is that $\mathbf{B}_1$ is continuous and that its divergence is zero,

$$\mathcal{E}_{\mathrm{M}} \le \mathcal{E}_{\mathrm{B}} = 2\pi M_s^2 V + \frac{1}{8\pi} \int \mathbf{B}_1^2 \, d\tau - \int \mathbf{B}_1 \cdot \mathbf{M} \, d\tau, \tag{7.3.46}$$

where the last integral is over the ferromagnetic body, and the one before it is over the whole space. As before, $V$ is the volume of the ferromagnetic body (or bodies). And, as in the case of the lower bound in eqn (7.3.43), the inequality becomes an equality if $\mathbf{B}_1 = \mathbf{B}'$ due to the actual magnetization distribution $\mathbf{M}$.

The constraint $\nabla \cdot \mathbf{B}_1 = 0$ is easily implemented by choosing $\mathbf{B}_1 = \nabla \times \mathbf{A}$, in which case the vector potential $\mathbf{A}$ is almost completely arbitrary, because it is very easy to take care of the requirement of continuity in any analytic model. It is always preferred to define some adjustable parameters in both the scalar potential $\Phi$ and the vector potential $\mathbf{A}$, and maximize $\mathcal{E}_{\mathrm{H}}$ and minimize $\mathcal{E}_{\mathrm{B}}$ with respect to these parameters. This technique ensures obtaining the best upper and lower bounds for any chosen functional form of these functions of space. With some intuition, or some luck, the upper and lower bounds may be close enough to make it unnecessary to go into the computations of the actual magnetostatic energy, and some such cases have been reported [288]. A different case will be given in section 10.5.1.

In this section, integrals over the ferromagnetic body (or bodies) and integrals over the whole space were treated on an equal basis, using even the same symbol for both. In practice there is a very big difference between these two integrals when it comes to numerical computations. Because of the long-range nature of the magnetostatic potentials, the integration outside the ferromagnetic body converges very slowly, and it is necessary to use many times the volume of the ferromagnet before the result can approximate an integration to infinity. It must always be borne in mind that when two expressions are identical mathematically, *e.g.* as are eqns (7.3.26) and (7.3.32), their computation should only converge *eventually* to the same numerical result for the same problem. It does *not* mean that they take the same time to compute the same result. Because of the slow convergence, such a numerical integration outside the ferromagnet has never been considered practical in any of the applications of this theorem reported [288] so far, with only one exception which will be discussed in section 11.3.4. Instead, the potential was always taken as a functional form for which at least the contribution to $\int (\nabla\Phi)^2 d\tau$ or $\int \mathbf{B}_1^2 d\tau$ from the part outside the ferromagnet could be carried out analytically. More details can be found in the references cited in [288], and it can only be added that there is a certain suggestion [290] for a rather general class of functions which can be used for this purpose, both for the scalar potential $\Phi$ and for the vector potential $\mathbf{A}$.

### 7.3.5 *Planar Rectangle*

It has already been mentioned in section 6.3 that the formal solution of eqn (6.3.48) involves integrations which can be very rarely carried out analytically. The main reason is that the numerator is a function of $\mathbf{r}'$ only, while the denominator involves $\mathbf{r} - \mathbf{r}'$. These expressions are difficult to mix, even for rather simple functions.

Consider in particular the case of a ferromagnetic body in the form of a prism, $-a \leq x \leq a$, $-b \leq y \leq b$ and $-c \leq z \leq c$. The potential due to the volume charge, namely the first integral in eqn (6.3.48), can be written in Cartesian coordinates as

$$U_{\text{volume}} = -M_s \int_{-c}^{c} \int_{-b}^{b} \int_{-a}^{a} \frac{\frac{\partial m_x(x',y',z')}{\partial x'} + \frac{\partial m_y(x',y',z')}{\partial y'} + \frac{\partial m_z(x',y',z')}{\partial z'}}{\sqrt{(x-x')^2 + (y-y')^2 + (z-z')^2}}$$

$$\times\, dx'\, dy'\, dz', \tag{7.3.47}$$

where $\mathbf{m}$ is defined in eqn (5.1.6). Integrating by parts, the first term with $\partial m_x/\partial x'$ with respect to $x'$, the second term with respect to $y'$, etc., it is seen that all the expressions between the limits $-a$ and $a$, etc., cancel the appropriate terms of the potential of the surface charge, namely the second integral in eqn (6.3.48). For the reader who has never tried this kind of exercise, I highly recommend following the details of the last statement, which is the best way to understand the meaning of the normal $\mathbf{n}$. At any rate, the result is that the *total* potential due to both surface and volume charge is

$$U(x,y,z) = M_s \int_{-c}^{c} \int_{-b}^{b} \int_{-a}^{a}$$

$$\times \frac{(x-x')m_x(x',y',z') + (y-y')m_y(x',y',z') + (z-z')m_z(x',y',z')}{[(x-x')^2 + (y-y')^2 + (z-z')^2]^{3/2}}$$

$$\times\, dx'\, dy'\, dz'. \tag{7.3.48}$$

If $\mathbf{m}$ is made out of (rather small) integral powers of $x'$, $y'$ and $z'$, all the integrations in this equation can be carried out analytically. For practically any other function of these variables it is impossible to do anything analytic unless the mixture in the denominator can somehow be transformed into a product of functions of $x$ and $x'$, and so forth.

No general transformation of this sort is known for the general, three-dimensional integral in eqn (7.3.48). However, a general transformation is known for the two-dimensional case, when $\mathbf{m}$ does not depend on $z$, which can be either because the sample is a very thin film, with $c \to 0$, or because the sample is very long in one dimension, and $c \to \infty$. In the two-dimensional case it is known from any undergraduate textbook that the potential of a unit charge is $\log(r/r_0)$, instead of the three-dimensional $1/r$ used in the derivation of eqn (6.3.48). Repeating the foregoing integration by parts for the rectangle $-a \leq x \leq a$, $-b \leq y \leq b$ leads to

$$U(x, y) = 2M_s \int_{-b}^{b} \int_{-a}^{a} \frac{(x - x')m_x(x', y') + (y - y')m_y(x', y')}{(x - x')^2 + (y - y')^2} \, dx' \, dy'. \tag{7.3.49}$$

In this case it is possible to use the well-known Laplace transform,

$$\int_0^\infty \cos[(y - y')t] e^{-|x - x'|t} \, dt = \frac{|x - x'|}{(x - x')^2 + (y - y')^2} \tag{7.3.50}$$

(and similarly for the second term), and rewrite eqn (7.3.49) *inside* the rectangle in the form

$$U_{\text{in}}(x, y) = 2M_s \int_0^\infty \left[ \int_{-b}^{b} \cos[(y - y')t] \left( \int_{-a}^{x} m_x(x', y') e^{-(x - x')t} \, dx' \right. \right.$$

$$\left. - \int_x^a m_x(x', y') e^{-(x' - x)t} \, dx' \right) dy' + \int_{-a}^{a} \cos[(x - x')t] \left( \int_{-b}^{y} m_y(x', y') \right.$$

$$\left. \left. \times \, e^{-(y - y')t} \, dy' - \int_y^b m_y(x', y') e^{-(y' - y)t} \, dy' \right) dx' \right] dt. \tag{7.3.51}$$

This expression was originally derived [291] as the limit $k \to 0$ of a periodic $z$-dependence of the form $\cos(kz)$, directly from the full three-dimensional potential in eqn (7.3.48). It should be noted that breaking the integrals for $x' > x$ and $x' < x$ (and similarly for $y'$) is due to the absolute value in eqn (7.3.50), so that the result presented in eqn (7.3.51) is valid only for the potential inside the rectangle. If either $x$ or $y$ is outside the ferromagnetic rectangle, this breaking down into two integrals has to be modified, and different expressions have to be fitted according to the quadrant outside the rectangle for which the potential is calculated. These distinctions are not usually necessary, because the potential in the rectangle, $-a \le x \le a$ and $-b \le y \le b$, is sufficient for calculating the magnetostatic energy.

Replacing a double integral by a triple one may not seem a good strategy at a first glance. However, the advantage of eqn (7.3.51) over eqn (7.3.49) is that the former contains trigonometric and exponential functions which are readily expressed as *products*, namely a function of $x'$ times a function of $x$, and similarly for $y$ and $y'$. In a product one is highly likely to be able to perform both integrations analytically, for a wide variety of functions $\mathbf{m}$, which cannot be handled in the form which contains $x - x'$ in eqn (7.3.49). For the magnetostatic energy in two dimensions there is a four-fold integration, which is transformed here into a five-fold one. But if four out of the five can be performed analytically, the numerical integration of the remaining integral over $t$ is much simpler than a four-fold numerical integration. This technique has indeed been found useful in the calculation of several cases, cited in [288].

The substitution in the energy will only be demonstrated here for the particular case of a magnetization which does not depend on $y$. In this case, after carrying out the integration over $y'$ in eqn (7.3.51),

$$U_{\rm in}(x,y) = 2M_s \int_0^\infty \left[ \frac{e^{-(b-y)t} - e^{-(b+y)t}}{t} \int_{-a}^{a} \cos[(x-x')t] m_y(x')\,dx' \right.$$
$$+ 2\frac{\cos(yt)\sin(bt)}{t} \left( \int_{-a}^{x} m_x(x') e^{-(x-x')t}\,dx' \right.$$
$$\left.\left. - \int_x^a m_x(x') e^{-(x'-x)t}\,dx' \right) \right] dt. \qquad (7.3.52)$$

According to eqns (7.3.26) and (6.1.2), the magnetostatic energy of such a one-dimensional magnetization structure in a rectangle is

$$\mathcal{E}_{\rm M} = \frac{1}{2}\int \mathbf{M}\cdot\nabla U\,dS = \frac{1}{2}M_s \int_{-a}^{a}\int_{-b}^{b} \left[ m_x(x)\frac{\partial U_{\rm in}}{\partial x} + m_y(x)\frac{\partial U_{\rm in}}{\partial y} \right] dx\,dy. \qquad (7.3.53)$$

Substituting from eqn (7.3.52), and carrying out the integrations over $y$,

$$\mathcal{E}_{\rm M} = 2M_s^2 \int_0^\infty \left[ \frac{2}{t}\sin^2(bt) \int_{-a}^{a} \left( \frac{2}{t}[m_x(x)]^2 - m_x(x) \right.\right.$$
$$\left. \times \int_{-a}^{a} m_x(x') e^{-|x-x'|t}\,dx' \right) dx + \frac{1-e^{-2bt}}{t} \int_{-a}^{a} m_y(x)$$
$$\left. \times \int_{-a}^{a} m_y(x') \cos[(x-x')t]\,dx'\,dx \right] dt. \qquad (7.3.54)$$

This expression can be simplified by noting that

$$\int_0^\infty \frac{\cos[(x-x')t]}{t}\left(1-e^{-2bt}\right) dt = 2\int_0^\infty \frac{\sin^2(bt)}{t} e^{-|x-x'|t}\,dt, \qquad (7.3.55)$$

because each of these expressions can be found in tables of integral transforms as being equal to

$$\frac{1}{2}\log\left(1+\frac{4b^2}{(x-x')^2}\right).$$

It is *not* advisable to use the latter expression before integrating over $x$ and $x'$ for specific magnetization configurations. However, eqn (7.3.55) allows combining the two expressions together. By substituting it and

$$\int_0^\infty \frac{\sin^2(bt)}{t^2}\,dt = \frac{\pi b}{2} \tag{7.3.56}$$

in eqn (7.3.54),

$$\frac{\mathcal{E}_{\mathrm{M}}}{2M_s^2} = \int_0^\infty \frac{1-e^{-2bt}}{t} \int_{-a}^{a}\int_{-a}^{a} \cos[(x-x')t][m_y(x)m_y(x')$$

$$- m_x(x)m_x(x')]dx'\,dx\,dt + 2\pi b \int_{-a}^{a} [m_x(x)]^2\,dx. \tag{7.3.57}$$

The advantage of this integral over one with $x - x'$ inside a logarithm should be quite obvious. This expression will be used in section 8.1 for the calculation of the magnetostatic energy of one-dimensional domain walls in thin films.

# 8

# ENERGY MINIMIZATION

## 8.1 Bloch and Néel Walls

The most popular case of minimizing all three energy terms (namely, the exchange, the anisotropy and the magnetostatic energies) is the study of the structure and energy of the wall between antiparallel domains in thin films. The Landau and Lifshitz solution described in section 7.2 assumes an infinite crystal, in which it is possible to get away with no magnetostatic energy contribution. If the crystal is finite, this wall structure contains a non-zero normal component of the magnetization on the surface, and the energy of the ensuing surface charge must be taken into account. Moreover, Néel recognized already in 1955 that the energy of this surface charge can be too large in the case of very thin films, which have more surface than volume. For this reason, Néel suggested a different structure for the wall in very thin films, in which the surface charge is replaced by a volume charge, and showed that the total energy could indeed be reduced by such a transformation.

This problem of a wall structure in thin films will be described here for the geometry shown in Fig. 8.1. A plate which is infinite in both the $x$- and $z$-directions has a thickness $2b$ in the $y$-direction. Two antiparallel domains have their magnetization along $\pm z$, which is also assumed to be an easy axis for a uniaxial anisotropy, and the wall between them occupies the region $-a \leq x \leq a$. The wall is assumed, in this section, to be one dimensional, namely $\mathbf{m}$ is assumed to be a function of $x$ only.

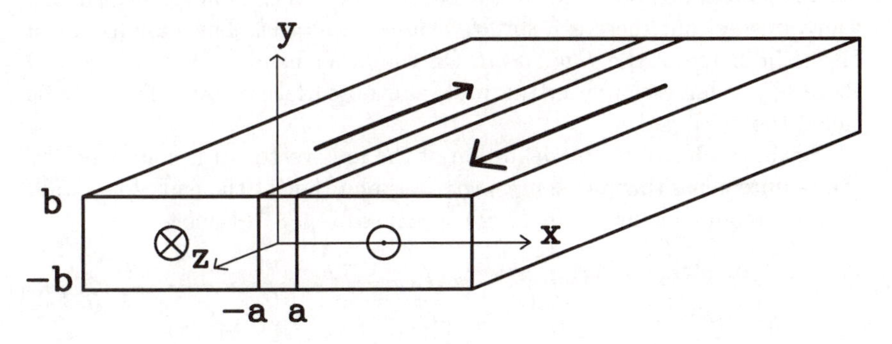


FIG. 8.1. The geometry of a domain wall in thin films.

One way to approach this problem is to use the solution for the bulk wall structure in eqn (7.2.17), and calculate its magnetostatic energy for the case of a finite $b$ shown in Fig. 8.1. Of course, it should be taken into account that the bulk structure may be modified for a finite thickness, and it is better to have a model with one or more parameters and minimize the total energy with respect to these parameters. The model should only tend to the structure of eqn (7.2.17) in the limit $b \to \infty$. However, the calculation of this energy term for this particular wall structure can only be done by a relatively complicated [292] numerical computation. Therefore, two methods have been used for resolving this difficulty. In one method certain approximations for the magnetostatic energy are introduced, and the other method uses functional forms for which the magnetostatic energy can be calculated analytically, and which *approximate* eqn (7.2.17) for a large $b$. Examples of both methods can be found in the literature cited in [288] and [292].

Here I choose to illustrate the problem by one of the models of the second type. It was first proposed by Dietze and Thomas in a paper cited in [288] and [292], then extended to more adjustable parameters by others. The original paper is in German, but it is not important for the reader to look it up, because the calculation of the magnetostatic energy given here uses a completely different method from the one given there. Only the result is the same. This model assumes that the $x$- and $y$-components of the unit vector $\mathbf{m}$ are

$$m_x(x) = \frac{q^2 \cos\phi}{q^2 + x^2}, \qquad m_y(x) = \frac{q^2 \sin\phi}{q^2 + x^2}, \tag{8.1.1}$$

where $q$ is an adjustable parameter, which essentially determines the wall *width.* Here $\phi$ is another parameter which is introduced in order to treat together the cases $\phi = 0$ (which makes $m_y = 0$) and $\phi = \pi/2$ (which makes $m_x = 0$). These cases were studied separately in the original paper of Dietze and Thomas, as well in all other models of a one-dimensional wall. It should be noted that in the case $\phi = \pi/2$ the volume charge in the wall vanishes, but there is a surface charge on $y = \pm b$. This case has been given the name of the *Bloch wall.* On the other hand, in the case $\phi = 0$ there is a volume charge in the wall, and no surface charge. This case is called the *Néel wall.*

For any value of $\phi$, the definition of the unit vector $\mathbf{m}$ is completed by the requirements that $m_x^2 + m_y^2 + m_z^2 = 1$, and that at the end of the wall, where the domains begin, $m_z(\pm\infty) = \pm 1$, see Fig. 8.1. Hence,

$$m_z(x) = \frac{x\sqrt{2q^2 + x^2}}{q^2 + x^2}. \tag{8.1.2}$$

The magnetostatic energy of this wall configuration can be calculated from eqn (7.3.57), for the particular case $a = \infty$, which is implied by eqns (8.1.1) and (8.1.2). For the integrations over $x$ and $x'$ it is only necessary to use

$$\int_{-\infty}^{\infty} \frac{\sin(xt)}{q^2 + x^2}\, dx = 0, \qquad \int_{-\infty}^{\infty} \frac{\cos(xt)}{q^2 + x^2}\, dx = \frac{\pi}{q}\, e^{-qt}, \tag{8.1.3}$$

and

$$\int_{-\infty}^{\infty} \frac{q^4}{(q^2 + x^2)^2}\, dx = \frac{1}{2}\left[\frac{q^2 x}{q^2 + x^2} + q \arctan\left(\frac{x}{q}\right)\right]_{-\infty}^{\infty} = \frac{\pi q}{2}. \tag{8.1.4}$$

Substituting all these relations in eqn (7.3.57), the magnetostatic energy per unit length in the $z$-direction is

$$\mathcal{E}_{\mathrm{M}} = 2b\pi^2 M_s^2 q \cos^2\phi + 2\pi^2 M_s^2 q^2 (\sin^2\phi - \cos^2\phi) \int_0^{\infty} \frac{e^{-2qt}}{t}\left(1 - e^{-2bt}\right) dt. \tag{8.1.5}$$

The remaining integration over $t$ is a well-known Laplace transform, which allows the whole expression to be written in an analytic closed form. However, in the study of domain walls it is customary to deal with the wall energy per unit wall area, denoted by $\gamma$, rather than with energy per unit wall length. In the case of Fig. 8.1, it is necessary to divide the wall energy per unit length by the film thickness, $2b$, to obtain the energy per unit wall area. Therefore,

$$\gamma_{\mathrm{M}} = \frac{\mathcal{E}_{\mathrm{M}}}{2b} = \pi^2 M_s^2 \left[q\cos^2\phi + \frac{q^2}{b}\left(\sin^2\phi - \cos^2\phi\right)\log\left(1 + \frac{b}{q}\right)\right]. \tag{8.1.6}$$

In particular, for the Bloch wall with $\phi = \pi/2$, this energy is proportional to $(q/b)\log(1 + b/q)$, which tends to zero for $b \to \infty$, and remains finite for $b \to 0$. For the Néel wall, with $\phi = 0$, the magnetostatic energy is proportional to $1 - (q/b)\log(1 + b/q)$, which tends to zero for $b \to 0$, and remains finite for $b \to \infty$. It is thus qualitatively clear that if there are no other types of walls, the Néel wall should exist for thin films (in which the energy of a surface charge is larger than that of a volume charge), and the Bloch wall should take over for thick films, for which the energy of a surface charge becomes smaller than that of a volume charge.

The other energy terms to be considered are the exchange and the anisotropy. The energy density for the former is given by eqn (7.1.4), which becomes, after substituting from eqns (8.1.1) and (8.1.2) and carrying out the differentiations with respect to $x$,

$$w_e = \frac{1}{2}C\left[\left(\frac{dm_x}{dx}\right)^2 + \left(\frac{dm_y}{dx}\right)^2 + \left(\frac{dm_z}{dx}\right)^2\right] = \frac{2Cq^4}{(q^2 + x^2)^2(2q^2 + x^2)} \tag{8.1.7}$$

where $C$ is the exchange constant defined in section 7.1. Note that this expression is independent of the parameter $\phi$, which was entered in the form $\sin^2\phi + \cos^2\phi$. The integration is obvious, and the exchange energy per unit wall area of this model is

$$\gamma_{\text{ex}} = \frac{\mathcal{E}_{\text{ex}}}{2b} = \frac{1}{2b}\int_{-b}^{b}\int_{-\infty}^{\infty} w_e dxdy = \frac{\pi C}{q}(\sqrt{2}-1). \tag{8.1.8}$$

Assuming that the anisotropy is uniaxial, whose easy axis is parallel to $z$, the anisotropy energy density is given by eqn (5.1.7), which will be used here for the case of a negligible $K_2$. For the particular configuration in eqn (8.1.1) this energy density is

$$w_u = K_1\left(m_x^2 + m_y^2\right) = K_1\frac{q^4}{(q^2+x^2)^2}. \tag{8.1.9}$$

Substituting in eqn (5.1.10), the anisotropy energy per unit wall area is

$$\gamma_{\text{a}} = \frac{\mathcal{E}_{\text{a}}}{2b} = \frac{1}{2b}\int_{-b}^{b}\int_{-\infty}^{\infty} w_u dxdy = \frac{\pi q}{2}K_1. \tag{8.1.10}$$

As was the case with the exchange term, this expression does not depend on $\phi$. Therefore, minimizing the total wall energy with respect to $\phi$ is achieved by minimizing the magnetostatic term only. And since $\partial\gamma_{\text{M}}/\partial\phi$ is proportional to $\sin\phi\cos\phi$, there are only the two solutions mentioned in the foregoing: the Bloch wall, $\cos\phi = 0$, which has a surface charge but no volume charge, and its total wall energy per unit wall area is

$$\gamma_{\text{B}} = \frac{\pi C}{q}(\sqrt{2}-1) + \frac{\pi q}{2}K_1 + \frac{\pi^2 M_s^2 q^2}{b}\log\left(1+\frac{b}{q}\right), \tag{8.1.11}$$

and the Néel wall, $\sin\phi = 0$, which has a volume charge but no surface charge, and its total wall energy per unit wall area is

$$\gamma_{\text{N}} = \frac{\pi C}{q}(\sqrt{2}-1) + \frac{\pi q}{2}K_1 + \pi^2 M_s^2 q\left[1 - \frac{q}{b}\log\left(1+\frac{b}{q}\right)\right]. \tag{8.1.12}$$

Note that the exchange energy term is trying to make the wall width, $q$, as large as it can, while the anisotropy energy term is trying to make $q$ as small as it can. This tendency is more general than the particular model discussed here, and fits the general, qualitative discussion in section 6.2.1. The role of the magnetostatic energy term is less obvious because of the dependence on the film thickness, $2b$. This feature is also rather typical, in that the magnetostatic energy term is usually quite complicated, and it is not easy to see its tendency and preferences. In the present case, the only obvious feature is that $\gamma_{\rm M}$ prefers a large $q$, if $b/q$ is constant. But this statement is not helpful because $b/q$ is not a constant. Even in this simple case, the only way to find out the role of $\gamma_{\rm M}$ is to minimize the total wall energy for different values of $M_s$, and try to see the tendency. There is no magnetic field in this calculation, which is just meant to find the static structure of a wall in zero applied field. It is not difficult to add an interaction with a field to this model, but the main effect of applying a field is to make the wall move somewhere else, which is a different problem.

The parameter $q$ is determined by minimizing the wall energy in either eqn (8.1.11) or eqn (8.1.12). It is achieved by equating to zero the derivative of the wall energy with respect to $q$, which leads to the transcendental equation

$$\frac{C}{q^2}(\sqrt{2}-1) = \frac{K_1}{2} + \pi M_s^2 \left[\frac{2q}{b}\log\left(1+\frac{b}{q}\right) - \frac{q}{q+b}\right], \tag{8.1.13}$$

for the Bloch wall, and to

$$\frac{C}{q^2}(\sqrt{2}-1) = \frac{K_1}{2} + \pi M_s^2 \left[1 - \frac{2q}{b}\log\left(1+\frac{b}{q}\right) + \frac{q}{q+b}\right], \tag{8.1.14}$$

for the Néel wall. These equations have to be solved for $q$ as a function of $b$, and then the energy can be calculated from eqn (8.1.11) or eqn (8.1.12), by substituting the computed $q$.

The solution of these equations is straightforward only in the limit $b \to 0$ for the Néel wall, or $b \to \infty$ for the Bloch wall. In both these cases the magnetostatic energy contribution vanishes, and the solution of either eqn (8.1.13) or eqn (8.1.14) is

$$q = \sqrt{\frac{2C}{K_1}(\sqrt{2}-1)}. \tag{8.1.15}$$

Substituting in eqn (8.1.11) or eqn (8.1.12),

$$\lim_{b\to 0} \gamma_{\rm N} = \lim_{b\to\infty} \gamma_{\rm B} = \pi\sqrt{2CK_1(\sqrt{2}-1)}. \tag{8.1.16}$$

This wall energy is $\pi(\sqrt{2}-1)/2 = 1.011$ times the energy of the Landau and Lifshitz wall, eqn (7.2.18), which has been obtained as a solution of the Euler equation of the problem. It means that eqn (7.2.18) is the absolute minimum for the energy of all possible one-dimensional wall structures in an infinite film thickness. A difference of only 1% from this absolute minimum, and *for any value of the physical parameters*, certainly makes the present model a very good approximation, at least for very thick films. Also, the

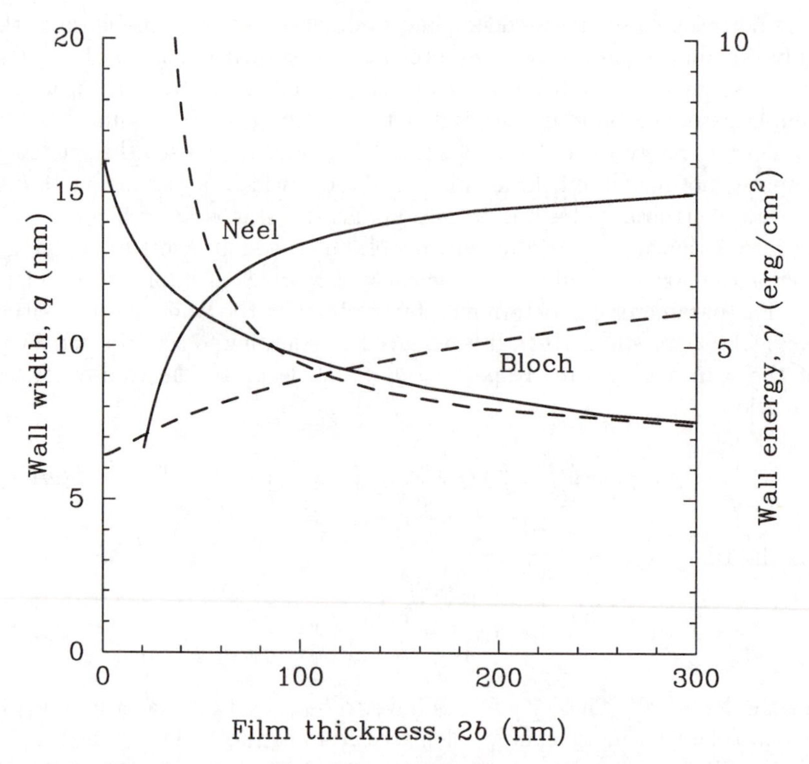


FIG. 8.2. The domain wall width, $q$ (dashed curves), and energy per unit wall area, $\gamma$, for Bloch and Néel walls in thin permalloy films.

wall width which can be defined by $q$ in eqn (8.1.16) is not significantly different from the width obtained from eqn (7.2.17).

For any finite film thickness eqns (8.1.13) and (8.1.14) have to be solved numerically, and a numerical solution can only be done for specific values of the physical parameters. As an example, I choose the parameters usually used in the study of permalloy films, namely $C = 2 \times 10^{-6}$ erg/cm, $K_1 = 10^3$ erg/cm$^3$, and $M_s = 800$ emu. For these values, the computed values of the wall width parameter $q$ are the dashed lines plotted in Fig. 8.2 as functions of the film thickness, $2b$. Once $q$ is known for either of these walls, its value can be substituted in eqn (8.1.11) or (8.1.12), for computing the wall energy per unit wall area, $\gamma_{\rm B}$ or $\gamma_{\rm N}$ respectively. The energy values thus computed are plotted as the full curves in Fig. 8.2.

Different one-dimensional models were published, and they all yielded very similar results, for the same values of the physical parameters. One of the obvious theoretical conclusions from Fig. 8.2, which has already been stated qualitatively in the foregoing, is that one should expect Néel walls in very thin films, then at a certain film thickness there should be a *sharp* transition to Bloch walls. This sharp transitions did not seem right, and several workers tried to work out a certain *mixed wall* around the transition between the Néel and the Bloch wall regions. None of these models worked, and they all collapsed in the same way that the present model in eqn (8.1.1) did. In the beginning, this model contained an extra parameter, $\phi$, which *could* have values for which the wall is partly Néel and partly Bloch. However, the energy minimization retained only the two values 0 and $\pi/2$, and did not allow any mixing. The same happened for any model which anybody tried. Later there was a general proof [292] that the same must happen to any one-dimensional model, and there can be no mixed wall in one dimension. This theorem does not necessarily invalidate certain semi-quantitative arguments [293] about the possibility of a mixed, but not strictly one-dimensional, wall.

Experimentally, the transition from a Néel to a Bloch wall is *not* sharp. It is possible to distinguish experimentally between the wall structure in thin films, identified as a Néel wall, and the wall structure in thick films (or in bulk materials), identified as a Bloch wall. However, between the regions where one or the other is observed, there is a certain region of film thicknesses in which a third type of wall is observed. This third type, which has been named the *cross-tie wall*, has [52, 294] a very complicated structure. It is definitely not a one-dimensional structure, because it has an obvious *periodicity* in the $z$-direction of Fig. 8.1. There have been several attempts [295, 296, 297, 298] to work out a theoretical model for the magnetization structure in this cross-tie wall, but none of them could produce satisfactory results. More recent computations [299] made a large advance towards the understanding of this wall structure, and compared favourably [300] with experiment. However, they have not really solved this problem completely, and the fine details of the cross-tie wall structure are not very well known yet.

The details of the Néel wall structure are not very well known either. In 1965, Brown tried to avoid the choice among the large number of the then-existing models for the Bloch and Néel walls. He thought that he could find the structure with the lowest possible energy, by a numerical minimization of all possible one-dimensional configurations, using a method which will be described in chapter 11. He and his student, LaBonte, solved [301] this problem for the Bloch wall in permalloy films, but they could not find such a structure for the Néel wall, because the computations just did not converge. It turned out that the Néel wall has a very long 'tail', which keeps changing with further iterations.

Later computations, reviewed in [295], found various *ad hoc* solutions for the problem of convergence of the tail, but these solutions only produce a converging result. They do not necessarily solve the physical problem, and it is not clear at all if these converging solutions actually present a physically valid wall structure. In the first place, the necessity for special tricks by itself should be regarded as a symptom of a deeper problem, which does not go away when the symptom is removed. Secondly, details of the computed wall structure do not quite agree [302] with experiment. Thirdly, the whole one-dimensional approach is based on the assumption that $m_y$ is zero everywhere, even though the magnetic field in this direction, $\partial U/\partial y$, is not zero, which somehow does not sound right, as noted [293, 295] already, together with some other details. Besides, there is too big a difference between the tails obtained for the same physical parameters, with a slightly different film thickness, as in curves $a$ and $b$ in Fig. 3 of [303], and it looks strange at best. On top of that, there is a general theorem [304] according to which *all* one-dimensional magnetization structures are unstable.

It is quite possible that these difficulties are not serious, and at least there is no experimental or theoretical proof that something is *basically* wrong with the theory of the one-dimensional Néel wall. The theorem about instability of all one-dimensional structures was never taken very seriously by anybody, not even by the authors of the original paper. For the particular case of the Landau and Lifshitz one-dimensional wall, they wrote [304] that it 'derives its justification from three-dimensional considerations implicit in the initial statement of the formally one-dimensional problem'. They had only a very mild criticism of other wall calculations. On p. 93 of his book [145], Brown still justified the Landau and Lifshitz wall calculations, but was more explicit in stating that the calculations of walls in thin films are essentially invalidated by this theorem. He wrote that they needed justification, without which they 'must be regarded as mere guesses'. However, this criticism was just ignored. Everybody else in those days regarded this theorem as a mere formality and a nuisance, and most people think so even today. They consider it as something equivalent to the mathematical proof that magnetism cannot exist in two dimensions, which does not prevent a theoretical study of two-dimensional systems, and its comparison with experiments on *nearly* two-dimensional samples, as discussed in section 4.5. They believe that these one-dimensional Néel walls, although formally wrong, are a very good approximation for the real three-dimensional wall structure. Therefore, no serious attempt has ever been made to check this point, and all that is known about Néel walls has not changed since the review [295] which listed the best one-dimensional models that were all generalizations of eqn (8.1.1) with more parameters. The results of an attempt [295] to allow $m_y$ to be a function of $y$, instead of just 0, were not very encouraging. The most recent numerical computations [299, 302] also start with an *a priori* assumption of dependence only on $x$ of Fig. 8.1.

In view of all the indications mentioned here, I do not consider this approach to be satisfactory. Although there is no clear-cut evidence for it, the solution may be in allowing *another* dimension. Since the structure of the cross-tie walls involves a handedness which changes periodically in $z$ of

Fig. 8.1, a real Néel wall may also have this periodicity. This calculation has not been tried yet.

## 8.2 Two-dimensional Walls

The theorem mentioned in the previous section, about the instability of all one-dimensional ferromagnetic configurations, also applies, in principle, to the case of the Bloch wall. Moreover, while the possibility of $z$-dependence stated in the previous section is only a speculation for the Néel wall, there is strong experimental evidence [52, 305, 306, 307] that such a periodic change of the handedness does exist in Bloch walls, even in bulk crystals. There is also a rather convincing argument [295, 308] that this periodic change reduces the wall energy, at least with respect to the one-dimensional Bloch wall. Still, the effect of this $z$-dependence on the wall energy, and on its structure in the other dimensions, has not been fully investigated yet. The usual assumption, which has never been justified in any way, is that the $z$-dependence is a minor perturbation, which affects only a small part of a long wall, and can be ignored without making a serious mistake.

However, for the Bloch wall there was also a rather common feeling that there must a way to reduce the magnetostatic energy by allowing a variation of the magnetization along $y$ of Fig. 8.1. Brown in particular used to go around advertising this idea, but neither he nor anybody else had a good model to try it on. The first published suggestion (cited in [295]) was a perturbation scheme of the one-dimensional wall, which has never been actually carried out. Then there were observations of *coupled* walls, which are the walls in a sample made of two ferromagnetic films, separated by a thin layer of a non-magnetic material. There is a strong magnetostatic interaction, affecting certain experimental results, between a wall in the upper layer which is just above one in the lower layer. Some references to both experiments and theory are given in [295], but the details are outside the scope of this book. It is sufficient to mention here that the theory of this phenomenon used two-dimensional models in which the magnetization in the walls was a function of both $x$ and $y$. The purpose was to form *closed loops* of the magnetization vector, which do not have a volume charge, and to hit the surfaces at small angles, thus reducing the surface charge.

It then occurred to me that the same model, in the limit of the thickness of the separating layer tending to 0, may be used for the wall structure in *one* film. The only difference between a single wall and a pair of coupled walls is that in the latter it is possible to draw closed loops whose radius shrinks to zero at the centre, where there is no ferromagnetic material. Such small-radius loops involve a very large exchange energy in a single layer, whose centre is also ferromagnetic. I solved this difficulty by defining a small transition region at the wall centre, $|x| < x_0$, in which the magnetization changed more gradually than in the coupled walls, thus avoiding the large exchange energy. This $x_0$ was left as a parameter with respect to which the energy was minimized. Two more adjustable parameters were suggested [309], but never tried.

Specifically, for the geometry of Fig. 8.1 the model [309] assumed that for $|x| < x_0$,

$$m_x = -\sin\left(\frac{\pi y}{2b}\right), \qquad m_y = \cos\left(\frac{\pi x}{2x_0}\right)\cos\left(\frac{\pi y}{2b}\right),$$
$$m_z = \sin\left(\frac{\pi x}{2x_0}\right)\cos\left(\frac{\pi y}{2b}\right), \tag{8.2.17}$$

while for $|x| > x_0$,

$$m_x = -\text{sech}\left[\frac{\pi}{2b}\left(|x| - x_0\right)\right]\sin\left(\frac{\pi y}{2b}\right),$$
$$m_y = \frac{x}{|x|}\text{sech}\left[\frac{\pi}{2b}\left(|x| - x_0\right)\right]\tanh\left[\frac{\pi}{2b}\left(|x| - x_0\right)\right]\cos\left(\frac{\pi y}{2b}\right), \tag{8.2.18}$$
$$m_z = \frac{x}{|x|}\sqrt{\tanh^2\left[\frac{\pi}{2b}\left(|x| - x_0\right)\right] + \text{sech}^4\left[\frac{\pi}{2b}\left(|x| - x_0\right)\right]\cos^2\left(\frac{\pi y}{2b}\right)}.$$

Note that

- This model obeys the constraint

$$m_x^2 + m_y^2 + m_z^2 = 1 \tag{8.2.19}$$

  everywhere.
- The magnetization is continuous everywhere, including at $x = \pm x_0$.
- It represents a wall in the sense that $m_z = \pm 1$ for $x = \pm\infty$.
- There is no surface charge, because $m_y = 0$ for $y = \pm b$, but there is a volume charge. Thus, this model does not go along with the rule argued in section 6.3, according to which a surface charge should be preferred over a volume charge in bulk materials.

The structure of this wall model is shown in Fig. 8.3, for the particular case of a permalloy film whose thickness is 2000 Å, with the value of $x_0$ which minimizes the energy for these particular parameters. In this figure, only $m_x$ and $m_y$ are plotted. The component $m_z$ is perpendicular to the plane of the plot, and its magnitude is large where the size of the plotted arrows is small, and vice versa.

The original publication [309] of this model did not contain this figure, because this method of plotting magnetization structures was only invented later, in the thesis of LaBonte. It shows how the volume charge is decreased by forming nearly closed loops, in which the head of each arrow nearly follows the tail of the one before it. This kind of magnetization structure

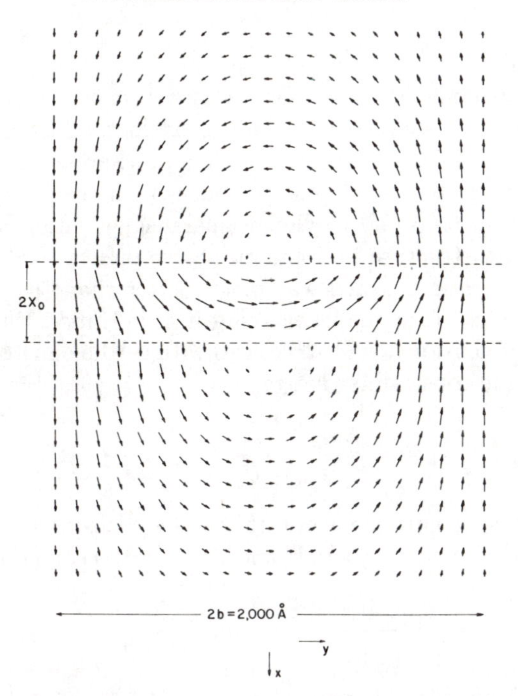


FIG. 8.3. The assumed structure of the first two-dimensional domain wall in permalloy films, as described by eqns (8.2.18) and (8.2.19).

is preferred by the magnetostatic energy, but involves work which must be done against the exchange energy, which prefers the magnetization to be aligned. It is especially noted in the figure that if the region with $x_0$ is removed, a circular vortex with a very small radius is formed at the centre, which is better avoided because it involves a very large exchange energy. The introduction of this extra transition region is an artificial, *ad hoc* solution originating from the adoption of a model for double walls, with no exchange at the centre. The physical system can find better ways to avoid this large exchange at the centre.

The exchange and anisotropy energies of this model were calculated analytically. The magnetostatic energy term was expressed [309] as a one-dimensional integral, which would be trivial to compute nowadays. However, computers in those days were not what they are now, and instead of computing this integral, it was only proved that the term which contained it was *negative*. Therefore, by dropping that term the magnetostatic energy was *increased* and the total wall energy thus calculated was an *upper bound* to the wall energy which can be obtained by such a wall structure. Such an upper bound was adequate to demonstrate the necessity of this second dimension, because even the upper bound for this wall energy was already considerably smaller than the computed [301] lowest possible energy for a one-dimensional wall.

In a way, this demonstration was a waste, because by the time it was published LaBonte was already concluding his doctoral thesis, in which he developed a method (which will be described in chapter 11) for numerical minimization of the total wall energy, made of the exchange, anisotropy and magnetostatic terms, of two-dimensional magnetization configurations. He computed the structure and energy of two-dimensional walls by this numerical minimization, which was much better than this crude model. The only usefulness of the model was that it gave Hubert the idea [248] to construct two-dimensional wall structures with zero magnetostatic energy. He introduced the constraints that $m_y = 0$ on $y = \pm b$ of Fig. 8.1 *and* that

$$\frac{\partial m_x}{\partial x} + \frac{\partial m_y}{\partial y} = 0 \tag{8.2.20}$$

everywhere. These conditions were enforced by choosing a scalar function $A(x, y)$, with $A = \text{const}$ on $y = \pm b$, and defining the components of $\mathbf{m}$ by

$$m_x = \frac{\partial A(x,y)}{\partial y}, \qquad m_y = \frac{\partial A(x,y)}{\partial x}, \tag{8.2.21}$$

with $m_z$ being defined by the constraint of eqn (8.2.19). To simplify matters, Hubert [248, 249] chose a certain functional form for $A(x, y)$, which contained certain adjustable *functions*. These functions were eventually defined numerically from point to point during the energy minimization process.

For his doctoral thesis, LaBonte solved the problem of the Bloch wall in permalloy films, but at that time he was still using symmetry considerations to reduce the computation time. He actually computed only the quarter $x > 0$ and $y > 0$ of the equivalent of Fig. 8.3, assuming that the rest of the wall can be obtained by taking the mirror images of this quarter, which is the same assumption used to make the model shown in Fig. 8.3. However, when he graduated, and went to work for CDC company, he had unlimited computer time at his disposal, which was unusual in those days. Therefore, he allowed the computer to look at all four quarters of the wall, and found [310] that the wall energy could be very much reduced by a structure which is *not* symmetric along $x$. When viewed from the domain on its right, the wall looks different than when viewed from the domain on its left. This result was unexpected, because one would tend to assume that there is no built-in directionality, and the wall cannot possibly tell which is right and which is left. However, it turns out that the asymmetric structure allows the magnetization to build nearly complete vortices, with a very small magnetostatic energy, *without* making the full circle at the centre, with its large exchange energy. It is a better solution than the *ad hoc* avoidance of that small circle by the $x_0$ introduced into Fig. 8.3.

The approximation of zero magnetostatic energy used by Hubert also led [248, 249] to an asymmetric wall, and its whole structure turned out to be very similar to that computed by LaBonte. The total wall energy calculated by these two methods was also very nearly the same, which shows that minimizing only the magnetostatic energy, as done by Hubert, is a good approximation to minimizing the total wall energy, as done by LaBonte. This result demonstrates that the magnetostatic energy is the leading energy term in sufficiently large samples, which is the conclusion already reached in section 6.2.2. It should be emphasized again that it is mainly the magnetostatic energy term which determines the complex structure of the magnetization in the wall (or in any other magnetization structure in bulk materials), while the exchange and anisotropy energy terms only play the role of small perturbations. However, once this structure is determined, the magnetostatic energy term computed from it becomes very small. In the computations of LaBonte [310] for permalloy films in the thickness range between 1000 and 2000 Å, the magnetostatic term varied between 5 % and 3 % of the total wall energy. For the thinner film with a thickness of 500 Å, the contribution of this term was 12 %, but at this thickness the minimization of the Bloch wall becomes doubtful, because experiments show that the cross-tie wall already takes over. Experiments on Ni platelets show [311] a strong dependence on $z$ of Fig. 8.1 in the Bloch wall, already in the thickness range approaching the occurrence of cross-tie walls, and an even more complex transition has been observed [312] in an Fe–Al alloy. It is reasonable to assume that this $z$-dependence also affects the $x$- and $y$-structure of the wall, and probably its energy too, in that thickness region, but the appropriate theory has not been worked out. In *this* region of film thickness it thus seems that the calculations of LaBonte and of Hubert are unreliable, and should be replaced by something else, in *three* dimensions, which is not known yet.

For thicker films, the asymmetric two-dimensional wall as computed by LaBonte or Hubert is in good agreement with electron microscope studies of these walls, at least to within the accuracy of these experiments. Actually, some wall asymmetry can already be seen in older pictures [313] which were published before these theories, but at that time this asymmetry was ignored. When more attention was paid to this detail, a very pronounced asymmetry, quite similar to the one predicted by the theory, was seen in 180° walls [314, 315, 316] in various materials. Such an asymmetry has also been seen [317] in 90° domain walls, which are outside the scope of this book. In these measurements electrons are shot *through* the film, and measure only the average of the magnetization in their path, namely the average along $y$ of Fig. 8.1. However, in some cases several passes are taken with the sample tilted [318, 319] at different angles. This technique allows a better look into the $y$-dependence of the magnetization, because several different averages are measured. The accuracy is not high, but to within this accuracy it seems that the computations of Hubert or LaBonte give a good description of the observed walls in this intermediate film thickness. It is not so for the thinner films, where the theory does not fit the experiment, as has already been mentioned. And it cannot be checked for very thick films, which electron cannot pass without being completely absorbed on the way. It may be worth mentioning that while each wall in a film is asymmetric, the sample as a whole does not have any directionality for this asymmetry, and walls with an opposite sense of asymmetry occur in the same sample. Sometimes the sense of asymmetry can be seen [320] to vary periodically in $z$ even within the same wall.

In as much as a plot of numerical results can be taken as a final solution, the wall structure in this intermediate film thickness is known. However, numerical results apply only to the particular physical parameters used for the computations, and if one wants to know the structure for a different exchange constant, or anisotropy constant, etc., the whole computation must be repeated. Also, details of the results cannot be published, and cannot be passed over to somebody who wants to use them as a start for another calculation, or to calculate a property which has not been included in the original computation. For such purposes it is much more convenient to have an *analytic approximation* of the magnetization structure with some parameters that can be fitted to any particular case. When a good approximation for the true minimal energy state can be specified over a wide range of the physical parameters, it is sufficient to minimize the energy with respect to the adjustable parameters for every specific case. It helps if the latter minimization is simpler than the original numerical computation of the whole structure. But even if it is not, it has at least the advantage of the possibility of specifying all the details of the structure for any particular case in terms of the numerical values of a few parameters. Moreover, it is possible to interpolate these parameters between computed values for a physical parameter (*e.g.* the anisotropy constant) and thus know the approximate structure of a wall for cases which have not been computed.

For the two-dimensional Bloch wall, there is such a model [321], with eight adjustable parameters, which is a very good approximation to both the structure and the energy of Bloch walls as computed by LaBonte. It is too complex for any analytic calculation of any of the energy terms, and the energy minimization must be done by the same numerical method as that of LaBonte [310], but the possibility of communicating the results in terms of the numerical values of the eight parameters is a large advantage. Attempts to make a *simpler* model have concentrated on cases for which the magnetostatic energy can be evaluated analytically, because it is this energy term which takes almost all the computational time in numerical minimizations. In the best of this class of models [322], the evaluation of the exchange and the anisotropy terms was done by a numerical integration of a one-dimensional integral for each. Such integration is a relatively easy computation, and the whole model is relatively simple to use, even if the definitions of the magnetization components take many lines. This model turned out [322] to be a sufficiently good approximation for comparing with electron microscopy data, but was not good enough for obtaining the finer details of the theoretical two-dimensional walls.

### 8.2.1 *Bulk Materials*

Above a certain film thickness, even the highest-voltage electrons cannot penetrate through the sample, and there is no way of knowing what the domain wall looks like. It is possible to shoot neutrons through the sample, but the accuracy of neutron diffraction is just sufficient to see the domains, not the details of the walls.

In the theoretical models for which the anisotropy energy term can be calculated analytically [309, 322], this energy term increases with increasing film thickness. This term is negligible for permalloy film whose thickness is around $10^3$ Å for which most of the studies have been carried out. However, for a much larger thickness this increase will make the anisotropy term larger than the other terms, and the *total* wall energy will start [309, 322] to increase with increasing thickness. Computations based on the model of Hubert also show [315] that, at least in one case, the total wall energy passes through a minimum, then starts to increase with increasing film thickness. It seems that it is going the way of the analytic models, namely the wall energy will keep increasing with increasing film thickness. Since the one-dimensional Bloch wall energy plotted in Fig. 8.2 decreases with increasing film thickness, there must be a certain thickness above which the energy of the two-dimensional wall will become larger than that of the one-dimensional wall. Therefore, the two-dimensional Bloch walls described here must cease to exist, and change into something else, above a certain film thickness, both for a cubic and for a uniaxial material.

For a long time it was taken for granted that at a sufficiently large thickness the wall will change into the one-dimensional Bloch wall of section 8.1, which eventually tends to the Landau and Lifshitz wall of section 7.2 in the bulk. For this reason the model of Jakubovics [321] was specifically designed to contain the one-dimensional Bloch wall as a particular case for certain values of the parameters, making sure that if there is a transition to this wall it will come out of the computations. Such a transition from the two-dimensional wall to the one-dimensional one was actually computed [321] for a large increase of the anisotropy constant. It could not be calculated for an increase in the film thickness, because the requirement of computer time and memory increases very rapidly with increasing film thickness, and all the computer resources are used up before such a transition is even approached.

The same difficulty of limited computer time and memory also applies to LaBonte-type computations. The thickest films studied theoretically so far [323] are a few $\mu$m thick iron, and in these ones the wall structure is still predominantly that of the thin films, with no way of seeing any possible transition towards a one-dimensional structure. Near the edge of the studied region there seems to be [323] a different kind of asymmetry, which *may* take over at a still larger film thickness, but it *still* has a slightly higher energy than that of the LaBonte-type thin film structure. Actually, the energy difference between these two structures is so small that the computer can be stuck for ever in the higher-energy state when the computations *start* from it. At this stage it is not clear if this other wall structure is indeed going to take over at still larger film thicknesses, and if it will eventually develop into something similar to the Landau and Lifshitz wall, or into something completely different. Some computations [324] were also carried out for 10 $\mu$m thick permalloy films, but they used a very rough grid, with the subdivision being an order of magnitude larger than in many other computations. Therefore, the results of these computations are unreliable. Besides, they yield [324] a magnetostatic energy term which is 21 % of the total wall energy. It is suspiciously larger than in many of the other two-dimensional wall computations, and seems to indicate an inadequate energy minimization.

By analysing the polarization of the electrons in a scanning electron microscope [55] it is possible to measure the magnetization of the last few atomic layers near the surface. Such experimental data clearly show [325, 326] that near the surface the domain wall looks like a Néel wall, in the sense that the magnetization there is nearly parallel to the surface. This result is not surprising, because in the two-dimensional wall discussed in the foregoing the magnetization also approaches the surface at a very small angle, see Fig. 8.3. At any rate, it should be clear to the reader by now that the magnetostatic energy will not allow any other approach to the surface, even if the crystal is very large, and the vast majority of the spins are very far from the surface. The working hypothesis in analysing such data [326] is that in a sufficiently thick sample, the wall is essentially the one-dimensional Bloch wall throughout *most* of the thickness, but when it approaches the surface it changes from the Bloch type to the Néel type, when the magnetization slowly turns around from the $y$- to the $x$-direction of Fig. 8.1. Detailed LaBonte-type computations [326, 327] both for iron and for permalloy support this picture, and are in good agreement with measured details of the surface part of the wall. They also prove that *all* the older measurements of domain wall *width* in bulk materials have measured only the width of the surface part of the wall, which is very different from the wall width in the bulk of the material. However, such computations are limited to relatively thin films, because of limited computer resources, already mentioned in the foregoing. The problem of what the wall really looks like inside bulk materials has not really been solved yet.

The wall energy in bulk materials is not known either. The Landau and Lifshitz result of section 7.2 is still often used for analysing domain configurations, but it is not clear if it is a good approximation. By using a certain analysis [328, 329], wall energies can be obtained from experimental data on domain widths, and this measurement is even often used for evaluating the exchange constant of the material. However, LaBonte-type computations for thin films involve no approximation, except for leaving out the third dimension, and must therefore be at least a reliable *upper bound* to the real wall energy. And yet, wall energies measured by this technique for thin permalloy films are [270] considerably *larger* than this theoretical upper bound. This discrepancy has never been accounted for, and it does seem to indicate that something is wrong with the analysis of the data in this technique, and casts some doubts on the values published for the bulk.

Finally, it will only be remarked that there is a vast experimental and theoretical literature on walls which are *not* straight lines, in particular walls in the form of a closed circle, around a circular domain, known as a *bubble domain.* For such a circle, even a one-dimensional wall such as the one in section 8.1 must be expressed in the two dimensions of the circle, and a two-dimensional wall as in [330] is too complicated to be discussed in this book. Some references can be found in [288].

## 8.3 Brown's Static Equations

Numerical computations as described in the previous section are relatively new, and are still limited to relatively simple cases. With present computers it is not even possible to find the lowest-energy configuration of a single wall, let alone a whole structure of domains separated by walls, or any other true three-dimensional magnetization distribution. For such problems it is still necessary to look for analytic solutions, or at least workable models. In as much as the wall energy is known, it is possible to compare the total energy of certain domain configurations, and find the one which has the lowest energy. This technique is the basis for what is known as the *domain theory* which has been used successfully for many cases. However, in principle it has two serious drawbacks.

The first one is that comparing the energies of different configurations always carries the risk of ignoring *another* configuration, which may have a still lower energy than all the ones being considered. If the basic structure can be taken from experiment, or if it is done by somebody with a high physical intuition, it may work out. But the probability of a correct guess is never very high, and many wrong results have been obtained by this method. It has already been seen in the previous sections how all sorts of one-dimensional models were compared with each other, till it turned out that the wall energy can be very much reduced when a variation is allowed in the second dimension. Many other examples also exist, and it is always a risk which must be borne in mind. In principle, any energy calculated for any particular model, with or without minimization with respect to some adjustable parameters, should be regarded as an *upper bound* for the actual energy minimum. The minimum cannot be larger than the energy of any special case, but there is always the possibility that the lowest-energy minimum is in a different configuration, which is not included in the assumed model. The estimation is really reliable only if a *lower bound* can also be found, in which case the true minimum must be between those bounds.

The second difficulty is that one is not always interested in the lowest-energy state, because of the hysteresis which is part of the study of ferromagnets. As seen already in the simple case discussed in section 5.4, the actual magnetization state may depend on the history of the applied field, and even though lower-energy states may exist, they are inaccessible due to an energy barrier between them and the present state. In such cases, comparing energies is meaningless.

For these reasons Brown set out to express the energy minimization rigorously, with an eye to performing it in a way that would take the hysteresis into account. A first step in this direction has already been demonstrated in the Landau and Lifshitz wall in section 7.2, where the energy minimization is done by solving the Euler differential equation which leads to the lowest possible energy minimum for the assumed form of the total energy. In that case there was no hysteresis, because no magnetic field was allowed, and the existence of the wall was *assumed, a priori.* Brown's idea was to have the most general Euler differential equation by a pure variational calculation, so that the existence of the wall (or of any other magnetization configuration) will be the *result* of the calculation, without having to assume it beforehand. It was this theory which Brown originally named *micromagnetics* although the name was later extended to mean any sort of calculation in which the atomic structure of matter is ignored, and the magnetization vector is taken as a continuous function of space.

Consider, therefore, a ferromagnetic body of any shape, in which the magnetization is any function of space. The total energy for this particular $\mathbf{m}(\mathbf{r})$ is made up of the exchange energy as in eqn (7.1.4), the anisotropy energy, the magnetostatic energy as in eqn (7.3.26), and an interaction with an applied magnetic field, $\mathbf{H}_a$, which is sometimes called the Zeeman energy term, namely

$$\mathcal{E} = \mathcal{E}_{\mathrm{e}} + \mathcal{E}_{\mathrm{a}} + \mathcal{E}_{\mathrm{M}} + \mathcal{E}_{\mathrm{H}} = \int \left\{ \frac{C}{2} \left[ (\nabla m_x)^2 + (\nabla m_y)^2 + (\nabla m_z)^2 \right] + w_{\mathrm{a}} \right.$$
$$\left. - \frac{1}{2} \mathbf{M} \cdot \mathbf{H}' - \mathbf{M} \cdot \mathbf{H}_a \right\} d\tau + \int w_{\mathrm{s}}\, dS, \qquad (8.3.22)$$

where the last term of the volume integral is the interaction of $\mathbf{M}$ with $\mathbf{H}_a$ as it comes out of the definition of $\mathbf{M}$, and as used many times in this book, although not in this form. The first integral is over the ferromagnetic body, and the second one is over its surface. Both the volume and the surface anisotropy energy density are left unspecified at this stage, but they are simple functions, which can be one or more of the cases specified in section 5.1. Magnetostriction can also be added, but it is neglected in this book, as mentioned in section 5.1, except for cases which can be written in the form of an anisotropy term, and are therefore included in $w_{\mathrm{a}}$.

This expression determines the energy if $\mathbf{m}(\mathbf{r})$ is known. The problem here is to determine $\mathbf{m}(\mathbf{r})$ so that this energy is a minimum. Brown [145] minimized this energy in several ways, the simplest of which is to consider a *small* variation of the magnetization vector around its value $\mathbf{m}_0$, bound by the constraint that the magnitude of $\mathbf{m}$ must be 1. The first two Cartesian coordinates can then be expressed as

$$m_x = m_x^{(0)} + \epsilon u, \qquad m_y = m_y^{(0)} + \epsilon v, \tag{8.3.23}$$

where $u$ and $v$ are *any* functions of space, and $\epsilon$ is small. The third component is determined by the constraint that $\mathbf{m}$ is a unit vector, and to a first order in $\epsilon$,

$$m_z = \sqrt{1 - \left(m_x^{(0)} + \epsilon u\right)^2 - \left(m_y^{(0)} + \epsilon v\right)^2} = \sqrt{m_z^{(0)2} - 2\epsilon\left(m_x^{(0)}u + m_y^{(0)}v\right)}$$

$$= m_z^{(0)}\left[1 - \epsilon\frac{m_x^{(0)}u + m_y^{(0)}v}{m_z^{(0)2}}\right] = m_z^{(0)} - \epsilon\lambda, \tag{8.3.24}$$

where

$$\lambda = \frac{m_x^{(0)}u + m_y^{(0)}v}{m_z^{(0)}}. \tag{8.3.25}$$

The variation of the exchange energy term due to this variation of $\mathbf{m}$ is

$$\delta\mathcal{E}_{\mathrm{e}} = \frac{C}{2}\int\left\{\left[\nabla\left(m_x^{(0)} + \epsilon u\right)\right]^2 + \left[\nabla\left(m_y^{(0)} + \epsilon v\right)\right]^2 + \left[\nabla\left(m_z^{(0)} - \epsilon\lambda\right)\right]^2\right.$$

$$\left. - \left(\nabla m_x^{(0)}\right)^2 - \left(\nabla m_y^{(0)}\right)^2 - \left(\nabla m_z^{(0)}\right)^2\right\} d\tau = \epsilon C\int\left\{\left(\nabla m_x^{(0)}\right)\cdot(\nabla u)\right.$$

$$\left. + \left(\nabla m_y^{(0)}\right)\cdot(\nabla v) - \left(\nabla m_z^{(0)}\right)\cdot(\nabla\lambda)\right\} d\tau\,, \tag{8.3.26}$$

to a first order in $\epsilon$. However, according to the divergence theorem, for any two functions $f$ and $F$,

$$\int(\nabla f)\cdot(\nabla F)\,d\tau = \int\left[\nabla\cdot(F\nabla f) - F\nabla^2 f\right]\,d\tau = \int F\frac{\partial f}{\partial n}dS - \int F\nabla^2 f\,d\tau, \tag{8.3.27}$$

where $\mathbf{n}$ is the normal to the surface. Using this relation three times in eqn (8.3.26) for the three products of this form which occur there, the variation of the exchange energy becomes

$$\delta\mathcal{E}_{\mathrm{e}} = \epsilon C \int \left[ u\frac{\partial m_x^{(0)}}{\partial n} + v\frac{\partial m_y^{(0)}}{\partial n} - \frac{m_x^{(0)}u + m_y^{(0)}v}{m_z^{(0)}}\frac{\partial m_z^{(0)}}{\partial n}\right] dS$$

$$- \epsilon C \int \left[ u\nabla^2 m_x^{(0)} + v\nabla^2 m_y^{(0)} - \frac{m_x^{(0)}u + m_y^{(0)}v}{m_z^{(0)}}\nabla^2 m_z^{(0)}\right] d\tau\,. \quad (8.3.28)$$

For the variation of the anisotropy energy it is sufficient at this stage to use eqn (8.3.24) in a first-order Taylor expansion,

$$\delta\mathcal{E}_{\mathrm{a}} = \int [w_{\mathrm{a}}(\mathbf{m}) - w_{\mathrm{a}}(\mathbf{m}_0)]\, d\tau$$

$$= \epsilon \int \left[ u\frac{\partial w_{\mathrm{a}}}{\partial m_x^{(0)}} + v\frac{\partial w_{\mathrm{a}}}{\partial m_y^{(0)}} - \frac{m_x^{(0)}u + m_y^{(0)}v}{m_z^{(0)}}\frac{\partial w_{\mathrm{a}}}{\partial m_z^{(0)}}\right] d\tau, \quad (8.3.29)$$

and similarly for the surface anisotropy term.

The variation of the magnetostatic energy term is *a priori* given by

$$\delta\mathcal{E}_{\mathrm{M}} = -\frac{1}{2}\int [(\mathbf{M} + \delta\mathbf{M})\cdot(\mathbf{H}' + \delta\mathbf{H}') - \mathbf{M}\cdot\mathbf{H}']\, d\tau, \quad (8.3.30)$$

where $\delta\mathbf{H}'$ is the field due to the small magnetization variation $\delta\mathbf{M}$. If this field had to be calculated by any of the methods used to calculate magnetostatic fields, this problem would have become hopelessly complicated. However, it is *not* necessary to calculate this field in order to evaluate the integral in eqn (8.3.30), because it is possible to use the *reciprocity theorem* of eqn (7.3.40), which ensures that in the present case,

$$\int \mathbf{M}\cdot\delta\mathbf{H}'\, d\tau = \int \mathbf{H}'\cdot\delta\mathbf{M} d\tau. \quad (8.3.31)$$

Substituting in eqn (8.3.30), and leaving out the *second*-order term,

$$\delta\mathcal{E}_{\mathrm{M}} = -\int \mathbf{H}'\cdot\delta\mathbf{M} d\tau, \quad (8.3.32)$$

which has the same form as the variation of the interaction with the applied field, $\mathbf{H}_a$. Using for both these terms the specific variation in eqns (8.3.23) and (8.3.24),

$$\delta\left(\mathcal{E}_{\mathrm{M}}+\mathcal{E}_{\mathrm{H}}\right)=-\epsilon M_s \int\left[H_x u+H_y v-H_z \frac{m_x^{(0)} u+m_y^{(0)} v}{m_z^{(0)}}\right] d\tau, \tag{8.3.33}$$

where

$$\mathbf{H}=\mathbf{H}_a+\mathbf{H}'. \tag{8.3.34}$$

At a minimum, the variation of the total energy, made out of all the above-mentioned terms, should vanish for *any choice* of $u$ and $v$. This requirement means that the coefficients of $u$ and $v$ in the volume integral should each vanish, and the same applies to the coefficients in the surface integral. Adding up all the appropriate terms, and omitting the index '0' which is not necessary any more, leads to two differential equations in the ferromagnetic body, and to two boundary conditions on its surface. The boundary conditions on the surface are

$$C\left(\frac{\partial m_x}{\partial n}-\frac{m_x}{m_z}\frac{\partial m_z}{\partial n}\right)+\frac{\partial w_{\mathrm{s}}}{\partial m_x}-\frac{m_x}{m_z}\frac{\partial w_{\mathrm{s}}}{\partial m_z}=0, \tag{8.3.35}$$

and

$$C\left(\frac{\partial m_y}{\partial n}-\frac{m_y}{m_z}\frac{\partial m_z}{\partial n}\right)+\frac{\partial w_{\mathrm{s}}}{\partial m_y}-\frac{m_y}{m_z}\frac{\partial w_{\mathrm{s}}}{\partial m_z}=0. \tag{8.3.36}$$

The two differential equations are

$$C\left(\nabla^2 m_x-\frac{m_x}{m_z}\nabla^2 m_z\right)+M_s\left(H_x-\frac{m_x}{m_z}H_z\right)-\frac{\partial w_{\mathrm{a}}}{\partial m_x}+\frac{m_x}{m_z}\frac{\partial w_{\mathrm{a}}}{\partial m_z}=0 \tag{8.3.37}$$

and

$$C\left(\nabla^2 m_y-\frac{m_y}{m_z}\nabla^2 m_z\right)+M_s\left(H_y-\frac{m_y}{m_z}H_z\right)-\frac{\partial w_{\mathrm{a}}}{\partial m_y}+\frac{m_y}{m_z}\frac{\partial w_{\mathrm{a}}}{\partial m_z}=0. \tag{8.3.38}$$

It looks as if $m_z$ plays a special role here, unlike $m_x$ or $m_y$, but it is only a matter of choice which two of the three components to use first in eqn (8.3.23). The symmetry can be seen if eqn (8.3.37) is multiplied by $m_y$, and subtracted from eqn (8.3.38) multiplied by $m_x$, which leads to a third equation of the same form,

$$C\left(m_y\nabla^2 m_x-m_x\nabla^2 m_y\right)+M_s\left(m_y H_x-m_x H_y\right)-m_y\frac{\partial w_{\mathrm{a}}}{\partial m_x}+m_x\frac{\partial w_{\mathrm{a}}}{\partial m_y}=0. \tag{8.3.39}$$

These three equations can be written together in a vector notation, as

$$\mathbf{m}\times\left(C\nabla^2\mathbf{m}+M_s\mathbf{H}-\frac{\partial w_{\mathrm{a}}}{\partial\mathbf{m}}\right)=0, \tag{8.3.40}$$

where $\partial w_{\rm a}/\partial \mathbf{m}$ is a notation for a vector whose Cartesian coordinates are $\partial w_{\rm a}/\partial m_x$, $\partial w_{\rm a}/\partial m_y$ and $\partial w_{\rm a}/\partial m_z$. The vector notation is easier for transforming into other coordinate systems, but it is somewhat misleading. It should be remembered that there are only two independent equations, and the third one is only a linear combination of the other two, due to the constraint $|\mathbf{m}| = 1$.

These equations are known as Brown's differential equations. They mean, as phrased in [331], that in equilibrium the torque is zero everywhere, and that the magnetization is parallel to an effective field,

$$\mathbf{H}_{\rm eff} = \frac{C}{M_s}\nabla^2\mathbf{m} + \mathbf{H} - \frac{1}{M_s}\frac{\partial w_{\rm a}}{\partial \mathbf{m}}. \tag{8.3.41}$$

Since $\mathbf{M}\times\mathbf{M} = 0$, any arbitrary vector proportional to $\mathbf{M}$ may be added to $\mathbf{H}_{\rm eff}$, without changing the result. In particular [331], there is no difference between using $\mathbf{H}$ and $\mathbf{B} = \mathbf{H} + 4\pi\mathbf{M}$.

Brown's equations have to be solved together with solving for $\mathbf{H}'$, which is part of eqn (8.3.34), by solving either the differential equations in section 6.1, or the integrals in section 6.3. The solutions of the whole set contain in principle all possible energy minima, but not only the minima. The condition that the variation vanishes is also fulfilled for energy maxima, and it is necessary to check each solution for being a maximum or a minimum.

There are also the boundary conditions of eqns (8.3.35) and (8.3.36), for which a linear combination can be added in the same way as to the differential equations. All three equations can then be written by a similar vector notation as

$$\mathbf{m}\times\left(C\frac{\partial \mathbf{m}}{\partial n} + \frac{\partial w_{\rm s}}{\partial \mathbf{m}}\right) = 0, \tag{8.3.42}$$

on the surface. In the particular case when the surface energy is as in eqn (5.1.11), namely if

$$w_{\rm s} = \frac{1}{2}K_s\left(\mathbf{n}\cdot\mathbf{m}\right)^2, \tag{8.3.43}$$

one should substitute in eqn (8.3.42)

$$\frac{\partial w_{\rm s}}{\partial \mathbf{m}} = K_s\left(\mathbf{n}\cdot\mathbf{m}\right)\mathbf{n}, \tag{8.3.44}$$

which is the form used by Brown [145], and by others [332]. Other, special cases, such as [146], will be ignored here. If there is no surface anisotropy, which is the assumption made in most of the theoretical calculations, the combination of eqn (8.3.42) with the *identity* $\mathbf{m}\cdot\partial\mathbf{m}/\partial n = 0$, which holds for any vector whose magnitude is constant, leads to $\partial\mathbf{m}/\partial n = 0$.

## 8.4 Self-consistency

Solving Brown's equations is not easy, but there are certain conclusions about the nature of the solutions which can be drawn right away. For example, consider the case $m_x = 0$ or $m_y = 0$, which is the basic assumption used in all the one-dimensional wall calculations in section 8.1. Substituting $m_y = 0$ in eqn (8.3.38) yields $H_y = 0$, because every one of the other terms vanishes, including $\partial w_\mathrm{a}/\partial m_y$, which is also proportional to $m_y$ for all the expressions of $w_\mathrm{a}$. In the absence of an applied field, as in the one-dimensional wall calculations, $H_y = 0$ means that $H'_y = 0$, which is *not* the case in those studies that take the average of $H'_y$. Therefore, the assumption $m_y = 0$ (and similarly for $m_x = 0$) cannot lead to a solution of Brown's equations, and cannot represent a true energy minimum.

This argument is rigorous, but it is not very useful, because it cannot be made quantitative. The fact that a particular function is not an *exact* representation of the actual physical state can hardly ever be a good reason to avoid it. Certain approximations are often inevitable, and after all some approximations have already been made on the way before deriving Brown's equations. The real question is not if $m_y = 0$ (or any other assumption of this sort) can be an absolute energy minimum, but if it may be a reasonably good approximation to this minimum. The above argument does not help answer this question, which requires some measure for how far the model is from the minimum.

A partial answer can be obtained when eqn (8.3.37) is multiplied by $m_x$, and eqn (8.3.38) is multiplied by $m_y$, then they are added together, and integrated over the volume of the ferromagnetic body. Part of the integrand can be transformed by the relation

$$\mathbf{m} \cdot \nabla^2 \mathbf{m} = \frac{1}{2} \nabla^2 m^2 - (\nabla m_x)^2 - (\nabla m_y)^2 - (\nabla m_z)^2 , \tag{8.4.45}$$

in which the first term vanishes as the derivative of a constant, and the rest is proportional to the exchange energy density. The total integral is then

$$-2\mathcal{E}_\mathrm{e} + \int \left[ \mathbf{M} \cdot \mathbf{H} - \mathbf{m} \cdot \frac{\partial w_\mathrm{a}}{\partial \mathbf{m}} + \frac{1}{m_z} \left( \frac{\partial w_\mathrm{a}}{\partial m_z} - C\nabla^2 m_z - M_s H_z \right) \right] d\tau = 0. \tag{8.4.46}$$

Comparing with eqn (8.3.22), it is seen that the total energy of the system may *also* be written as

$$\mathcal{E} = \frac{1}{2} \int \left[ 2w_\mathrm{a} - \mathbf{m} \cdot \frac{\partial w_\mathrm{a}}{\partial \mathbf{m}} - \mathbf{M} \cdot \mathbf{H}_a \right] d\tau + \int w_\mathrm{s} \, dS$$
$$+ \frac{1}{2} \int \frac{1}{m_z} \left[ \frac{\partial w_\mathrm{a}}{\partial m_z} - C\nabla^2 m_z - M_s H_z \right] d\tau. \tag{8.4.47}$$

This energy expression does not contain the exchange energy, and does not contain most of the magnetostatic energy term, only part of which is included in $H_z$ of the last term. It is thus much easier to compute than the integral in eqn (8.3.22). However, it cannot be used as a substitute for eqn (8.3.22), because it applies only to magnetization structures which fulfil Brown's equations. What it *can* be used for is as a measure of how close a particular model, or a particular minimization under constraints, is to the true energy minimum which is a solution of Brown's equations. For a good model, the energy computed from eqn (8.4.47) must be approximately equal to that computed from eqn (8.3.22). If these energies are very different, the model is a bad approximation, as has been found when this criterion was first applied [333] to the then-used models of Néel walls. The two values of this wall energy differed by an order of magnitude at certain film thicknesses, which showed that the models used for this wall calculation were very bad approximations, at least in that range of film thicknesses.

It is thus possible to use the difference, or ratio, of the energies computed from eqn (8.4.47) and from eqn (8.3.22) as a *quantitative* measure for the validity of the model or the assumptions used in any micromagnetic calculation. If these numbers are very different, this calculation is wrong and must be discarded. If they are reasonably close to each other, the calculation has a good chance of being correct and self-consistent, and a good approximation to the real energy minimum. It is only a chance, because this criterion is only a necessary, not a sufficient, condition for the calculation to be correct. In the first place, Brown's equations are also fulfilled by the maxima, and not only the energy minima. Also, the boundary conditions have *not* been used in the above derivation of eqn (8.4.47), and a solution of Brown's equations which does not fulfil the boundary conditions is not necessarily an energy minimum. However, a solution obtained by any sort of energy minimization is not very likely to be close to a real *maximum*, and at any rate, it is always better to have a necessary condition for eliminating *some* wrong cases than to have no criterion at all, and have no idea if the calculation has any meaning at all. It should be particularly noted that eqn (8.4.47) contains *second* derivatives in the term with $\nabla^2 m_z$, and as such is very sensitive to the details of the magnetization structure, which makes this criterion of self-consistency quite effective.

This criterion was first suggested [333] for a particular case of eqn (8.4.47) which applies only to a 180° domain wall in a film which has a uniaxial anisotropy. In this particular case, there is neither applied field nor surface anisotropy, and the first two terms in the first line of eqn (8.4.47) cancel each other, so that the whole first line of this equation vanishes. It was then extended to a cubic symmetry [323], and to a 90° domain wall, first in one dimension [334] and then in two dimensions [335]. A non-zero applied field [336] and a moving wall [337] were also considered. Other self-consistency tests also exist, which either are not useful [295] or apply only to the specific case [334] of one dimension.

## 8.5 The Dynamic Equation

The time-dependence of the magnetization can be obtained directly from the quantum-mechanical expression for a precession of the magnetization in a magnetic field, by considering the terms in the brackets of eqn (8.3.40) as an *effective* magnetic field. Other methods can also be used [145] to derive the same result, which is

$$\frac{d\mathbf{M}}{dt} = -\gamma_0\,\mathbf{M}\times\boldsymbol{\mathcal{H}}, \qquad \boldsymbol{\mathcal{H}} = \frac{C}{M_s}\nabla^2\mathbf{m} + \mathbf{H}_a + \mathbf{H}' - \frac{1}{M_s}\frac{\partial w_{\mathrm{a}}}{\partial \mathbf{m}}, \tag{8.5.48}$$

where $t$ is the time,

$$\gamma_0 = \frac{g|e|}{2m_e c} \tag{8.5.49}$$

is the gyromagnetic ratio, already mentioned in section 5.2, and $g$ is the 'Landé factor', already mentioned in section 2.1. In some sense, Brown's static equations can be considered as a particular case of eqn (8.5.49), giving the static equilibrium when there is no change in time. The boundary conditions here are the same as in the static case, namely eqn (8.3.42).

This equation represents an undamped precession of the magnetization, which can continue for ever. However, actual changes of the magnetization are known from experiment to decay in a finite time. As is the case with the anisotropy in section 5.1, the damping cannot be derived theoretically from basic principles, and is just added as a phenomenological term. One way to add it is to modify eqn (8.5.48) into

$$\frac{d\mathbf{M}}{dt} = -\gamma_0\,\mathbf{M}\times\left(\boldsymbol{\mathcal{H}} - \eta\frac{d\mathbf{M}}{dt}\right), \tag{8.5.50}$$

where $\eta$ is a phenomenological damping parameter.

This form of the equation is due to Gilbert. It is actually equivalent to an older form of Landau and Lifshitz [145, 338], which can be derived as follows. First $\mathbf{M}\cdot$ is applied to both sides of eqn (8.5.50). The right hand side vanishes, and therefore $\mathbf{M}\cdot d\mathbf{M}/dt = 0$. This result ascertains that $dM^2/dt = 0$ from which it follows that $M^2$ remains a constant during the motion, and this constant can be identified with $M_s^2$. Then $\mathbf{M}\times$ is applied to both sides of eqn (8.5.50), using the general rule for a cross product of a cross product and $\mathbf{M}\cdot d\mathbf{M}/dt = 0$. The result is

$$\mathbf{M}\times\frac{d\mathbf{M}}{dt} = -\gamma_0\,\mathbf{M}\times(\mathbf{M}\times\boldsymbol{\mathcal{H}}) - \gamma_0\,M_s^2\eta\frac{d\mathbf{M}}{dt}. \tag{8.5.51}$$

Substituting the right hand side of this equation for $\mathbf{M}\times d\mathbf{M}/dt$ in eqn (8.5.50), and rearranging, leads to

$$\frac{d\mathbf{M}}{dt} = -\gamma_0' \mathbf{M} \times \boldsymbol{\mathcal{H}} + \lambda \mathbf{M} \times (\mathbf{M} \times \boldsymbol{\mathcal{H}}) \,, \tag{8.5.52}$$

where

$$\gamma_0' = \frac{\gamma_0}{1 + \gamma_0^2 \eta^2 M_s^2} \qquad \text{and} \qquad \lambda = \frac{\gamma_0^2 \eta}{1 + \gamma_0^2 \eta^2 M_s^2} \,. \tag{8.5.53}$$

Equation (8.5.52) is the older form of Landau and Lifshitz, which some still prefer to use, besides other forms [339] which also exist. It can be seen from the derivation here that the two forms are *mathematically* equivalent [145, 338] if the physical constants are modified according to eqn (8.5.53). The physical interpretation is not very different either, if the damping is small. For a large damping there are some practical reasons [340] to prefer the Gilbert form of eqn (8.5.50).

# 9

# THE NUCLEATION PROBLEM

It is very difficult to solve *any* non-linear differential equation, and it is even more difficult to choose the appropriate physical case among the various solutions which such an equation may have. Therefore, before trying any solution of Brown's differential equations, it seemed desirable to define first the possible branch on which the required solution may be. For this purpose a solution was first sought of a set of *linear* differential equations, to be defined in section 9.1, which became known as the nucleation problem.

This problem was misunderstood by many from the very beginning, and even more so later, when micromagnetics became popular among the researchers of digital recording. Neither the purpose nor the techniques or the results of the nucleation problem seem to have been properly followed, and there is a large number of wrong quotations and misrepresentations of this problem in the literature. The older papers are only too often just presumed to say the exact opposite of what they really do, which may be due to the fact that the writers of these papers reported only certain details, assuming that the main assumptions were well known. They were actually known to the small number of specialists working on them at the time, because practically everybody else just ignored them, but they were unknown, and therefore misinterpreted, when the field was revived later.

I will try to clarify here some of the wrong concepts that have been attached to the studies of nucleation. However, even before defining what this problem is, it is worth mentioning what it is *not*. The linearization of the equations is not an approximation, and should not be confused with the physically completely different problem of the *approach to saturation* in a material which contains point, line or plane imperfections. The latter calculations, which are outside the scope of this book, are mathematically similar to the nucleation problem, because they use the same linearized differential equations (although with different boundary conditions). They certainly played an important and a crucial role [145, 268] in the initial development of micromagnetics in general. But they have nothing to do with the nucleation discussed in this chapter.

## 9.1 Definition

Let a ferromagnetic body be first put in a magnetic field which is large enough to saturate it in this field's direction. Let this field be then reduced *slowly*, to avoid dynamic effects. If necessary, the field is decreased to zero,

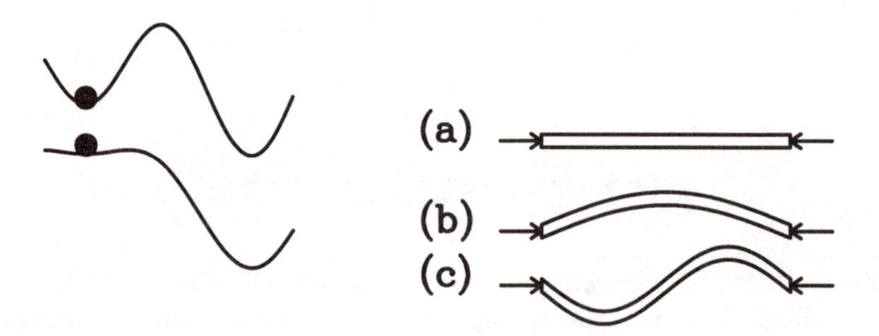


FIG. 9.1. Two mechanical analogues to the nucleation problem.

then increased slowly in the opposite direction. At some stage during this process, the state of saturation along the original direction of the applied field must stop to be stable, and some change must start to take place, because (after all) the sample must eventually be saturated in the opposite direction. The field at which the original saturated state becomes unstable, and any sort of a change in the magnetization configuration can *just start*, is called the *nucleation field*. The name is somewhat misleading, because [341] 'nucleation' seems to imply that something happens at a particular point, around a certain *nucleus*, whereas in the present context, this term is used for something which may happen all over the crystal. Brown argued all his life against this use of the word, but he had nothing better to offer, and the name stuck and was used by everybody, including Brown himself. The important point is that this nucleation has an unambiguous meaning, which is as defined by the foregoing process. It must be emphasized that the definition contains the history of the applied field, which must be the case in any definition involving the hysteresis which is part of ferromagnetism.

The concept of nucleation field is analogous to the critical force which is necessary to bend a beam in the experiment shown schematically on the right hand side of Fig. 9.1, which is partly based on [341]. If an elastic beam is pushed from both sides by a force (represented by the arrows), as shown in (a), nothing happens at first. With an *increasing* force, a 'critical' value is reached, at which the beam suddenly buckles to a particular shape, which is an eigenfunction of a certain differential equation, as shown in (b). In the analogous case of a ferromagnetic crystal nothing happens when the field is first *reduced* till a certain value of the field — the nucleation field — is reached, when the magnetization suddenly 'buckles', or changes in another way; and this change is also an eigenfunction of certain differential equations, as will be seen in the following. It must be particularly emphasized that during this first stage, when nothing happens, there may well be states whose energy is *lower* than that of the saturated state, but they are not accessible to the system. This possibility has already been encountered in the case of the Stoner–Wohlfarth model in section 5.4, and is further illustrated by the second mechanical analogue on the left hand side of Fig. 9.1. In the beginning the saturated magnetization state is the only energy minimum, and is analogous to the little ball placed at a *single* depression of a smooth surface. When the field passes zero, the saturated state in the *opposite* direction already has a lower minimum, and the situation is analogous to the ball placed on the surface shown in (a): there *is* already a lower pit, but the ball cannot roll there, because of the energy barrier in between. This ball can roll down only when the surface is further distorted into removing the barrier, as in (b), which is analogous to the nucleation for magnetization reversal to start. Therefore, just choosing a particular magnetization as a function of space, and proving that its energy is lower than that of the saturated state (or any other state, for that matter) is completely meaningless, and does *not* prove that the magnetization will actually choose that state. It must be shown that the lower-energy state is accessible to the system, and that the situation is not as in Fig. 9.1 (a).

It should also be noted that the surface on which the ball is placed in Fig. 9.1 is actually a representation of a multi-dimensional surface in the function space, and in the actual case of magnetization reversal there are many possible paths for this ball. Therefore, showing that a particular path is blocked by a barrier does not mean anything either, because there may be a way around it. In principle, *all* possible functions must be considered.

For the elastic beam shown in Fig. 9.1, all the harmonics of a given solution are usually also solutions of the differential equation with its boundary conditions. An example is shown schematically in (c), and there is generally a whole set of such solutions. In the case of this beam, all these higher-order solutions have larger eigenvalues, namely need a larger applied force, than the basic solution of Fig. 9.1 (b). In this case, none of the other solutions have any physical meaning, as can be seen from the following argument. Suppose it takes a certain force $F_1$ for the beam to buckle as in (b), and suppose that it theoretically takes a force $F_2 > F_1$ to create the deviation of (c). In order to apply the force $F_2$, it is necessary to pass through the application of the force $F_1$, at which time the beam already changes into the configuration (b). By the time $F_2$ is reached, the initial conditions (a) do not exist any more, and a theoretical transition from (a) to (c) at the force $F_2$ cannot be realized. The higher-order solutions may apply in *other* cases, *e.g.* if the beam is *clamped* at its centre in such a way that (b) cannot take place, or if the larger force is applied very fast, when dynamic conditions may prevent the formation of (b). For the nucleation problem as formulated here, the higher harmonics have no physical meaning, because they cannot be achieved. By the same token, in the case of magnetization only the *largest* nucleation field has a physical meaning. If something starts to reverse at a field $H_1$, the reversal will continue on that path, and another nucleation which can theoretically take place at $H_2 < H_1$ does not have the initial conditions of a saturated state any more, and has no physical meaning. For any case of nucleation it is thus necessary to find only the largest possible eigenvalue of the appropriate differential equations.

The nucleation process, as a *start* of the reversal, is studied by *linearizing* Brown's equations defined in section 8.3, namely by leaving out of these equations the higher-than-linear powers of the magnetization components in the directions perpendicular to the applied field. This linearization is *not* an approximation. It is only a manifestation of the requirement of a *continuity*. Every change of the magnetization structure must *start* with a *small* change. Therefore, everything is linear in the beginning. The idea is that once the correct eigenfunction for the nucleation is known, it is going to be possible to study the rest of the process by solving the non-linear equations for the case which starts with this particular eigenfunction, and not move in the dark through all sorts of mathematically possible solutions which have no physical meaning.

The linearization of Brown's equations will not be done here for the most general case. Instead, the following restrictive assumptions will first be made.

1. The applied magnetic field, $\mathbf{H}_a$, is homogeneous, and is parallel to an easy axis of either cubic or uniaxial anisotropy. This assumption was made in *all* studies of nucleation published so far. In principle, a theory could also be developed for other magnetic fields, but it has never been tried. The second part follows, because it can be shown that it is impossible to reach saturation in any finite, homogeneous field in a direction which is not an easy axis. If $z$ is chosen as the direction of the magnetic field, it is readily seen that to a first order in $m_x$ and $m_y$,

$$-\frac{\partial w}{\partial m_x} + \frac{m_x}{m_z}\frac{\partial w}{\partial m_z} = -2K_1 m_x, \qquad (9.1.1)$$

   where $w$ stands for either $w_u$ of eqn (5.1.7) or $w_c$ of eqn (5.1.8). In either case, $K_2$ is *not* neglected, as it has been in some other calculations in this book. In spite of the presentation in [342], there is just no first-order term in the expressions with $K_2$, and it does not enter the *nucleation* problem. A similar expression applies for $m_y$.

2. The sample is an ellipsoid, and the field is applied parallel to one of its major axes. The first part is inevitable if a homogeneous field is assumed, because only inside an ellipsoid is the demagnetizing field homogeneous, see section 6.1.3. The second part is not essential, and is only introduced here for the sake of simplicity. Some calculations *have* been reported [343, 344, 345, 346] for a field applied at an angle to a major axis, but only for a very limited number of cases, which will be ignored here. According to section 6.1.3, the demagnetizing

field at the saturated state, before nucleation, is a constant, and is parallel to $z$. Therefore, the total field of eqn (8.3.34) which has to be used in Brown's equations is

$$H_x = -\frac{\partial U}{\partial x}, \qquad H_y = -\frac{\partial U}{\partial y}, \qquad H_z = H_a - N_z M_s - \frac{\partial U}{\partial z}, \quad (9.1.2)$$

where $N_z$ is the demagnetizing factor in the $z$-direction, in the saturated state before nucleation, and $U$ is the potential due to the starting deviation from saturation, and is of the order of $m_x$ and $m_y$.

3. The material is homogeneous, and has no surface anisotropy. Only a few cases with a non-zero surface anisotropy [146, 347, 348], or with certain inhomogeneities [349, 350, 351, 352], were ever studied, and both will be ignored here. As mentioned at the end of section 8.3, the boundary conditions are in this case

$$\frac{\partial m_x}{\partial n} = \frac{\partial m_y}{\partial n} = 0 \qquad (9.1.3)$$

on the surface.

Using all these assumptions in eqns (8.3.37) and (8.3.38), and omitting all terms which are higher than linear in $m_x$ or $m_y$ or $U$, *including* a term such as $m_x U$, which is also second order when $U$ is of the order of $m_x$ or $m_y$, yields the differential equations

$$\left[C\nabla^2 - 2K_1 - M_s\left(H_a - N_z M_s\right)\right] m_x = M_s \frac{\partial U_{\text{in}}}{\partial x} \qquad (9.1.4)$$

and

$$\left[C\nabla^2 - 2K_1 - M_s\left(H_a - N_z M_s\right)\right] m_y = M_s \frac{\partial U_{\text{in}}}{\partial y} \qquad (9.1.5)$$

inside the ferromagnetic body. Most of the earlier studies started with an extra term of the form $g_{12} m_x m_y$ in the anisotropy energy density, which added another term to each of these equations. But then this term was taken to be zero at a later stage, and at any rate it is not usually a part of any anisotropy energy.

These equations have to be solved together with eqn (9.1.3) as the boundary conditions, and simultaneously with the equations and boundary conditions, (6.1.4) to (6.1.6) which define the potential $U$. For the linearized case of taking only first-order terms in $m_x$ and $m_y$, the differential equations are

$$\nabla^2 U_{\text{in}} = 4\pi M_s \left(\frac{\partial m_x}{\partial x} + \frac{\partial m_y}{\partial y}\right), \qquad \nabla^2 U_{\text{out}} = 0, \qquad (9.1.6)$$

with the boundary conditions

$$U_{\text{in}} = U_{\text{out}}\,, \qquad \frac{\partial U_{\text{in}}}{\partial n} - \frac{\partial U_{\text{out}}}{\partial n} = 4\pi M_s \mathbf{m}\cdot\mathbf{n}\,, \tag{9.1.7}$$

on the surface, as well as the regularity of the potential at infinity. All these equations have to be solved for all possible eigenvalues of the applied field, $H_a$, and then the largest of them has to be chosen. As has been explained in the foregoing, only the largest allowed value for $H_a$ has a physical meaning.

## 9.2 Two Eigenmodes

Brown wrote this set of linearized equations [353, 354] already in 1940, for the sake of another problem. Only 17 years later he formulated them as the nucleation problem [355] and realized that two analytic solutions could be written right away, both for a sphere and for an infinite circular cylinder. It actually turned out that one of them could be generalized to any ellipsoid, and the other could be generalized to an ellipsoid of revolution, namely one which has two equal axes. These two modes will be described here first, in this more general form, before addressing the problem of other possible eigenfunctions, which Brown [355] presented at the time as a gap in the theory. As has already been mentioned in the previous section, only the *largest* eigenvalue has a physical meaning, so that any one eigenfunction does not mean very much.

### 9.2.1 *Coherent Rotation*

If both $m_x$ and $m_y$ are constants, eqn (9.1.3) is fulfilled. The volume charge is 0, and eqn (9.1.6) for the potential becomes $\nabla^2 U = 0$, both inside and outside. It thus reduces to the problem of a homogeneously magnetized ellipsoid discussed in section 6.1.3, which leads to a homogeneous field inside, with

$$\frac{\partial U_{\text{in}}}{\partial x} = N_x M_s m_x \quad \text{and} \quad \frac{\partial U_{\text{in}}}{\partial y} = N_y M_s m_y, \tag{9.2.8}$$

where $N_x$ and $N_y$ are the appropriate demagnetizing factors. Equations (9.1.4) and (9.1.5) are in this case

$$\left[\frac{2K_1}{M_s} + H_a + (N_x - N_z)M_s\right] m_x = \left[\frac{2K_1}{M_s} + H_a + (N_y - N_z)M_s\right] m_y = 0. \tag{9.2.9}$$

These equations are all that is left in this case from the whole set of the linearized equations, and nucleation can take place once they are fulfilled.

To remind the reader of what this calculation is all about, the start was a saturated state $m_x = m_y = 0$ at a large positive $H_a$, which was later reduced through 0, and started increasing in the opposite direction. The nucleation of a magnetization reversal becomes possible when $H_a$ reaches such a value that allows a (small) deviation from the saturated state. It can happen when these equations can be fulfilled *for the first time* with either $m_x \neq 0$, or $m_y \neq 0$. Therefore, this nucleation is achieved when the applied field $H_a$ reaches such a value that makes one of the square brackets of eqn (9.2.9) pass through zero. If the ellipsoid has a symmetry for revolving around $z$, which means that $N_x = N_y$, the eigenvalue is the same for rotation in the $x$- or in the $y$-direction. If they are not the same, the rotation will be towards the *longer* axis, because it has a larger (less negative) eigenvalue than a rotation in the other direction. Suppose the longer axis is $x$, namely $N_x < N_y$; then the nucleation is for $m_y = 0$ and takes place when $H_a$ reaches the *nucleation field* value of

$$H_n = -\frac{2K_1}{M_s} + (N_z - N_x)M_s. \tag{9.2.10}$$

In this mode the magnetization rotates in the same angle everywhere through the ellipsoid, and it is therefore known as the coherent rotation mode. The name 'rotation in unison' was also tried [356] for a while, but it did not catch on. Actually, it is just the Stoner–Wohlfarth model, studied in section 5.4 for the more general case of a field applied at an angle to the easy axis. In the present case the field is parallel to the easy axis, which is $\theta = 0$ in the notation of that section. In section 5.4 it was assumed that there was only a crystalline anisotropy, with no shape anisotropy, which essentially means a sphere. For that case it was seen that at a zero angle, nothing happens till the field reaches the value of the first term in eqn (9.2.10) here. Then in section 6.1.3, the second term of the nucleation field was introduced for the case of an ellipsoid without anisotropy, namely with $K_1 = 0$. Here, eqn (9.2.10) is for the combination of *both* anisotropy terms, but only for the case when the easy axes of both are parallel to the applied field. In this case, their values are just added together, at least during the *start* of the deviation from saturation.

The Stoner–Wohlfarth model, which started as a *model*, namely as a postulated structure of the magnetization in space, has thus been shown to be a *mode*, namely an eigenfunction of Brown's equations. As such, it is a real energy minimum, and not just an arbitrary configuration for comparing energies as in the domain theory, which Brown tried to avoid, see section 8.3. It is not the end of the road yet, because this mode will actually be used by the physical system only in cases in which it is the mode with the least negative nucleation field. In order to find out if it is, all the other modes must be investigated, and compared with this mode.

### 9.2.2 *Magnetization Curling*

Another mode which Brown [355] found to be a solution of the linearized set of equations applies only to an ellipsoid of revolution, or at least nobody has tried to generalize it to any other ellipsoid. In cylindrical coordinates $\rho$, $z$ and $\phi$, what has become known as the curling mode is the solution of

Brown's equations for which

$$m_x = -F(\rho, z)\sin\phi, \quad m_y = F(\rho, z)\cos\phi, \quad U_{\text{in}} = U_{\text{out}} = 0. \qquad (9.2.11)$$

It is an arbitrary set of constraints on the solution, whose only justification is that it *works*, namely that such a solution does exist for any ellipsoid of revolution. In order to see that it does, it is obviously sufficient to substitute these constraints in eqns (9.1.3) to (9.1.7), and see that there is a solution to the constrained set. Substituting thus eqn (9.2.11) in eqns (9.1.4) and (9.1.5), it is seen that they are *both* fulfilled if

$$\left[C\left(\frac{\partial^2}{\partial\rho^2} + \frac{1}{\rho}\frac{\partial}{\partial\rho} - \frac{1}{\rho^2} + \frac{\partial^2}{\partial z^2}\right) - 2K_1 - M_s\left(H_a - N_z M_s\right)\right] F(\rho, z). \qquad (9.2.12)$$

Equation (9.1.6) is obviously fulfilled, and so is eqn (9.1.7), even though the latter takes a little thinking about to see that eqn (9.2.11) actually leads to $\mathbf{m}\cdot\mathbf{n} = 0$ on the surface of any ellipsoid of revolution. The point is that according to this definition of $F$, this $F$ is actually the component of $\mathbf{m}$ in the direction of the coordinate $\phi$, namely $m_\phi$. This component is parallel to the surface in any body which has a cylindrical symmetry, in particular an ellipsoid of revolution. Therefore, the only equation which is still left from the original set is eqn (9.1.3). Substitution in it yields the boundary condition

$$\frac{\partial F}{\partial n} = 0, \qquad (9.2.13)$$

on the surface, where $\mathbf{n}$ is the normal to the surface.

The assumption of eqn (9.2.11) has thus reduced the three-dimensional problem to a two-dimensional one, which is not difficult to solve in the ellipsoidal coordinate system. However, since these coordinates may be too advanced for some readers, the solution will be expressed first for the two cases of a sphere and an infinite circular cylinder, originally studied, almost simultaneously, both by Brown [355] and by Frei *et al* [356]. There is, however, a big difference between the two which has been forgotten during the years, and which is worth emphasizing.

Brown started from the differential equations, and guessed a particular solution for a sphere and for an infinite cylinder. Frei *et al* started from a particular functional form for the magnetization, and compared its energy with that of some other *models*. In their paper, they presented the curling mode as an arbitrary *model* for a sphere and for an infinite cylinder, because they did not know that it was the same solution of Brown's equations, published by Brown a few months earlier. Since Brown did not give a *name* to this mode, and since people usually feel that when something is given a name they understand it better, the paper of Frei *et al* [356], in which the *name* 'curling' was first invented, became much better known and cited than the paper of Brown [355]. In his talks and publications, Brown kept emphasizing the strange coincidence of the same solution being worked out simultaneously and independently in two places, but he never mentioned this difference between a reversal mode that obeys his equations, and a mere model, probably because it was so obvious to him, and to everybody else at the time. The unintentional result was that too many people were left with the impression that [355] was just the same as [356], and was not worth reading. Thus, too many books and reviews give a schematic picture of what the curling 'looks like', usually only in an infinite cylinder, but not the mathematical definition of the function, as given here. And papers have been, and still are, published with all sorts of models for magnetization reversal in which they compare theirs with the 'curling model', as *e.g.* [357]. It is impossible to convince them that the curling is not a model, and cannot be treated on the same level as their arbitrary models. It is a nucleation reversal mode, which is a solution of Brown's equations, and as such can only be compared with other reversal modes, such as coherent rotation, or the other modes discussed in section 9.4.

9.2.2.1 *Infinite Cylinder* For an infinite cylinder the normal $\mathbf{n}$ to the surface is parallel to the coordinate $\rho$, and it is only necessary to consider $F$ which does not depend on $z$. Eqn (9.2.12) is the well-known differential equation for the Bessel functions. It has two solutions, one of which is not regular at $\rho = 0$, and cannot be used as a solution. The other one is

$$F \propto J_1(k\rho), \tag{9.2.14}$$

which is a solution of eqn (9.2.12) provided

$$Ck^2 + 2K_1 + M_s H_a = 0, \tag{9.2.15}$$

where the term with $N_z$ has been omitted, because $N_z = 0$ for an infinite cylinder, see section 6.1.3. The boundary condition (9.2.13) is also fulfilled, if

$$\left(\frac{dJ_1(k\rho)}{d\rho}\right)_{\rho=R} = 0, \tag{9.2.16}$$

where $R$ is the radius of the cylinder.

Equation (9.2.16) has an infinite number of solutions, out of which only the *smallest* one has to be considered, because the larger ones lead to a more negative nucleation field, which can never take place if a less negative one exists. Let $q_1$ be the *smallest* root of

$$\frac{dJ_1(q)}{dq} = 0 \tag{9.2.17}$$

(which is $q_1 \approx 1.8412$). Then the nucleation field for this mode is, according to eqn (9.2.15),

$$H_n = -\frac{2K_1}{M_s} - \frac{Cq_1^2}{R^2 M_s}. \qquad (9.2.18)$$

Comparing with eqn (9.2.10), and taking into account that for an infinite cylinder $N_x = 2\pi$ and $N_z = 0$, it is seen that *if there is no other mode*, magnetization reversal in an infinite cylinder should start by coherent rotation if $R < R_c$, and by curling if $R > R_c$, where

$$R_c = \frac{q_1}{M_s}\sqrt{\frac{C}{2\pi}}, \qquad (9.2.19)$$

because it is always the largest $H_n$ which counts, as explained in section 9.1. This statement is not true if there is a third mode with a still larger eigenvalue for $H_a$. This possibility will be further discussed in section 9.4. Frei *et al* [356] introduced the *reduced radius*,

$$S = \frac{R}{R_0}, \qquad \text{with} \qquad R_0 = \sqrt{\frac{C}{2M_s^2}}, \qquad (9.2.20)$$

and this notation was later used in many papers on micromagnetics. In this notation, the turnover from coherent rotation to curling is at the reduced radius

$$S_c = \frac{R_c}{R_0} = \frac{q_1}{\sqrt{\pi}} \approx \frac{1.8412}{\sqrt{\pi}} \approx 1.039. \qquad (9.2.21)$$

9.2.2.2 *Sphere* The second case considered by Brown [355] and by Frei *et al* [356] was that of a sphere. For this case, the cylindrical coordinates $\rho$ and $z$ are changed to the spherical coordinates $r$ and $\theta$, with $\phi$ kept the same. In these coordinates, eqn (9.2.12) transforms into

$$\left[C\left(\frac{\partial^2}{\partial r^2} + \frac{2}{r}\frac{\partial}{\partial r} + \frac{1}{r^2}\frac{\partial^2}{\partial \theta^2} + \frac{\cos\theta}{r^2\sin\theta}\frac{\partial}{\partial\theta} - \frac{1}{r^2\sin^2\theta}\right)\right.$$
$$\left. -2K_1 - M_s\left(H_a - \frac{4\pi}{3}M_s\right)\right] F(r,\theta) = 0\,, \qquad (9.2.22)$$

because $N_z = 4\pi/3$ for a sphere. One of the solutions of this equation is

$$F \propto j_1(kr)\sin\theta, \qquad (9.2.23)$$

where $j_1$ is the spherical Bessel function, which can also be expressed in terms of the trigonometric functions,

$$j_1(x) = \frac{\sin x}{x^2} - \frac{\cos x}{x}. \qquad (9.2.24)$$

It is seen to be a solution, provided

$$Ck^2 + 2K_1 + M_s\left(H_a - \frac{4\pi}{3}M_s\right) = 0. \tag{9.2.25}$$

Actually, Brown [355] also considered other solutions of the same equation, but they had a smaller (*i.e.* more negative) nucleation field than the one in eqn (9.2.25), and as such are of no interest. The whole eigenvalue spectrum is left to be discussed in section 9.4.

The boundary condition (9.2.13) is fulfilled if

$$\left(\frac{dj_1(kr)}{dr}\right)_{r=R} = 0, \tag{9.2.26}$$

where $R$ here is the radius of the sphere. This equation has an infinite number of solutions, out of which only the *smallest* one has to be considered. Let $q_2$ be the *smallest* root of

$$\frac{dj_1(q)}{dq} = 0 \tag{9.2.27}$$

(which is $q_2 \approx 2.0816$). Then the nucleation field for this mode is, according to eqn (9.2.25),

$$H_n = -\frac{2K_1}{M_s} - \frac{Cq_2^2}{R^2M_s} + \frac{4\pi}{3}M_s. \tag{9.2.28}$$

Comparing with eqn (9.2.10), and taking into account that $N_z = N_x$ for a sphere, it is seen that *if there is no other mode*, magnetization reversal in a sphere should start by coherent rotation if $R < R_c$, and by curling if $R > R_c$, where

$$R_c = \frac{q_2}{M_s}\sqrt{\frac{3C}{4\pi}}, \tag{9.2.29}$$

which is rather similar to the expression for an infinite cylinder. In the notation of Frei *et al* [356], the turnover from coherent rotation to curling in a sphere is at the reduced radius

$$S_c = \frac{R_c}{R_0} = q_2\sqrt{\frac{3}{2\pi}} \approx 2.0816\sqrt{\frac{3}{2\pi}} \approx 1.438. \tag{9.2.30}$$

9.2.2.3 *Ellipsoid of Revolution* Both early studies of curling [355, 356] speculated that this result could be extended to a prolate spheroid, for which the curling nucleation field should be

$$H_n = -\frac{2K_1}{M_s} - \frac{Cq^2}{R^2M_s} + N_zM_s, \tag{9.2.31}$$

where $R$ is the semi-axis of the ellipsoid in a direction perpendicular to the field direction $z$, and $q$ is a parameter whose value is between $q_1$ and $q_2$. By solving eqns (9.2.12) and (9.2.13) in terms of the ellipsoidal harmonics, it was shown that eqn (9.2.31) is indeed the nucleation field for this mode, both for a prolate [358] and for an oblate [359] spheroid. The parameter $q$ is a geometrical factor, which depends only on the ratio of the ellipsoidal axes, and does not depend on the properties of the material. For a prolate spheroid the value of $k = q^2/\pi$ is plotted as a function of the ratio of the axes in Fig. 6 of [358], and it is indeed a monotonically decreasing function of the aspect ratio of the ellipsoid. It varies between the limits of $q_2 = 2.0816$ for the sphere with an aspect ratio of 1, and $q_1 = 1.8412$ for the infinite cylinder, with an infinite aspect ratio. For an oblate spheroid $q$ keeps increasing monotonically with decreasing aspect ratio, from the value $q_2$ for the sphere, to the value $q_3 = 2.115$ in the limit of an infinite *plate* with an aspect ratio tending to 0. The change in the whole region of an oblate spheroid is, thus, very small, and a constant value of $q \approx 2.1$ is correct to within 1%. Or, it is possible [359] to use $q^2/\pi \approx 1.4$, which is correct to within 2% for any oblate spheroid.

Comparing eqn (9.2.31) with eqn (9.2.10), it is seen that for any ellipsoid of revolution, nucleation must be by coherent rotation for small radii, and by curling for larger ones, and the 'critical' radius for changing between these two modes is

$$R_c = \frac{q}{M_s}\sqrt{\frac{C}{N_x}}, \qquad \text{or} \qquad S_c = q\sqrt{\frac{2}{N_x}}, \tag{9.2.32}$$

provided there is no third mode whose nucleation field is larger.

The curling mode is *not* really limited to cases of circular symmetry. For the case of a prism, which is infinite in the $z$-direction, but has a *rectangular* cross-section in the $xy$-plane, the curling is defined [359] as the reversal mode for which $m_x$ is an even function in $x$ and an odd function in $y$, and $m_y$ is an even function in $y$ and an odd function in $x$. The potential for this mode is *not* zero, and neither is the magnetostatic energy, but it also yields a nucleation field which has a term with an essentially $1/R^2$ dependence.

## 9.3 Infinite Slab

It has been shown in section 9.1 that the only solution of Brown's linearized equations which has any physical meaning is the one which has the largest nucleation field. The two modes described in the previous section do not mean very much unless the whole eigenvalue spectrum is analysed, and it is shown that all the other possible modes have a smaller eigenvalue. The first of such studies of the whole spectrum was for the case of an infinite cylinder. It showed [360] the existence of a third mode, but eliminated all possibilities of a fourth one. This case is atypical, and its results are not as conclusive as other cases, which will be discussed in the next section. Its algebra is also rather complicated for a start. Therefore, the principle of covering the whole eigenvalue spectrum will be demonstrated here by a detailed study of an infinite plane, as presented in [361]. This case is also atypical, and its results are rather ambiguous and of no particular interest [362], as usually happens whenever an infinity is assumed in magnetostatic problems. However, the *method* is the same as in the more complicated cases, and it is easier to understand this method by considering this simple case first.

Therefore, consider a plate which is infinite in both the $x$- and the $y$-directions, and extends over the finite range $-c \leq z \leq c$ in the direction of the applied field. In an infinite material, any non-divergent function of space must be a *periodic* function. With an appropriate consideration for the symmetry properties of eqns (9.1.4)–(9.1.6), this periodicity means that the most general solution can be written in the form

$$m_x = A(z)\sin(kx - x_0)\cos(ny - y_0), \tag{9.3.33}$$

$$m_y = B(z)\cos(kx - x_0)\sin(ny - y_0), \tag{9.3.34}$$

$$U = u(z)\cos(kx - x_0)\cos(ny - y_0), \tag{9.3.35}$$

where $k$ and $n$ are real numbers, and $A$, $B$ and $u$ are functions which have to be determined. Substituting in eqns (9.1.4)–(9.1.6), and noting that $N_z = 4\pi$ for an infinite plate, because $N_x = N_y = 0$,

$$\left[C\left(\frac{d^2}{dz^2} - k^2 - n^2\right) - 2K_1 - M_s\left(H_a - 4\pi M_s\right)\right] A(z) = -kM_s u_{\text{in}}(z), \tag{9.3.36}$$

$$\left[C\left(\frac{d^2}{dz^2} - k^2 - n^2\right) - 2K_1 - M_s\left(H_a - 4\pi M_s\right)\right] B(z) = -nM_s u_{\text{in}}(z), \tag{9.3.37}$$

$$\left[\frac{d^2}{dz^2} - k^2 - n^2\right] u_{\text{in}}(z) = 4\pi M_s\left[kA(z) - nB(z)\right], \tag{9.3.38}$$

$$\left[\frac{d^2}{dz^2} - k^2 - n^2\right] u_{\text{out}}(z) = 0. \tag{9.3.39}$$

The boundary conditions are obtained by substituting the same equations, (9.3.33)–(9.3.35), in eqns (9.1.3) and (9.1.7). The first two conditions are

$$\left(\frac{dA}{dz}\right)_{z=\pm c} = \left(\frac{dB}{dz}\right)_{z=\pm c} = 0. \tag{9.3.40}$$

The other boundary conditions are easier to incorporate if it is noted first that eqn (9.3.39) has only two possible solutions, one of which diverges at infinity. Its only solution which is regular at infinity is

$$u_{\text{out}} = V_{\pm} e^{\sqrt{k^2+n^2}(c \mp z)}, \tag{9.3.41}$$

where the upper sign applies to the region $z > c$ and the lower sign applies to the region $z < -c$, and where $V_+$ and $V_-$ are two integration constants. Substituting this solution in the rest of the boundary conditions, they are seen to be

$$u_{\text{in}}(\pm c) = V_{\pm}\,, \qquad \text{and} \qquad \left(\frac{du_{\text{in}}}{dz}\right)_{z=\pm c} = \mp\sqrt{k^2+n^2}\,V_{\pm}\,, \tag{9.3.42}$$

which completes the reduction of the partial differential equations to a set of ordinary ones, in one dimension.

The coherent rotation mode is the particular case $k = n = u_{\text{in}} = u_{\text{out}} = 0$, with either $A = 0$ and $B =$ const, or $B = 0$ and $A =$ const, according to the definition in section 9.2.1. It can be verified by substitution that this case is indeed a solution of all the foregoing equations, and that the nucleation field is the same for rotation in the $y$-direction or in the $x$-direction, and is

$$H_n = -\frac{2K_1}{M_s} + 4\pi M_s, \tag{9.3.43}$$

which is a particular case of eqn (9.2.10), for $N_x = 0$ and $N_y = 4\pi$. But this eigenvalue is degenerate not only with respect to the direction of the rotation. If the infinite slab is taken as the limit of an oblate spheroid for which $R \to \infty$, eqn (9.2.31) for the nucleation by *curling* also tends to the same value as in eqn (9.3.43). The point is that an infinity is never well-defined in magnetostatic problems, and it must be specified the limit of which shape this infinity is. In a finite body, a coherent rotation involves doing work against the magnetostatic forces due to the surface charge on the side towards which the magnetization rotates, but does not involve any work against exchange, because all the spins are aligned parallel to each other. On the other hand, the curling mode does work against exchange forces, because there is a spatial dependence of the magnetization, but does not involve any work against magnetostatic forces because there is neither surface nor volume charge. In this atypical case of an infinite slab, the exchange contribution to the curling vanishes because the radius in infinite, while the magnetostatic contribution to the coherent rotation vanishes because there is no surface in the direction of rotation when the plate extends to infinity in the $xy$-plane. In this case, the only barrier is due to the anisotropy energy, which is *the same* for both modes, and the eigenvalue turns out to contain only the $K_1$ term. However, it is obvious that the vanishing of one term is not the same as that of the other, and the real physical limit depends on the way of approach to this infinity, as often happens in many problems in magnetism. In spite of all the arguments in [361], the conclusion that only coherent rotation takes place in this plate is only due to the separation of variables in Cartesian coordinates, which implies approaching the infinity as the limit of a growing square plate. If it is approached as the limit of a growing oblate spheroid, the mode at infinity is the curling mode.

Before proceeding, it should also be mentioned that the argument about the periodicity is not strictly correct, although it has been used in other studies, in particular for the case [360] of an infinite circular cylinder. In principle it is possible to imagine some sort of a *localized* mode, which does not spread all over the slab, and such a mode need not be periodic. It is not possible to build such a mode by the separation of variables technique, as used in writing eqns (9.3.33)–(9.3.35), but this shortcoming does not necessarily rule out the possible existence of a localized mode. This problem will be further discussed in the next section. Here it is sufficient to say that the inadequacy of the separation variables into a function of $x$ times a function of $y$, etc., is another manifestation of the infinite dimension of the sample. The problem is *not* encountered in any finite ellipsoid, for which all the possible modes can be written as series in the spheroidal wave function, and it is not necessary to superimpose any extra assumption which is equivalent to the present assumption of periodicity. If the infinity is approached as an appropriate limit of a finite particle, the results are better defined than they are when the start is from a particle which is infinite in one or more dimension. As has been mentioned already, there is no meaning to infinity in magnetism, and it is always necessary to specify in which way this infinity is approached.

Leaving this problem of infinity for the meantime, and accepting eqns (9.3.36)–(9.3.38) as the most general case, such a set of three second-order differential equations should have a solution with six arbitrary integration constants. Therefore, any solution which has such six constants is the most general one. In particular, if it is shown to be a solution, it is sufficient to take a solution of the form

$$A(z) = \sum_{i=1}^{6} A_i e^{\mu_i z}, \quad B(z) = \sum_{i=1}^{6} B_i e^{\mu_i z}, \quad u_{\text{in}}(z) = \sum_{i=1}^{6} U_i e^{\mu_i z}, \quad (9.3.44)$$

where $\mu_i$ are six complex numbers, provided six out of the 18 constants $A_i$, $B_i$ and $U_i$ are independent of the others. Substituting eqn (9.3.44) in the differential equations (9.3.36)–(9.3.38), it is seen that the conditions for it being a solution are

$$\left[C\left(\mu_i^2 - k^2 - n^2\right) - 2K_1 - M_s\left(H_a - 4\pi M_s\right)\right] A_i + kM_s u_i = 0, \quad (9.3.45)$$

$$\left[C\left(\mu_i^2 - k^2 - n^2\right) - 2K_1 - M_s\left(H_a - 4\pi M_s\right)\right] B_i + nM_s u_i = 0, \quad (9.3.46)$$

$$\left(\mu_i^2 - k^2 - n^2\right) u_i = 4\pi M_s \left(kA_i - nB_i\right). \quad (9.3.47)$$

For each value of $1 \leq i \leq 6$ these are three homogeneous equations in $A_i$, $B_i$ and $u_i$, and the condition for them to have a non-zero solution is that the determinant of the coefficients vanishes. This determinant is a third-order polynomial in $\mu_i^2$, and as such should have six (complex) roots for $\mu_i$. For each of these roots, eqns (9.3.45)–(9.3.47) can be used to solve for two out of the three constants $A_i$, $B_i$ and $u_i$ in terms of the third one, thus leaving six arbitrary integration constants, which means that eqn (9.3.44) is the most general solution, and contains all possible modes. These six constants, with the two additional ones $V_\pm$, should now be evaluated by the requirement of fulfilling the eight equations for the boundary conditions in eqns (9.3.40) and (9.3.42). There are eight homogeneous equations for determining these eight constants, and the condition for a non-zero solution is that the determinant of the coefficients vanishes. Equating this determinant to zero then yields the allowed values for the applied field $H_a$, and these are the eigenvalues of the problem.

Such an algebra is not trivial, but it is straightforward in principle. Moreover, the present case of an infinite plate is particularly simple in that it can all be carried out analytically. The third-order determinant can be factorized [361] into a quadratic and a linear equation in $\mu_i^2$, and all six roots can be written in a closed form. The result is [361] that all other modes have a more negative nucleation field than the coherent rotation mode, and as such can be ignored as being physically unattainable.

The details of these other solutions will not be given here, because it is simpler to employ a technique [362] which has also proved a very powerful tool in other cases. The method is based on calculating an upper bound to the nucleation field of a certain mode, namely a value which is proved to be larger than or equal to the true nucleation field of that mode. If this upper bound is found to be smaller (*i.e.* more negative) than that of another mode, the actual nucleation field of the first mode (which is not larger than its upper bound) is certainly smaller than that of the second mode. And since only the largest nucleation field has a physical meaning, any mode which is shown to have a smaller nucleation field than that of another mode is of no interest, and may be safely left out. Even when two modes have the *same* nucleation field one of them may usually be left out.

One way to calculate an upper bound is to drop a *positive* energy term, thus decreasing the energy barrier, and making the reversal easier than it really is. In the present case it is convenient to obtain an upper bound to a set of modes by dropping the magnetostatic energy term, which is known to be a non-negative term, see section 7.3.2. It is clear from the derivation of Brown's equations in section 8.3 that dropping the magnetostatic energy term is equivalent to writing $u_{\text{in}} = 0$ on the right hand side of eqns (9.3.36) and (9.3.37), and ignoring eqns (9.3.38) and (9.3.42), *before* the substitution of eqn (9.3.44). The differential equations for $A(z)$ and $B(z)$ are then identical, and either one of them may be used for the upper bound. For example, with $A(z) = 0$ the most general solution is

$$B(z) = B_1 e^{\mu z} + B_2 e^{-\mu z}, \tag{9.3.48}$$

with the two arbitrary constants $B_1$ and $B_2$, which is a solution provided

$$C\left(\mu^2 - k^2 - n^2\right) - 2K_1 - M_s\left(H_a - 4\pi M_s\right) = 0. \tag{9.3.49}$$

Substituting eqn (9.3.48) in the boundary conditions, eqn (9.3.40), yields

$$\mu\left(B_1 e^{\mu c} - B_2 e^{-\mu c}\right) = \mu\left(B_1 e^{-\mu c} - B_2 e^{\mu c}\right) = 0. \tag{9.3.50}$$

These two equations have a common non-zero solution if and only if the determinant of the coefficients of $B_1$ and $B_2$ vanishes, namely

$$\mu\left(e^{2\mu c} - e^{-2\mu c}\right) = 0, \tag{9.3.51}$$

whose most general solution is

$$\mu = \frac{m\pi i}{2c}, \tag{9.3.52}$$

where $m$ is an integer. Substituting in eqn (9.3.49), the upper bound for the nucleation field is

$$H_n = 4\pi M_s - \frac{2K_1}{M_s} - \frac{C}{M_s}\left(\frac{m^2\pi^2}{4c^2} + k^2 + n^2\right). \tag{9.3.53}$$

The least negative of these eigenvalues is the one for which $m = k = n = 0$, and for this mode eqn (9.3.53) is the same as eqn (9.3.43) of the coherent rotation.

It has thus been proved that an upper bound for all other modes is larger than or equal to the true nucleation field of the coherent rotation mode. Therefore, for this case of an infinite plate, the coherent rotation is the mode which has the least negative nucleation field. It has *not* been proved that there is no other mode which has the *same* nucleation field as that of the coherent rotation, and there is no justification to the claim in [361] that this calculation proves that only coherent rotation can take place in such a plate. In fact, it has already been demonstrated in the foregoing that the curling mode does have the same nucleation field as the coherent rotation, and this degeneracy leaves some ambiguity as to which mode will take place in a real physical situation. The main idea of linearizing the equations was to know from which mode to start a numerical solution of Brown's non-linear equations, out of the very many possibilities. For this purpose, degeneracy of the nucleation modes is undesirable, because it allows more than one possibility to proceed from. However, in this respect the case of an infinite plate is not representative, because there is less ambiguity in any finite ellipsoid. The assumption of an infinity is problematic anyway, because of the necessity to assume a periodicity, already mentioned in the foregoing.

## 9.4 The Third Mode

Using similar methods to those outlined in the previous section, it was first proved for a sphere [358] and later for any oblate spheroid [359] that the coherent rotation and the curling are the only possible nucleation modes. Curling takes place above a certain size, and coherent rotation below it, where the turnover from one to the other is given by eqn (9.2.32). There cannot possibly be any competition from a third mode, because for all other modes the nucleation field is more negative than for these two. The only ambiguity is encountered in the limit of an infinite plate, for which the curling and the coherent rotation tend to the same eigenvalue, as discussed in the previous section. But even for that limit, there is no third mode, if the infinity is approached from an oblate spheroid with $R \to \infty$.

For a prolate spheroid, there *may* be a third mode, which will be named *buckling*, for lack of a more appropriate name. The name buckling was first applied to a particular *model*, suggested [356] together with the *model* for curling for the particular case of an infinite cylinder. Its nucleation field was later found [360] to be a good approximation to that of a third nucleation *mode* that was shown to exist in an infinite cylinder, and which was given the same name. Actually, this mode turned out to be always easier than the coherent rotation in an infinite cylinder [360], so that in such a cylinder there can only be buckling below a certain size, and curling above it, without any third possibility. The first study [359] of the whole eigenvalue spectrum of a finite prolate spheroid showed that coherent rotation could be the easiest mode in some region of size and elongation, but did not rule out the possibility that the buckling would take over in another range. It did rule out, however, the possibility of a fourth mode, so that it could be definitely stated that none but these three modes should be considered for any prolate spheroid.

A later evaluation [363] put more specific limits on the possible modes, as reproduced in Fig. 9.2 here. It plots regions in which modes *may* be allowed for given elongation and radius of a prolate spheroid, where the radius is the semi-axis $R$ in the direction perpendicular to the applied field, and is plotted in terms of the reduced radius defined in eqn (9.2.20). For an elongation of $4.6:1$ or less, there can only be either curling or coherent rotation, as shown in the figure, and as is the case for all oblate spheroids. For higher elongation, there is no final result yet. This third mode has not

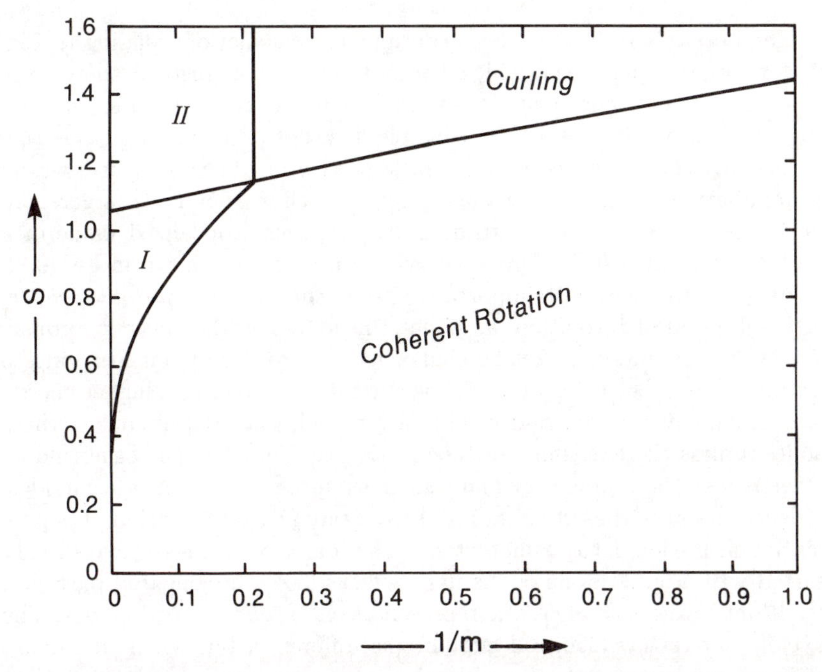


FIG. 9.2. The possible nucleation modes in a prolate spheroid with a reciprocal elongation (minor to major axis) $1/m$, and a reduced radius (along the minor axis) $S$, defined in eqn (9.2.20). Only curling or coherent rotation is possible in the regions so marked, while either buckling or coherent rotation may take place in region I, and buckling or curling in region II. Copied from [363].

been completely ruled out, and it *may* take place in the regions as marked in Fig. 9.2. It is not likely to make any significant contribution, when even for the most elongated ellipsoid, namely the infinite cylinder, the buckling mode is not very different from the coherent rotation, for small radii before the curling takes over. There is some uncertainty in this conclusion, because of the infinity which is never a good assumption in magnetism. In the study of the infinite cylinder [360] it was assumed that the eigenmode should be *periodic* in the $z$-direction. Therefore, in this case, as in the study of the infinite plate in section 9.3, the possibility of a different, *localized* mode is still left open. But unlike the case of an infinite plate, the infinite cylinder cannot be approached from a *known* solution for a finite ellipsoid. The best one can say is that if there is a localized mode, the third mode, named 'buckling' in Fig. 9.2, must be similar to that one, and not to the known [360] buckling in the infinite cylinder. If there is no localized mode in an infinite cylinder, it is reasonable to assume that the buckling mode is only a manifestation of the infinity, and does not exist in any finite ellipsoid.

The best way to find out is to continue the calculations leading to Fig. 9.2 for larger elongations, which has not been done. Instead there were some attempts to find a model for such a mode in an infinite cylinder. None of them worked, which in principle does not prove anything one way or the other. The most recent attempt [364] is wrong because it uses wrong approximations, neglecting an energy term which is [365] much larger than the terms which are taken into account. However, a localized mode of a similar nature may still be possible, within the bounds shown in Fig. 9.2.

At any rate, the most important part is that *there is no other mode*, for an ellipsoid of revolution, and that this statement has been rigorously proved and the proof cannot be challenged. There is no point in trying to postulate any other reversal mode, because it must lead to a higher energy barrier, namely a more negative nucleation field, than at least one of these modes, unless there is some mistake in the calculation of the other mode. Nevertheless, there were very many such attempts to look for other modes, evidently because the presentation of this statement in the original papers was not clear enough to be understood. The confusion seems to have already started with Fig. 3 of one of the first reviews [366] on elongated particles, which put together a schematic representation of *four* reversal modes. The text of that review [366] did explain the difference, but when this figure was copied to many reviews and books, it was used out of context. It then led some people to believe that there are actually four *models* for reversal, which may be used on an equal basis, not paying attentions to their different geometries. These models are the coherent rotation, buckling and curling, mentioned in the foregoing, plus a fourth one called *fanning*.

Historically, the magnetization fanning model was the first attempt to calculate any form of a non-coherent magnetization reversal, in order to explain why the Stoner–Wohlfarth model did not agree with experiment on certain materials. It came at the time when General Electric was developing the production of elongated fine particles for what was later sold under the commercial name of Lodex magnets, and it was noted that these particles were shaped [366] more or less like peanuts. Therefore, this shape was approximated [367] by a linear 'chain of spheres', which touch each other at a point so that there is no exchange interaction between spheres, but there is a magnetostatic interaction between them. For this case, a model was proposed [367] in which the magnetization rotates coherently in each of the spheres, but the angle of rotation may be different for the different spheres. This model was called 'non-symmetric fanning'. However, computers were not available in those days, and computation of the different angle by hand was rather elaborate. Therefore, it was found adequate [367] to study only an *approximation* called 'symmetric fanning', in which there is only one angle, with one half of the spheres rotating at that angle, and the other half at minus the same angle. Even this approximation was shown [367] to be easier to reverse than by coherent rotation, which is not surprising for this particular geometry. Obviously, the non-symmetric fanning, which is equivalent to the buckling mode in a cylinder, must be even easier than the symmetric fanning, because the energy is minimized over more parameters, with the symmetric fanning being a particular case of the more general minimization. This mode was studied in more detail for a chain of only *two* spheres with a uniaxial anisotropy whose easy axis is parallel to the chain axis [368] (which is actually just an *additive* to the nucleation field in *all* modes), and the nucleation field for magnetization curling in a chain of any length has been evaluated [344] by a perturbation scheme. The problem of the whole set of modes in such a chain of spheres has never been fully solved, but it seems that the result should be very similar to that of an ellipsoid, namely that the reversal is by curling above a certain radius, and by non-symmetric fanning for a smaller radius, with very little chance for any other mode. In any case, the reversal modes for an ellipsoid, and those for a chain of spheres, are for different body shapes. They cannot be compared with each other, or mixed together in any other way.

Nevertheless, it was quite popular for some time to compare [369] the values 'predicted by each of the known mechanisms of fanning, buckling and curling' with some experimental results, in order to find out which of these 'mechanisms' takes place in a given experiment. No attention was paid to the different geometries involved, and actually there was not even an attempt to define any particular geometry, even when specific pictures of the sample were available. These transmission electron microscope (TEM)photos showed [369] particles of a rather irregular shape, which could certainly not be described as ellipsoids. They look more like distorted ellipsoids, but they can much *less* be approximated by the picture of a chain of spheres. However, the real shape was not even mentioned in the quasi-theoretical interpretations, which considered it as a part of the adjustable parameters, stating, for example, that [370] certain experimental results 'lay in the range consistent with the chain-of-spheres and prolate ellipsoid models'! The discussed modes were not very well-defined either, with more attention paid to the name than to any sort of a mathematical definition, and nucleation fields used in these comparisons were, more often than not, those of curling in an *infinite cylinder* [369, 371]. With this approach there is little wonder that too many workers felt free to invent and propose all sorts of new reversal models [372] 'in addition to the existing models of coherent rotation, fanning, buckling and curling', and demanded these new models to be treated on the same footing as the 'existing' ones.

A different particle shape can be easily imagined to support modes which are different from those mentioned here. But no new shape was assumed in these studies, and for the same geometry other modes cannot take place if properly calculated. It should be especially noted that it is not sufficient for a reversal mode to have a lower energy than the saturated state. States with lower energy do exist as soon as the applied field reverses its direction, but they are not necessarily accessible, as explained in section 9.1. One should be careful to go via energy minima on a well-defined path, and not to use approximations, which can lead to large errors, as they have in these modes. These new modes essentially assumed the same chain of spheres, but considered a vagueness about the geometry as an excuse for poor approximations in the calculations. The 'quasi-curling' and 'quasi-buckling' [370] were particularly ill-defined, and were hardly more than just names, it being stated that 'calculations appropriate to these cases are very difficult'. The 'novel reversal mechanism' [372], named 'flipping', could be the same as what used to be called the non-symmetric fanning in a chain of spheres, if done properly and with no extra approximations or inconsistencies, such as a total thickness, $T$, of the 'fully developed wall' in the chain, which is allowed to be larger than the total number of spheres.

In an invited talk at a conference [373] I tried to point out these distorted concepts, but at least Knowles was not convinced by these arguments. He published [374] a 'reply' which stated that the irregular shape of the particles invalidated all the results for an ellipsoid, and allowed him to legislate curling out of existence, and to choose other modes at will. The reader will hopefully understand that this approach leads nowhere. There are more appropriate studies of chains of spheres with anisotropy [375], or of spheres which are cut before joining together [376] so that they touch each other over a rather wide area, and not only at one point, and thus resemble better the peanut-shaped particles. There is also a theory for a chain of disks [377], and one for a chain of oblate spheroids [378], which are even better approximations for this shape. Yet none of them has encountered any new mode, and they all come back to the 'old' modes. In order to conclude this discussion it will only be mentioned that even the angular dependence which was so emphasized in [374], as well as that of [371], was later accounted for [379] by a completely different approach. The data of Knowles were fitted to a Stoner–Wohlfarth model, with a cubic magnetocrystalline anisotropy superimposed on the shape anisotropy of a prolate spheroid, and when the demagnetizing field was introduced, it led [379] to an 'excellent agreement'. Other possibilities [368, 380, 381] have also been considered. It does happen quite often [382] that the same experimental data fit different theories.

## 9.5 Brown's Paradox

The main reason for the (futile) search for other reversal modes is that in spite of all the rigour in the evaluation of the nucleation fields, the results do *not* agree with experiment, in particular for the case of bulk materials. Although the nucleation field is a theoretical concept which is not usually measured, it is not even necessary to continue the calculation beyond the nucleation point in order to see that this theory cannot agree with experiment, and that something is basically wrong with it. The nucleation field

$H_n$ is defined in section 9.1 as the field at which some change *just starts* in the previously saturated state. The coercivity (or coercive force) $H_c$ is defined in section 1.1 as the field at which half the magnetization has been reversed, which at any rate means that a *large* change has already occurred in the previously saturated state. Therefore, $H_c$ can only be reached *after* going through $H_n$ in the sequence described in section 9.1, or they can be at most the same field in the case when there is a big jump of the magnetization at the field $H_n$. Noting that the definition of $H_n$ is for an applied field that starts positive, and goes through zero to negative values, while $H_c$ is defined as a *positive* quantity, the above statement means $H_c \geq -H_n$.

Suppose first that the ferromagnetic body is a large ellipsoid. Small particles will be discussed later in this section. The main conclusion from all the studies in sections 9.2 and 9.4 is that for a large enough body, the curling mode takes over, whether it is coherent rotation or buckling for the smaller radii. Therefore, according to eqn (9.2.31)

$$H_c \geq \frac{2K_1}{M_s} + \frac{Cq^2}{R^2 M_s} - N_z M_s. \tag{9.5.54}$$

The right hand side *may* be negative, when nucleation occurs already at positive fields. For example, for iron at room temperature $2K_1/M_s = 560\,\mathrm{Oe}$, and $4\pi M_s = 21\,600\,\mathrm{G}$. If the ellipsoid is a sphere, with $N_z = 4\pi/3$, the first and the last term of eqn (9.5.54) add up to $-6640\,\mathrm{Oe}$. The middle term is positive, but it is certainly negligible for a sufficiently large $R$, and the right hand side of eqn (9.5.54) is negative. In such a case all that the inequality states is that the *positive* $H_c$ is larger than a negative number, which is an empty statement. Effectively it means that the calculation of $H_n$ is inadequate to tell anything about $H_c$ in such a case. Starting from a large positive field, a reversal already nucleates at a positive applied field, before a zero field is reached. It is then necessary to solve the non-linear Brown's equations for smaller positive fields, after nucleation, and follow the solution down to negative fields till $-H_c$ is reached. No statement about the theoretical value of $H_c$ is valid before this calculation is carried out.

However, if the iron crystal is a very elongated prolate spheroid instead of a sphere, $N_z$ can be much smaller (it tends to 0 for an infinite cylinder), and the whole $N_z M_s$ term may become negligibly small. In this case the right hand side of eqn (9.5.54) is positive. Since the middle term is positive anyway, it can be stated that for very elongated iron bodies $H_c \geq 560\,\mathrm{Oe}$. For iron *whiskers*, which are very elongated particles indeed, with diameters of several $\mu$m and a length of the order of 1 cm, the experimental value for $H_c$ is usually 0.1 Oe or less, which is a very large discrepancy. It is known in the literature as *Brown's paradox*, or Brown's coercivity paradox.

Iron is given here as a representative of a class of materials which are called *soft magnetic materials* and defined as materials for which $K_1 \ll$

$\pi M_s^2$. The other extreme is materials for which $K_1 \gg \pi M_s^2$, and this class is known by the name of *hard magnetic materials.* There is no law of nature which prevents materials from being in between these extreme cases, but such materials are not being produced or investigated, because they do not have any practical application. Soft materials are used where the coercivity is preferred to be as small as possible, *e.g.* in motors or transformers, where a high permeability and low losses are required. Hard materials are used in applications which require the magnetization to be fixed for a long time after the crystal has been magnetized, such as in permanent magnets. Recording materials, such as $\gamma$-$Fe_2O_3$, are in the latter category, but for them it is preferred that the magnetizing field for writing the data should not be large. Therefore, they are made with a $K_1$ which is rather large, but not too large. Such materials are sometimes referred to as *semi-hard magnetic materials.*

In hard materials, Brown's paradox is even more outstanding than in soft ones, because it applies to any ellipsoid, and not only to elongated ones. For example, in $BaFe_{12}O_{19}$ at room temperature $2K_1/M_s = 16\,600$ Oe, and $4\pi M_s = 4500$ G. The largest possible value for $N_z$ in any ellipsoid is $4\pi$, for an infinite plate with the field perpendicular to the plate. Even for this value, and just noting that the middle term in eqn (9.5.54) is *positive* without checking what its value may be, this inequality says that $H_c \geq 12\,000$ Oe. The experimental value for $BaFe_{12}O_{19}$ particles of the order of $1\,\mu$m is 3000 Oe. Similar discrepancies are encountered in practically any magnetic material, when the crystal size is large enough.

It should be particularly noted that this discrepancy, or paradox, can be formulated without any of the calculations in the previous sections, and would have applied even if all that algebra was wrong. The value of the middle term in eqn (9.5.54) has not even been used here, except for its being positive, so that all the details of the curling mode do not enter the argument. At least the same discrepancy would have applied if the coherent rotation was the easiest mode, because it is in the term with $K_1$ which is common to *all* reversal modes. Other modes may even make this discrepancy worse by having additional terms, such as the $N_x$ term in eqn (9.2.10) for the coherent rotation, but the term with $K_1$ is certainly always there. For this reason, inventing new modes cannot help remove the paradox, even if all the discussion in section 9.4 was wrong. Actually, the problem was already known *before* all the foregoing calculations of nucleation fields. Already in 1945 Brown [383] noted that the barrier for nucleation of any sort of a reversal is *at least* that of the anisotropy energy, which is the first term in eqn (9.5.54), just because the exchange and magnetostatic energy terms are always positive. Even this anisotropy term by itself is too large, and the experimental coercivity is considerably smaller than it, in all bulk ferromagnets. Moreover, even the values of the coercivity need not be used in order to realize the paradox. Equation (9.5.54), or already eqn

(9.2.31), implies a *negative* nucleation field, but domains can be observed already in *zero* applied field, see section 4.1. It has been shown in section 6.2 that the existence of these domains is favourable energetically, but a lower energy is not a sufficient condition for the domains to enter. As has already been emphasized in section 9.1, the existence of a lower-energy state is not sufficient for the system to be able to reach that state.

The reasons for this paradox are rather well understood, *qualitatively*, and will be listed separately for hard and for soft materials. However, before going into these details it must be emphasized that the discrepancies are too large for being taken lightly. Until the theory is modified to take into account quantitatively the effects which cause this paradox, everything discussed in this chapter can only be applied to fine particles. None of this study of nucleation can serve any useful purpose when it comes to crystals which are large enough to support a subdivision into many domains. For them it is possible to get away with a theory which ignores all this chapter and the question of how domains enter into the crystal, and takes it for granted that they are there if their existence reduces the energy. Theories which only compare energies of various configurations, such as the one in section 6.2, or that of domain wall structures in chapter 8, work very well in this region, and can be used to interpret all sorts of experimental data. A particularly nice example is the shape of the so-called Néel spikes, which are formed [384] near non-magnetic inclusions. Once their general shape is assumed, an energy minimization leads [385] to all the fine details of their structure, with a perfect fit to experiment. Such theories are still being done [386], and they are actually inevitable, as long as the present theory cannot be extended to take care of the defects which will be specified in the following. Still, this situation does *not* justify discarding this chapter altogether, much less discarding the whole theory of micromagnetics, as has been suggested on dubious grounds, discussed in [382]. The nucleation theory does agree with experiment for small particles, and the modifications which are necessary to make it agree with experiment for larger particles are known in principle, and may be worked out in detail sometime.

### 9.5.1 *Hard Materials*

When optically transparent plates of $BaFe_{12}O_{19}$ are saturated in a large field, then the field is reduced, domains appear already at a positive applied field. However, it has been noted [387] that these domains do not appear anywhere in the sample. They seem to emerge radially from a well-defined 'nucleation centre' which is always at the same spot in a given crystal, for various cycling of the field. In *some* crystals, no domains were observed at zero applied field, and they only appeared up to several hours [388] after the field had been switched off. In some cases, domains nucleated only after a negative field of $-1000$ to $-2000$ Oe was applied [389]. Obviously, these observations seem to indicate that the domains nucleate before the theoretical value is reached only at those points in the crystal where there is some sort of a defect, which may be, for example, an impurity atom, or a dislocation. The nature of these defects has not been determined, except for one case in which the nucleation centre could be identified [389] with a crack in the crystal. Similar nucleation centres were produced [390] in MnBi films by pricking them with a non-magnetic needle. Some domains nucleate [391] at the *edges* of a plate.

These experiments may mean that the perfect parts of the crystal obey the nucleation theory, and that nothing would have nucleated if it were not for these imperfect spots. Of course, once the domains nucleate at one of these centres, there is no extra energy barrier, and it is very easy for them to spread all over the crystal, when the state of subdivision into domains has a lower energy than the saturated state even for the perfect parts of the crystal. A model was proposed [392] in which the nucleation centres were *assumed* to be dislocation lines, and their effect was assumed to be a high local stress that could effectively be taken as a *local lowering* of the anisotropy constant, $K_1$. Nothing specific can be done without at least some indication of the amount of this reduction, and the size over which it may extend [393], but these parameters are not known. It is also not possible to know what the domains actually look like in the early stages of their formation, and an attempt to study this stage [394] in one material could only report that the initial growth of the domains proceeds too rapidly for observation. For these reasons, such a calculation, or its modifications [395, 396], can only be described as a semi-quantitative evaluation. Instead of the dislocation lines, the defects may actually be planar [397], but in either case it seems [392] that such a mechanism can resolve Brown's paradox in hard materials, at least for particles which are not much larger than the size at which domains start to be energetically favourable, although a more quantitative theory is still needed. There are also some experiments on milling and annealing hard materials, whose results are [112, 113, 114, 398] in qualitative agreement with this picture for the role of dislocations.

For much larger crystals, the role of dislocations is reversed, and there are both theoretical [399, 400, 401] and experimental [402, 403, 404] indications that the coercivity of bulk materials increases with increasing number of defects, probably because they hold the domain walls, and do not let them move freely when the field is changed. There have been several attempts [405] to separate the mechanisms of nucleating a domain and of moving its wall, but the difference has not been [406] very well established. Or, as concluded in a review [407], 'more work is needed to make interpretations unambiguous'. The problem is particularly complicated by experiments which do not start with a sufficiently large applied field for driving the domains away, and report measurements which are actually minor loops. Sometimes they are presented as such [403], but sometimes they are not, as in several examples listed in [373, 382, 392]. To repeat just one case, in certain MnBi crystals the domains completely disappeared [408] at an applied field of 5000 Oe, and reappeared when the field was reduced to a smaller, but still positive, value. When such a crystal was once put in a field of 20 000 Oe, the domains disappeared and never appeared again with any cycling of the field. In some other cases, it has been stated that the 'nucleation field depends on the value of previously applied positive field and the crystal imperfection' [409], so that the nucleation thus measured has obviously nothing to do with the nucleation as defined in section 9.1, or with Brown's paradox. Similar observations [410, 411, 412], and others cited in [407], show that many experiments do not start from saturation. On the whole, the magnetization reversal in bulk hard magnets can only be said to depend on crystalline defects, which are not included in the theory, and that Brown's paradox will be resolved when they *are* included.

### 9.5.2 *Soft Materials*

If a long iron whisker is held in a sufficiently large magnetic field, and an opposite field is applied to a small part of it, it is possible to study the reversal of *that* part of the whisker, while the rest of it is held saturated parallel to its long axis. By picking the signal from the reversing part, it is possible to determine the field at which the reversal just starts, namely the nucleation field, for different *parts* along the whisker. This experiment [413] and its later modification [414] obtained nucleation fields which were quite close to the theoretical value of −560 Oe (for a very long iron crystal at room temperature) at some parts of selected whiskers. In other parts of the same whisker, less negative values of $H_n$ were measured, obviously because the crystal was less perfect in those regions. Unlike hard materials, for which the nature of the defects at the nucleation centres is not known, for soft materials it has been established that reversed domains nucleate where the surface is rough. The first demonstration [414] of this conclusion was an electropolishing of the whisker, which resulted in a complete change of the whole nucleation pattern. A more direct proof was an observation of the surface of the whisker by an optical microscope. It [415] showed a good correlation between the volume of the surface defects and the difference between the theoretical and experimental $H_n$ in that vicinity.

Surface roughness must be important in any ferromagnet, but in a soft material the magnetostatic energy is particularly large by definition. There is thus a large effect due to the surface charge created at the points where the magnetization of the saturated state is not parallel to the surface, which must be the case where the surface is not smooth. This charge gives rise to a demagnetizing field in that region, which reduces locally the energy barrier, thus allowing domains to nucleate there. And it is quite easy to be convinced [392] that a local hill or a local valley has approximately the same effect. A similar demagnetization, due to a similar surface charge, should also occur near voids and inclusions inside the crystal, which have also been demonstrated [416, 417] to interact with walls. Nucleation at such internal defects is also possible, especially in less perfect samples. Even in whiskers, which are particularly good crystals, there were some cases where a local reduction in $|H_n|$ could not be assigned to any surface imperfection [415] and must have been due to an internal void. The opposite never occurred, and whenever a major surface defect could be seen on at least one of the four surfaces of a whisker, there was always a minimum in $|H_n|$ there.

This total dependence on the fine details of the surface in a large crystal must seem strange at first sight, especially to somebody who is used to thinking in terms of the theories in the first few chapters of this book. In those theories, a sufficiently large crystal (and sometimes even quite small ones) are just assumed to extend to infinity, with no surface at all. The 'natural' approach is that the surface can have a strong effect only for small crystals, but it is 'expected to be less important, the larger the crystal' as I wrote [418] in one of my earlier papers, before I understood the nature of the problem. The point which must be remembered is that for sufficiently large crystals the multi-domain state has a lower energy than that of the single-domain one, but theoretically the domains cannot nucleate, before a certain negative field is reached, because of an energy barrier on the way. Once this barrier is lifted at any point in the crystal or on its surface, it is easy for them to propagate all over the crystal, as seen experimentally [419], and because their existence reduces the energy even in the perfect parts. Therefore, any measurement of the whole crystal will measure the nucleation property of the *worst* point [382], *i.e.* the point at which the crystal is farthest from being perfect. The effect is the same as in pulling a chain, with a steadily increasing force. The whole chain breaks at the force which is sufficient to break its *weakest* link, even if all the other links are much stronger. The only way to measure the properties of other links is to pull them one at a time, while holding the others from being pulled. And the only way to find the true nucleation field of the perfectly smooth whisker is to measure one region at a time, while holding the other parts from letting the domains come in there, by keeping them in a saturating field, as is done indeed [413, 414, 415] in the experiment of De Blois.

In this sense it can be said that the experiment of De Blois proves that the nucleation theory does agree with experiment for the perfect crystals assumed in the theory, and there is no paradox. For less perfect crystals, the theory should be modified to take into account the effect of imperfections, which has not been done yet. More details were revealed in two extensions of this experiment. In one [420] a local field was applied to different regions of a thin film, and the shape of the developing domain was observed. In another [421] whiskers under stress were studied by the same technique of De Blois, and the nucleation field at 'good' points was found to become more negative with increasing stress, which is equivalent to increasing the anisotropy constant, $K_1$. There is a first theoretical step towards a better analysis of the data in the less perfect parts of the whisker in the experiment of De Blois [422], and a statistical correlation [423] between the probability of finding a defect and the values of $H_n$ measured by De Blois. But none of these studies has been carried far enough for a quantitative theory of imperfect crystals.

As is the case in hard materials discussed in the foregoing, there are also theories for soft materials which ignore nucleation, and try to analyse experimental data on different measurements, in particular coercivities, of the crystal as a whole. They assume that the domains are already there, and consider pinning of their walls by crystalline defects. For these theories there is not really much difference between hard and soft materials, except for the numerical parameters which are used. Some theories still use the assumption of one wall pinned to one defect [424], while others consider the statistical aspects [425, 426] of many defects, randomly distributed in each domain. Various models have been proposed [427, 428, 429, 430] for the pinning. They all use rather rough approximations.

There is one point on which the experiment of De Blois cannot give a clear answer, and which therefore remains obscure. It is the question of whether the edge (or tip) of the whisker, which cannot be accessed by the technique of De Blois, behaves differently than other regions. This question has already been discussed in [392] in connection with some suggestions for resolving Brown's paradox by the argument that real crystals are never saturated to start with, and some unseen domains remain near the edges, where the demagnetizing field (for non-ellipsoidal shapes) is very large. Of course, there is a large amount of evidence, some of which has already been mentioned in the foregoing, that the field used in many experiments is not sufficient to saturate the sample, and these published results confuse the issue. But there are those who claim that no saturation is possible *in principle* for some bodies (such as a prism or a plate), with a sharp corner at the edge, which is a different matter altogether. My argument at the time [392] was that on an atomic scale, a sharp corner does not have any more meaning than a rounded one, see Fig. 3.1. Therefore, the approximation of an ellipsoid is at least as good as that of a prism, and a saturation in a finite field should be possible in principle for real particles, although higher fields than are usually considered adequate may have to be used. My view is still the same, but the opposite is just as legitimate, and there are those who prefer to consider a prism, which takes an infinite field to saturate, and for which there can be no nucleation, see also section 10.5.3. On this basis, there is a model [431] for domains that enter from the corner of a plate, even in *hard* materials.

### 9.5.3 *Small Particles*

All the foregoing explanations of *why* the theory does not agree with most experiments do not change the fact that the theory presented in this chapter is not useful for most practical cases. Therefore, this theory was considered for a long time to be just a curiosity, which could at most interest some pure theorists, or may be applied to extremely unusual kinds of experiments. And in spite of all the great hopes of Brown and others in the beginning, of having one theory that can explain everything, it is undeniably a complete failure for bulk materials. It should be obvious from the above analysis that until a big improvement is incorporated, this theory can at most be applied to small particles, below the value for which subdivisions into domains reduce the total energy. If the nucleation is not that of reversed domains, the exact shape of the crystal becomes less important, and both the effect of crystalline defects and the probability of their existence are very much reduced. For such cases, the nucleation theory has a chance to work well.

In a way it can be said that it works indeed in a relatively narrow size region of 'small particles'. This region is best defined from a plot, such as Fig. 1 of [190], of the remanence and coercivity *vs.* particle size. These properties have low values due to superparamagnetism in the smaller particles, and due to subdivision into domains in the larger ones, with a maximum in between. In the vicinity of this maximum, nucleation theory usually works, and since the smallest particles are eliminated, it means in most cases that the theory of the curling mode agrees with experimental data. Examples of such agreement have been listed in reviews [270, 359, 373, 392], and it will only be repeated here that the coercivity of very elongated nickel particles was found [432] to be quite well approximated by a *linear function* of $1/R^2$, at two temperatures. A linear function of $1/R^2$ was also observed [433] in the coercivity of alumite, although the particles were not ellipsoids, and even in *cubes* of cobalt-doped $\gamma$-$Fe_2O_3$ [434]. The latter fit was originally presented [434] as 'perhaps fortuitous', but it now seems to be a real part of a pattern. This example, and others, prove that the theory works on the whole in this size range, and may at most need some slight modifications, when the finer details are taken into account. The data suggest, as a rough criterion, that the theory works well if the middle term of eqn (9.5.54) is larger than the first one, and breaks down when the first term becomes large. This rule is demonstrated by the experimental coercivity [435] of whiskers. It is close to the theoretical value for curling nucleation at small radii, but when the $K_1$ term becomes dominant, the data just keep going down, orders of magnitude below the theoretical value.

However, such a comparison of $H_n$ with $H_c$ does not really show an agreement, and it is more accurate to say that in this size region there is no big discrepancy, than to say that it fits. The agreement [436] between other properties, calculated from the curling mode, and the experimental data, is also semi-qualitative. Such fits used to be considered good enough when it was difficult to make small particles, and when experiments were done on a large number of particles together, which involved the crucial factor of size distribution. And even with the more recent controlled dispersion

[437], there is still the difficulty of interactions among the particles. This problem has never been solved, and is still studied [438] for the case of a regular array of particles, or by certain averaging [439] schemes. The former theory should certainly be compared with experiments [440, 441] on such an artificial, regular array. But it is unrealistic when interpreting experiments on disordered powders, as is demonstrated by more rigorous calculations [442, 443, 444] on *two* interacting dipoles, which do not fit the local-field concept. Such calculation can probably be a good basis for studying particles which have [445] an odd shape, but they can obviously not be extended to interactions within a random ensemble of particles.

Nowadays, this problem of interactions can be evaded instead of being solved, because more and more measurements are performed on a single particle [369, 446, 447, 448, 449, 450, 451, 452, 453, 454, 455, 456]. For these experiments, the semi-quantitative comparison of $H_n$ with $H_c$ is not good enough any more, and it should be possible to try a more detailed comparison. In particular, a good theory is needed for the coercivity in ellipsoids, preferably somewhat distorted ellipsoids. It should be noted that the coercivity is not only an important parameter in its own right. It is one of the *very few* parameters from which the whole hysteresis curve can be [457, 458] constructed.

The original idea was to continue with the solution of the non-linear equations, once the nucleation is determined, and it is possible to identify the branch on which to proceed. It was not done because the disagreement with experiment for bulk materials gave the impression that there was no point in continuing, when already $H_n$ was wrong. Small particles were not available experimentally at that time, and the whole field was considered impractical. The comeback was when the size of recording particles became small enough, and the old theory suddenly fitted many experimental data. But at that time many of the original papers were already forgotten, or misunderstood, and the new theoretical approach did not proceed the way it could have. There was too much effort put into finding other modes that would fit better, and too little effort put into necessary modifications in the curling mode that could be applied to real particles. And in particular there is still no attempt to do what Brown meant to be the next step to start with, namely to find out what happens *after* nucleation. In the beginning, there is no change while the field is reduced from a large positive value, through zero, and down to a certain negative value, when something nucleates. After that, the linear equations are not valid any more, and the next step should be to solve the non-linear equations for *more negative* fields, and follow the rest of the magnetization hysteresis. In this solution, the one that is to be chosen out of many possibilities is the one which tends to the nucleation eigenmode when $H_a \to H_n$. Experiments are ready now for a good theory, but this part is still missing.

A recent attempt to do just that part [459] for a particular case missed the main point, and after finding the nucleation field, $H_n$, these authors solved the non-linear equations for $|H| < |H_n|$. They [459] found that the energy of the curling mode, in this range of field, is larger than that of the uniformly magnetized state, which is not surprising. All it means is that nothing will happen in the range $|H| < |H_n|$, which is essentially the definition of the nucleation field in section 9.1. They should have solved for $H < -|H_n|$, in order to find out what happens *after* nucleation. There are indeed in the literature some works in which the energy is calculated for $|H| < |H_n|$, but they [460, 461] look for the energy barrier, which is a different problem, or try to find a *local* mode, as mentioned in section 9.4, which is also a different problem.

A solution of the non-linear equations was only tried for an infinite cylinder [360], which is an atypical case. In that case it was found that there was one jump from saturation along $+z$ to one along $-z$, and $H_c = -H_n$. For a sphere [462] and for a finite cylinder [463, 464] there are only some *approximations* to the true curling mode after nucleation. The behaviour of an infinite plate was only studied [465] at the first stage after nucleation. Nothing has been done for a more general ellipsoid, and it is still needed.

# 10

# ANALYTIC MICROMAGNETICS

In this chapter and the next one, various topics in micromagnetics, outside the nucleation problem, will be described. The subdivision into analytic and numerical studies in the two chapters is rather artificial, and quite arbitrary. In most cases there is no real distinction between analytic solutions and numerical ones, and many physical problems use a mixture of both. Nevertheless, it seems desirable from a didactic point of view to keep them as separate chapters.

## 10.1 Ferromagnetic Resonance

The basic equation governing this resonance is eqn (8.5.48), or rather one of its modifications as either eqn (8.5.50) or eqn (8.5.52), because there is always damping in real systems. The experimental setup always involves an application of a large DC field $H_a$ which holds the magnetization almost parallel to its direction, $z$. It means that the components perpendicular to $z$ are rather small, and may be taken to a first order only, as is the case with the nucleation. Besides the DC field, there is also an AC field, at a given frequency, $\omega$, which 'tickles' the magnetization into a periodic motion at this frequency, with a small amplitude. One is then looking for a resonance of *this* motion at the frequency $\omega$, when the applied DC field $H_a$ passes through the appropriate value which corresponds to $\omega$ being the frequency of one of the natural oscillations of the system.

If the AC field is sinusoidal, its time-dependence can be expressed by a factor $e^{i\omega t}$, and the same factor can then be inserted into the steady state solution of $m_x$ and $m_y$. The linearization of the equations for small $m_x$ and $m_y$ is mathematically the same as for the nucleation problem in section 9.1, for the same assumptions used there, namely that the sample is an ellipsoid, that the field is applied along one of its major axes, which is also an easy axis for either a uniaxial or a cubic anisotropy, etc. For this case, and when the damping of eqns (8.5.50) or (8.5.52) is omitted for simplicity, it is seen that the equations of motion for the *amplitudes*, namely when the factor $e^{i\omega t}$ is omitted, are

$$\left[\frac{C}{M_s}\nabla^2 - \frac{2K_1}{M_s} + N_z M_s - H_a\right] m_x - \left(\frac{i\omega}{\gamma_0}\right) m_y = \frac{\partial U_{\text{in}}}{\partial x}, \qquad (10.1.1)$$

and

$$\left[\frac{C}{M_s}\nabla^2 - \frac{2K_1}{M_s} + N_z M_s - H_a\right] m_y + \left(\frac{i\omega}{\gamma_0}\right) m_x = \frac{\partial U_{\text{in}}}{\partial y}, \qquad (10.1.2)$$

where all the notations are the same as in section 8.5.

The boundary conditions are the same as in the case of nucleation, and on the whole the nucleation problem may be regarded as a particular case of the resonance problem, for the particular value $\omega = 0$. There is, however, one big difference in that the nucleation has a physical meaning only for the mode which has the largest eigenvalue, as discussed in chapter 9. In the case of the resonance, *all* the modes can be excited in principle, if the conditions are right, and the existence of one mode does not eliminate any of the others. And many different modes have indeed been studied experimentally in the same sample.

In spite of the similarity, resonance modes have been studied without paying much attention to the relation to the nucleation problem, before *and* after this point had been discussed [341] by Brown. In some ways, the theory of ferromagnetic resonance is more general than the equations given here, because it sometimes includes other terms, such as a surface anisotropy, *e.g.* in [158, 466, 467]. It also has to take into account a more general form of Maxwell's equations than is used in this book, because the dynamic effects of eddy currents and skin depth are important [468] at the high frequencies used in these experiments, whereas they are negligible for the static nucleation. In most cases, however, the geometry is limited to that of an infinite plate [469], for which the solution of the differential equations is just made out of sinusoidal variation, as in section 9.3 here. Besides these sinusoidal modes [470], the theory of resonances recognizes only the coherent rotation mode (known as the 'uniform mode' in this context), in which both $m_x$ and $m_y$ are constants. For the non-coherent modes in an ellipsoid, known as the *magnetostatic modes*, the exchange energy has been left out [471] by writing $C = 0$ in eqns (10.1.1) and (10.1.2), which makes them algebraic instead of differential equations. This approximation is justified as long as the samples studied experimentally are rather large. A rough estimation [472] showed that the neglected exchange term was indeed negligible for a sphere of the size used then, but smaller spheres were made [473] a short time after that estimation. These particles indeed support new resonance modes [474], which depend on the particle size, and which have not been observed in somewhat larger particles.

In such small spheres, the exchange energy becomes dominant, and it is necessary to solve the differential equations in the same way as in chapter 9. As a first step, this problem was solved for very small spheres, for which the magnetostatic term is negligible compared with the exchange energy. For the case of a cylindrical symmetry, the whole problem can be solved analytically, and the result is [472] the same as the curling mode. In particular, there should be a term proportional to $1/R^2$ in the resonance field. It has been suggested [472] to call these modes *exchange modes*, but this name was used [475] immediately after for a different geometry. It should be particularly emphasized again that higher modes can also be excited, and that their resonance can also be observed, unlike the case of the nucleation field for which only the least negative eigenvalue has a physical meaning. For the usual sinusoidal variation in a plate, the higher modes are the harmonics, with integral multiples of the frequency. For the curling exchange modes, the higher modes are obtained from the larger roots of eqn (9.2.26), which should make them easy to distinguish. This study, however, does not take into account the case of an intermediate size range, for which *both* magnetostatic and exchange energies may be important. In this case, these exchange modes are not completely separated, and their mixing can make them less easy to distinguish. Also, the modes which are not cylindrically symmetric, and which have not been calculated, may still appear in the *same* experiment.

The theory of these modes has also been extended [476] to include, for example, damping, which is left out in eqns (10.1.1) and (10.1.2). But it does not address the case of mode mixing when the magnetostatic energy is not completely negligible, nor does it address the possible modes which do not have a cylindrical symmetry. The latter must enter at still smaller size, when a curling configuration will involve a very large exchange energy, and may overlap the other modes. Before sufficiently small particles could be made, this theory was only a mathematical exercise. But now that such particles seem to be available, these possible gaps in the theory should be investigated.

## 10.2 First Integral

It has been mentioned several times in this book that the assumption of a one-dimensional magnetization configuration, in an infinite crystal, is quite risky and may lead to serious errors. However, this assumption is made anyway in many calculations, whether justified or not. In some of these cases there is no other theory, and they cannot be just ignored.

Besides the domain wall discussed in chapter 8, one-dimensional models have been used in an attempt to find the effect of planar crystalline defects on nucleation and coercivity. The physical properties in a certain region were assumed to be different from those in the rest of the material, and Brown's equations were solved separately in the perfect and in the imperfect regions, and then matched together. This problem was first solved [477] for a defective region in which only the anisotropy constant was different from that in the bulk, and later extended [478] to a modification of the exchange constant $C$ and the saturation magnetization, $M_s$, besides the anisotropy constant $K_1$. The former was later used [479] to explain the coercivity of a hard material. The *same* model was revived for the case of a wall which is assumed to be already in the defective region, instead of its being nucleated there. The coercivity in this case is determined by the pinning of that wall to the defect, which is mathematically the same problem as in the different physical case of the nucleation. Results were reported for a region in which both $C$ and $K_1$ were modified [480, 481], and for a case in which only $C$ was changed, as different functions [482] of space. There is also a case of a film [483] with a variable thickness, for which the mathematics is still essentially the same.

With the new interest in multiple films, the *same* one-dimensional model has been used for studying strongly coupled films. In this case there are again two regions (which are the films of different compositions) with the physical constants $C$, $M_s$ and $K_1$ being different for each of the regions. Because of the coupling, the solution in one region should pass smoothly to that in the other region, so that the mathematical problem is identical to that of [478], although these models are always reinvented without paying attention to the previous work. Results have been reported for two films each of which is essentially saturated, with only a transition layer between them being a function of space [484], and for a case of full variation over each of the two films [485], as well as such a variation for two films with an antiferromagnetic [486] coupling. Details will not be given here for any of these cases, and it will only be remarked that the assumption of one dimension may be too restrictive for describing the physical situation in many of the problems to which such models are applied. Even a large film ends *somewhere*, and it has been noted [487] that the effects of the edges may sometimes be very large and can invalidate the one-dimensional approach.

An example is shown schematically in Fig. 10.1. In (a) the magnetization is parallel to the film plane, creating a charge on the surface. When the hard material 'A' is magnetized to the right, the field due to its surface charge points to the left, and the soft material 'B' can be in a negative field when the applied field is positive. In (b) the films are not continuous and one material 'penetrates' through the other one (which *can* happen in practice). The applied field and the magnetization are perpendicular to the film, but the surface charge of 'A' can still create a field at 'B' in an opposite direction to that of the applied field. Such cases can give a magnetization curve which is qualitatively different [487] from the one calculated by neglecting these effects. Surface roughness can also [488] cause a similar demagnetization.

Nevertheless, such calculations do exist in which this assumption of one dimensionality is made. And once it is made, one may as well take advantage of a particular property of one-dimensional micromagnetics which is only little known, although it can facilitate the rest of the calculation very considerably. Before specifying this theorem, it should be noted that in a true one-dimensional case, *i.e.* when $\mathbf{M}$ and $U$ are functions of only (say) $x$, eqn (6.1.4) becomes

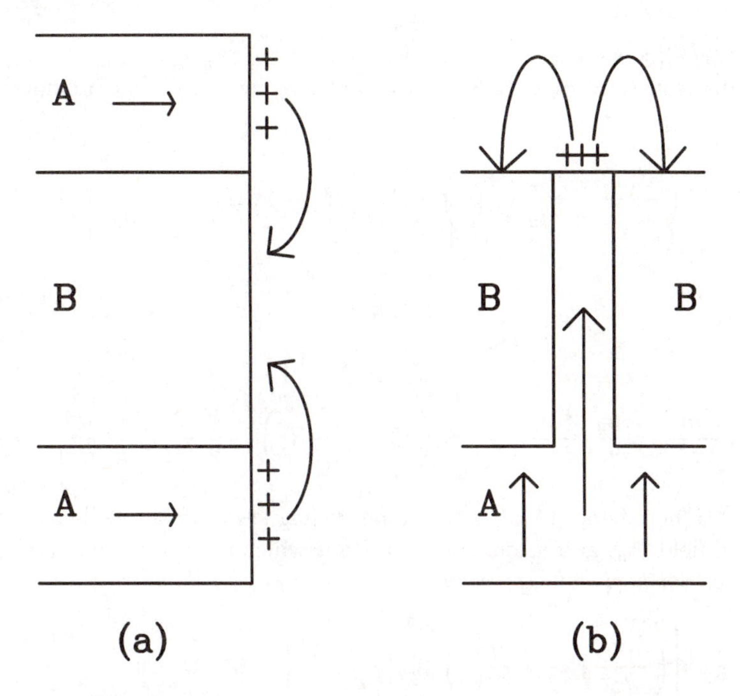


FIG. 10.1. The edge of a multilayer film made of a hard material 'A' and a soft material 'B'. Surface charge is shown schematically for the layers magnetized parallel (a) or perpendicular (b) to the film plane. Copied from [487].

$$\frac{d^2U_{\text{in}}}{dx^2} = 4\pi M_s \frac{dm_x}{dx}, \tag{10.2.3}$$

which is readily integrated to

$$\frac{dU_{\text{in}}}{dx} = 4\pi M_s m_x. \tag{10.2.4}$$

The constant of integration depends on the geometry, but it can always be absorbed into the demagnetizing factor. The whole term in eqn (10.2.4), when substituted in eqn (8.3.34), and then used in Brown's equations (8.3.37)–(8.3.39), has the form of a uniaxial anisotropy term. Therefore, the whole magnetostatic energy may be left out *in a true one-dimensional calculation*, and included only as a modification of the anisotropy. This feature did *not* appear in the domain walls in thin films discussed in chapter 8, because in that case the calculation was not a true one-dimensional one, with the magnetostatic energy of the surface along $y$ superimposed on the assumption of no dependence on $y$. However, here the term of eqn (10.2.4) will be written separately, and not included in the anisotropy, in order to emphasize its existence.

Substituting in eqns (8.3.37)–(8.3.38), for the case when $\mathbf{m}$ does not depend on $y$ or $z$,

$$C\left(\frac{d^2m_x}{dx^2} - \frac{m_x}{m_z}\frac{d^2m_z}{dx^2}\right) + M_s\left(H_x - 4\pi M_s m_x - \frac{m_x}{m_z}H_z\right) - \frac{\partial w_\mathrm{a}}{\partial m_x} + \frac{m_x}{m_z}\frac{\partial w_\mathrm{a}}{\partial m_z} = 0, \tag{10.2.5}$$

and

$$C\left(\frac{d^2m_y}{dx^2} - \frac{m_y}{m_z}\frac{d^2m_z}{dx^2}\right) + M_s\left(H_y - \frac{m_y}{m_z}H_z\right) - \frac{\partial w_\mathrm{a}}{\partial m_y} + \frac{m_y}{m_z}\frac{\partial w_\mathrm{a}}{\partial m_z} = 0, \tag{10.2.6}$$

where $\mathbf{H}$ now contains at most some demagnetizing factors besides the applied field $\mathbf{H}_a$, and as such is a *constant* which does not depend on $x$.

Consider now the expression

$$A = \frac{1}{2}C\left[\left(\frac{dm_x}{dx}\right)^2 + \left(\frac{dm_y}{dx}\right)^2 + \left(\frac{dm_z}{dx}\right)^2\right] + \mathbf{M}\cdot\mathbf{H} - w_\mathrm{a} - 2\pi M_s^2 m_x^2. \tag{10.2.7}$$

If this expression is differentiated with respect to $x$, one should use

$$\frac{dw_\mathrm{a}}{dx} = \frac{\partial w_\mathrm{a}}{\partial m_x}\frac{dm_x}{dx} + \frac{\partial w_\mathrm{a}}{\partial m_y}\frac{dm_y}{dx} + \frac{\partial w_\mathrm{a}}{\partial m_z}\frac{dm_z}{dx}, \tag{10.2.8}$$

because $w_\mathrm{a}$ depends on $x$ only via the components of $\mathbf{m}$. Substituting for the second derivatives of $m_x$ and $m_y$ from eqns (10.2.5) and (10.2.6), it is seen that the coefficient of $d^2m_z/dx^2$ is proportional to

$$m_x\frac{dm_x}{dx} + m_y\frac{dm_y}{dx} + m_z\frac{dm_z}{dx} = \frac{1}{2}\frac{d}{dx}(m^2),$$

which is zero because $m^2 = 1$. Also, the *same* factor multiplies $H_z$, and $\partial w_\mathrm{a}/\partial m_z$, and on the whole it is seen that

$$\frac{dA}{dx} = 0, \qquad i.e. \qquad A = \text{constant}, \tag{10.2.9}$$

for any function $\mathbf{m}$ which fulfils eqns (10.2.5) and (10.2.6). In other words, $A$ is a *first integral* of Brown's equations in one dimension.

This expression was first proved [489] to be first integral for the particular case of a uniaxial anisotropy, and then generalized [270] for any anisotropy. It has not been generalized to more than one dimension, but it applies as written here to *all* published one-dimensional models of the different physical problems, with constant $C$, $K_1$ and $M_s$, *including* the cases of these constants changing abruptly from one value to another in different regions. It does not apply only to the cases in which one or more of these constants changes continuously over a certain region of space, such as in [477] and [482], although it should not be difficult to generalize it for these cases as well. In some of the published cases this first integral has been rediscovered for the particular case under study. In others, it has been ignored, even though its use could have simplified the calculations. In particular, when the differential equations are solved numerically, it is easier and more accurate to solve a first-order equation than a second-order one, and it helps if one of the constants of integration can be determined directly from the physical problem, instead of having to be adjusted during the computations. Sometimes the use of this first integral may even lead [490] to a complete analytic solution, making the numerical computations unnecessary.

The same first integral has also been used [334, 491] as a self-consistency test in the computations of certain one-dimensional domain walls. For the remanent state, $H = 0$, another first integral was also found [489] for the case of a uniaxial anisotropy. Together with the present $A$, it allows a complete analytic solution of all one-dimensional problems in zero field and this anisotropy. It has not been generalized to any other case.

## 10.3 Boundary Conditions

Matching together the solutions in different regions, as mentioned in the previous section, usually means that both the magnetization and its normal derivative are continuous on the surface between the two types of materials. This condition is quite obvious where $M_s$ is the same on both sides of that surface, because the exchange is very strong over short range, see section 6.2.2, and this energy term prefers neighbouring spins to be parallel to each other.

When $M_s$ is not the same for the two materials, the same purpose is served if both the *direction* of the magnetization vector and its normal derivative are continuous on the boundary. In one-dimensional calculations, as in the previous section, this requirement means that the angle between the magnetization and the $x$-axis is continuous and smooth. This condition has been used in almost all the calculations of this sort, except for [485] in which an angle discontinuity was assumed, and was defined in terms of a certain unknown parameter measuring the exchange coupling at the interface. This approach was criticized in [487] as being too drastic a change. It may, however, be necessary to take into account the possibility that the exchange coupling at the interface between two materials being different from that within each of the materials, may add some extra term to the boundary condition there.

The continuity problem is not limited to thin films. It is also encountered in other types of heterogeneous materials, such as magnetic alloys in which regions of different chemical compositions of the alloy may exist [492, 493, 494] in the same sample. Another example is the so-called 'cobalt-modified' $\gamma$-$Fe_2O_3$, in which particles of this ferric oxide are coated by a layer of cobalt ferrite. The theory in this case [349, 350] also assumed a continuity of the magnetization direction and its derivative on the boundary between the $\gamma$-$Fe_2O_3$ and the $CoFe_2O_4$. Again, it does not prove that other boundary conditions should not be used.

One possibility to modify the boundary conditions between such two materials [492] is to postulate a kind of surface integral, of the general form of an exchange integral, which is supposed to manifest the different exchange on that surface. For the case of no other surface anisotropy term, such a postulate leads to the boundary conditions

$$\frac{C_1}{M_1^2}\mathbf{M}_1 \times \frac{\partial \mathbf{M}_1}{\partial n_1} - K_{12}\mathbf{M}_1 \times \mathbf{M}_2 = 0, \tag{10.3.10}$$

$$\frac{C_2}{M_2^2}\mathbf{M}_2 \times \frac{\partial \mathbf{M}_2}{\partial n_2} - K_{12}\mathbf{M}_2 \times \mathbf{M}_1 = 0, \tag{10.3.11}$$

on the surface which separates $\mathbf{M}_1$ (with exchange constant $C_1$) and $\mathbf{M}_2$ (with exchange constant $C_2$). Here $n_1$ and $n_2$ are the normals from either side of the interface, and $K_{12}$ is a parameter of the theory.

These boundary conditions have the advantage that they reduce to the conventional ones of eqn (8.3.42) for the limit of a boundary between a ferromagnet and a non-ferromagnet ($\mathbf{M}_2 = 0$), in the absence of a surface anisotropy. They have the disadvantage [270] that they do *not* reduce to some trivial continuity requirement in the limit $\mathbf{M}_1 = \mathbf{M}_2$ when the boundary is just an arbitrary surface *inside* the ferromagnet. In this limit $\mathbf{M}_1 \times \mathbf{M}_2 = 0$, and eqn (10.3.10) or (10.3.11) leads to $\mathbf{M} \times \partial\mathbf{M}/\partial n = 0$, which is an impossible requirement for every arbitrary surface inside the ferromagnet. It is non-physical to have a special form for only the boundary surface, which makes it impossible to adopt these boundary conditions. The physical problem, however, is still there, and more appropriate boundary conditions may still have to be developed.

## 10.4 Wall Mass

The motion of a real wall through an imperfect material is quite complicated, and outside the scope of this book. Discussion is limited here to the case of an ideal, *straight* wall (unlike the bubble wall, which is [495] a different problem), moving in a perfect crystal with a perfectly smooth surface. Even in this case, the wall structure cannot be the same as that of a stationary wall, because of two reasons. One is the effect of the applied field which drives the wall, and the other is the gyromagnetic effect, as expressed by the dynamic equation in section 8.5. Only the second one will be described here, and only for an undamped, uniform motion.

All early work on this problem, such as [496], started from the specific assumption of a one-dimensional wall configuration. Even in the work [497] which could be readily extended to three dimensions, and in its extensions [498, 499, 500], the actual examples were those of a one-dimensional wall. In these works it was noted that $\mathcal{H}$ of eqn (8.5.48) was actually the variation of the energy density, $w$, which should also be quite clear from the derivation of Brown's equations in section 8.3. Therefore, if $\mathbf{M}$ is replaced by its direction, which can be expressed by its polar and azimuthal angles, $\theta$ and $\phi$, eqn (8.5.48) is actually

$$\frac{d\theta}{dt}\sin\theta = -\frac{\gamma_0}{M_s}\frac{\delta w}{\delta\phi} \quad \text{and} \quad \frac{d\phi}{dt}\sin\theta = \frac{\gamma_0}{M_s}\frac{\delta w}{\delta\theta}, \tag{10.4.12}$$

where $\delta$ designates the variational derivative. In the particular case of a uniform motion at a velocity $v$ in the $x$-direction, the derivative with respect to the time may be expressed as a derivative with respect to $x$, according to

$$\frac{d\theta}{dt} = -v\frac{d\theta}{dx} \quad \text{and} \quad \frac{d\phi}{dt} = -v\frac{d\phi}{dx}. \tag{10.4.13}$$

In this particular case it is seen [501] that eqn (8.5.48) can be rewritten as

$$\frac{\delta}{\delta\theta}(w - w_1) = \frac{\delta}{\delta\phi}(w - w_1) = 0, \tag{10.4.14}$$

where

$$w_1 = \frac{M_s v}{\gamma_0}\cos\theta\frac{d\phi}{dx}. \tag{10.4.15}$$

This result means that the dynamics of a uniform motion can be taken into account by minimizing the integral of $w - w_1$ instead of the minimization of the integral of $w$ in the static case.

Consider specifically the case of a moving domain wall, which has the same statics as in chapter 8, and with the same geometry as defined by Fig. 8.1. It will also be assumed that the wall structure does not depend on $z$. In this case, the foregoing conclusion means that the energy per unit wall area of a uniformly moving wall can be taken as that of the stationary wall, plus a *dynamic term* [501],

$$\gamma_{\mathrm{D}} = -\frac{1}{2b}\int_{-b}^{b}\int_{-\infty}^{\infty} w_1 dx dy, \tag{10.4.16}$$

where $w_1$ is defined in eqn (10.4.15). Changing from $\phi$ and $\theta$ to the more conventional Cartesian coordinates of $\mathbf{m}$, this dynamic energy term can be written as

$$\gamma_{\mathrm{D}} = -\frac{M_s v}{2b\gamma_0} \int_{-b}^{b} \int_{-\infty}^{\infty} \frac{m_z}{m_x^2 + m_y^2} \left( m_y \frac{\partial m_x}{\partial x} - m_x \frac{\partial m_y}{\partial x} \right) dx dy. \quad (10.4.17)$$

Such a minimization of the integral of $w - w_1$ is equivalent to minimizing the Lagrange function, defined as the potential energy *minus* the kinetic energy, in mechanics. Therefore, it is convenient to define a *wall mass* $m_{\mathrm{wall}}$, so that the kinetic energy is equal to $\frac{1}{2} m_{\mathrm{wall}} v^2$, and write this wall mass per unit wall area as

$$m_{\mathrm{wall}} = \frac{2\gamma_{\mathrm{D}}}{v^2}, \quad (10.4.18)$$

after minimizing the Lagrange function. This expression is not necessarily independent of $v$, and the mass may depend on the velocity. It is often found [502, 503] that the behaviour can be approximated by

$$m_{\mathrm{wall}} = \frac{m_{\mathrm{D}}}{\sqrt{1 - (v/v_\infty)^2}}, \quad (10.4.19)$$

at least for rather low velocities. The mass $m_{\mathrm{D}}$ in the limit $v \to 0$ is known as the Döring mass, after Döring who predicted the existence of such a mass already in 1948. It should be emphasized that the wall mass in thin films is a real entity, and experiments on wall motion indeed show [504] a behaviour similar to that of a particle with an inertial mass.

When written in the form of eqn (10.4.17) it is clear [501] that this kinetic energy, and therefore also the mass, is *identically zero* for all one-dimensional domain wall models of section 8.1, because either $m_x$ or $m_y$ is identically zero in all of them, which makes the integrand always zero. Many workers managed to obtain non-zero values of mass from these one-dimensional wall models, but it was only because they did *not* use eqn (10.4.17). In particular, Schlömann [499] noted that eqn (10.4.15) was also 'another possible choice' for $w_1$, which he preferred to write in a different form. There are, of course, other forms to write the Lagrange function, but the point is that they should all lead to the same wall structure and energy, after proper minimization. If they do not lead to the same result, one should not just choose among them, but realize that the different results mean that it is not an energy minimum. The logic is the same as in the self-consistency checks in section 8.4, and indeed it should be clear from chapter 8 that even the static one-dimensional wall is *not* a proper approximation for a minimal energy wall structure in films. Still, a recent look into some relations between the differential equations [505] goes back to the picture of an essentially one-dimensional wall.

It can also be seen from eqn (10.4.17) that the kinetic energy and the mass will both vanish whenever $m_x$ is an odd function of $y$, while $m_y$ and $m_z$ are even functions of $y$. This symmetry is found in all the two-dimensional stationary walls in zero applied field, and it must therefore be concluded that an additional asymmetry in $y$ is added when the wall moves. Such an asymmetry is indeed found in the study of Hubert [249] of a domain wall in a non-zero field. It is also found in the computations of moving wall structures which will be described in the next chapter. For this reason, models of moving walls have been constructed [503, 506] with this kind of asymmetry in the $y$-direction.

## 10.5 The Remanent State

An argument presented in section 6.2 was meant to convince the reader that the total energy of sufficiently large particles is reduced by subdivision into domains in zero applied field, while for a small particle the exchange is too strong to allow it, and the particle should remain a uniformly magnetized 'single domain'. The study of superparamagnetism in section 5.2 is based on an even stronger assumption, that sufficiently small particles are *always* uniformly magnetized. This assumption is actually a little *too* strong for this purpose, because superparamagnetism is not always due to a coherent rotation of the magnetization. It has been demonstrated [180] that under certain circumstances, the thermal fluctuations can excite a back-and-forth magnetization reversal by the curling mode.

This qualitative argument can be made quantitative, giving the size under which a particle is a single domain. Sometimes the procedure is to compare the energy of the uniformly magnetized state with that of some chosen spatial configurations of the magnetization, as done in *e.g.* [283, 346]. But as in the case of Brown's equations for the magnetization process, this comparison is not sufficient, because there is always the risk of overlooking a state whose energy is still lower. Thus, for example, the first modern study of this problem in a sphere compared the [252] energies of all the configurations which could be obtained by cutting the sphere into slices. But it then turned out [253] that a still lower energy can be obtained by dividing the sphere into cylindrical domains, which cannot be expressed by that slicing. It is therefore necessary to take into account *all* possible magnetization configurations, and to do it by a rigorous calculation with no approximations. This problem, which Brown named 'the fundamental theorem', can be stated [507] as claiming that the state of lowest energy of a ferromagnetic particle, whose size is less than a certain critical size, is one of uniform magnetization. It also goes without saying that above this critical size, the lowest-energy state is one of non-uniform magnetization, although it does not necessarily have to be a subdivision into domains. There can be states of non-uniform magnetization just above the critical size, which are not the fully developed domain configurations, with the latter coming in only at a still larger size. The proof of this theorem will be given here for the simple geometry of a sphere, before discussing other geometries for which the problem is still controversial.

### 10.5.1 *Sphere*

The total energy of the different configurations considered here is the same as in eqn (8.3.22), except for $\mathcal{E}_{\mathrm{H}}$ which is omitted here, because the present calculation is for $\mathbf{H}_a = 0$. The integrations are over a sphere whose radius is $R$. It will also be assumed here, as in the calculations of Brown [507, 508], that the surface anisotropy is zero. Experimentally, this anisotropy is not zero in many cases, such as [200, 201, 202], see also section 5.3.

In the *single-domain* state, namely when the sphere is uniformly magnetized in a direction parallel to an easy axis of the anisotropy energy, $\mathcal{E}_{\mathrm{e}} = \mathcal{E}_{\mathrm{a}} = 0$, and the total energy is the magnetostatic term. The latter has already been calculated for a uniformly magnetized sphere in (6.1.14). Therefore, for this state,

$$\mathcal{E}_{\mathrm{uniform}} = \mathcal{E}_{\mathrm{M}} = \frac{8\pi^2}{9} R^3 M_s^2. \tag{10.5.20}$$

The energy of *all* the other possible states has to be compared with this expression.

For the energy of those other states, lower and upper bounds for $\mathcal{E}_{\mathrm{M}}$ are calculated, according to the general idea of the technique described in section 7.3.4, although the particular trick used here for the lower bound is not mentioned there. For finding a lower bound, the constraint of eqn (7.1.7) is replaced by the *weaker* constraint,

$$\int \left(m_x^2 + m_y^2 + m_z^2\right) d\tau = v = \frac{4\pi}{3} R^3, \tag{10.5.21}$$

where the integration is over the sphere, and $v$ is the volume of the sphere. This constraint allows functions of space which are not allowed by the stronger constraint of eqn (7.1.7). It means that the search for a minimum is done in a larger group. Therefore, a minimum found for the weaker constraint may be due to a function which does not belong to the original group, in which case this minimum is *lower* than the lowest minimum in the original group. It cannot be higher, because the weaker constraint also covers all the functions which are allowed by the stronger one.

For calculating the lower bound, the anisotropy energy is also omitted. Whether cubic or uniaxial, the anisotropy energy term is always *positive*, and it is legitimate to omit a positive energy term for calculating a lower bound, because such omission decreases the total energy. Energy minimization under constraint is carried out by the standard use of Lagrangian multipliers, leading to the three differential equations

$$\left[C\nabla^2 + \lambda\right] m_\alpha = \frac{4\pi}{3} M_s^2 \langle m_\alpha \rangle, \quad \text{for} \quad \alpha = x,\ y,\ \text{or}\ z, \tag{10.5.22}$$

where $\lambda$ is a constant Lagrangian multiplier, and

$$\langle m_x \rangle = \frac{1}{v}\int m_x d\tau, \quad \langle m_y \rangle = \frac{1}{v}\int m_y d\tau, \quad \langle m_z \rangle = \frac{1}{v}\int m_z d\tau. \quad (10.5.23)$$

The boundary conditions are

$$\frac{\partial m_x}{\partial r} = \frac{\partial m_y}{\partial r} = \frac{\partial m_z}{\partial r} = 0, \qquad \text{on} \quad r = R. \quad (10.5.24)$$

The constant $\lambda$ and any other integration constants must be adjusted so that the constraint in eqn (10.5.21) is satisfied.

Multiplying the equation with $\alpha = x$ of eqn (10.5.22) by $m_x$, the one with $\alpha = y$ by $m_y$, and the one with $\alpha = z$ by $m_z$, adding, integrating over the sphere, and using the divergence theorem and eqns (10.5.21) and (10.5.24), the energy of each of the solutions of these equations can be written as

$$\mathcal{E}_{\text{non-uniform}} = \frac{1}{2}\lambda v, \quad (10.5.25)$$

see also section 8.4. Comparing with eqn (10.5.20), the ratio of the energy of any of the solutions of these differential equations to that of the uniformly magnetized state is

$$\mathcal{R} = \frac{3\lambda}{4\pi M_s^2}. \quad (10.5.26)$$

Integrating now each of the equations in (10.5.22) over the sphere, and using eqn (10.5.24) and the definitions in eqn (10.5.23),

$$\left(\lambda - \frac{4\pi}{3}M_s^2\right)\langle m_x \rangle = \left(\lambda - \frac{4\pi}{3}M_s^2\right)\langle m_y \rangle = \left(\lambda - \frac{4\pi}{3}M_s^2\right)\langle m_z \rangle = 0. \quad (10.5.27)$$

If $\langle m_x \rangle \neq 0$, or $\langle m_y \rangle \neq 0$, or $\langle m_z \rangle \neq 0$, this equation means that

$$\lambda = \frac{4\pi}{3}M_s^2, \quad (10.5.28)$$

which according to eqn (10.5.26) means that the energy of such states is *equal* to that of the uniformly magnetized state. It can thus be concluded that the energy can be smaller than that of the uniformly magnetized state only if

$$\langle m_x \rangle = \langle m_y \rangle = \langle m_z \rangle = 0. \quad (10.5.29)$$

With the substitution from eqn (10.5.29), the equations in (10.5.22) are linear and homogeneous differential equations, whose solution is well known. One can write for example for $m_z$ the most general regular solution in the form

$$m_z = A j_n(k_{n,\nu} r/R) Y_{n,\mu}(\theta, \phi), \quad (10.5.30)$$

where $A$ is a constant which has to be adjusted to satisfy eqn (10.5.21), $j_n$ is a spherical Bessel functions, $Y_{n,\mu}$ is a spherical harmonic, and $k_{n,\nu}$ is the

$\nu$-th solution of $dj_n(x)/dx = 0$. The latter condition is necessary to satisfy eqn (10.5.24), and it is seen that all the equations which involve $m_z$ are fulfilled, with $\lambda$ of eqn (10.5.22) being equal to one of the eigenvalues

$$\lambda_{n,\nu} = \frac{Ck_{n,\nu}^2}{R^2}, \tag{10.5.31}$$

for all allowed values $n$ and $\nu$. However, according to eqn (10.5.25), the energy increases with $\lambda$, and for the lowest-energy minimum the smallest $\lambda$ should be taken, which is

$$\lambda = \frac{Cq_2^2}{R^2}, \tag{10.5.32}$$

where $q_2 \approx 2.0816$ is the *smallest* root of eqn (9.2.27). Note that for this solution the other components must be $m_x = m_y = 0$, because any other solution of eqn (10.5.22) is not compatible with the same $\lambda$. A similar solution is possible for $m_x$ or $m_y$ instead of $m_z$, but its energy is the same. Therefore, the smallest energy for all possible functions which solve this lower-bound problem is given by eqn (10.5.32).

Substituting in eqn (10.5.26), it is seen that $\mathcal{R} > 1$, namely the lowest energy is that of the uniformly magnetized state, if

$$\frac{Cq_2^2}{R^2} > \frac{4\pi}{3}M_s^2. \tag{10.5.33}$$

It has thus been proved that the lowest-energy state is one of a uniform magnetization for a sphere whose radius $R$ fulfils

$$R < R_{\rm co} = \frac{q_2}{M_s}\sqrt{\frac{3C}{4\pi}} \approx \frac{1.017\sqrt{C}}{M_s}. \tag{10.5.34}$$

It should be noted that even without an upper bound, this result already proves the statement that the uniformly magnetized state is the one which has the lowest energy, for 'sufficiently small' spheres. For a quantitative evaluation of how small 'sufficiently small' is, an upper bound is also needed.

For calculating an upper bound, it is first noted [508] that for any real numbers $m_x$, $m_y$ and $m_z$, whose squares add up to 1,

$$m_x^2m_y^2 + m_y^2m_z^2 + m_z^2m_x^2 = (m_x^2 + m_y^2) - (m_x^4 + m_x^2m_y^2 + m_y^4) < m_x^2 + m_y^2. \tag{10.5.35}$$

Therefore, a cubic anisotropy energy is always *smaller* than a uniaxial anisotropy with the same $K_1$. Since it is always legitimate to *increase* the total energy in calculating an upper bound, any upper bound which is calculated for a uniaxial anisotropy, of the form $w_{\rm a} = K_1(m_x^2 + m_y^2)$, is also valid for the case of a cubic anisotropy with the same value of $K_1$.

The technique for calculating an upper bound in this case is based on restricting the energy minimization to a particular class of functions, or even to one particular function. Such a restriction may eliminate the spatial variation for which the energy is the lowest minimum, so that the lowest energy of the special restricted class is *larger* than the real minimum. It cannot be smaller, and can at most be equal to the real minimum, when the lowest-energy state happens to be included in the restricted class. In this respect, a computation such as in [252], which minimizes the energy of all magnetization configurations that can be obtained by slicing the sphere, is also an upper bound, because it considers a particular *class* of functions. For the present problem, Brown [507, 508] considered two kinds of functions. One is a rough imitation of the curling mode,

$$m_\rho = 0, \qquad m_z = 1 - \left(\frac{\rho}{R}\right)^2, \qquad m_\phi = \sqrt{1 - m_z^2}, \tag{10.5.36}$$

where $\rho$, $\phi$ and $z$ are the *cylindrical coordinates.* The second one is a rough imitation of a two-domain structure, with a wall between them, taken as

$$\begin{aligned} m_x &= -1, & m_y &= 0, & &\text{for} \quad z < -h \\ m_x &= \sin\left(\tfrac{\pi z}{2h}\right), & m_y &= \cos\left(\tfrac{\pi z}{2h}\right), & &\text{for} \quad -h < z < h \\ m_x &= 1, & m_y &= 0, & &\text{for} \quad z > h \end{aligned} \tag{10.5.37}$$

where $h$ is a parameter with respect to which the energy is minimized.

An upper bound to the energy can be calculated analytically [508] for each of these particular functions. After comparing this energy with that of eqn (10.5.20), the result is that the lowest-energy state is that of a non-uniform magnetization if $R > R_{c_1}$, when eqn (10.5.36) is used, or if $R > R_{c_2}$, when eqn (10.5.37) is used, where

$$R_{c_1} = \frac{4.5292\sqrt{C}}{\sqrt{4\pi M_s^2 - 5.6150 K_1}}, \tag{10.5.38}$$

provided the expression under the square root is positive, and

$$R_{c_2} = \frac{9}{8(3\sigma - 2)M_s^2}\sqrt{C(K_1 + 8\pi\sigma M_s^2)}, \qquad \sigma = 0.785398. \tag{10.5.39}$$

Since two separate functions are used, the smaller of the two radii may be used, and it is possible to state that a sufficient condition for a non-uniform magnetization state to have the lowest energy is that $R > \min\,(R_{c_1}, R_{c_2})$.

Equation (10.5.38) is useful only for small values of $K_1$. It is meaningless and invalid if $K_1$ is so large that the expression under the square root becomes negative. Even when this expression is still positive, but small, this equation is not useful because it leads to a very large $R_{c_1}$. Therefore, eqn (10.5.38) is applicable only to soft materials, with $\pi M_s^2 \ll K_1$, leaving only eqn (10.5.39) for larger values of $K_1$. In practice it turns out that the latter is not very useful either for very large values of $K_1$, and can lead to an $R_{c_2}$ which is orders of magnitude larger than $R_{c_0}$. Knowing that the turnover from a uniform to non-uniform state is somewhere between $R_{c_1}$ and $R_{c_0}$, is as good as knowing this transition to within 10% or so for very small $K_1$, because these radii are that close together. For such cases it is not necessary to calculate the actual critical radius. This purpose is not achieved, however, for very large values of $K_1$, for which the present calculation leaves an uncertainty in the order of magnitude.

There are some indications that the lower bound may be pretty close to the exact value, while both expressions for the upper bound lead to too large bounds, and should be replaced by better ones. This conclusion may be drawn from certain computations, to be described in section 11.3.2, on one case of a uniaxial anisotropy [253] and two cases [509] of a cubic anisotropy. In all three cases, the computed value was much closer to the lower bound than to the upper bound of Brown. In principle, the results of these computations are also upper bounds, because they apply a certain constraint, and do not really minimize the energy of *all* possible functions. They can only be presented as good approximations to the true, three-dimensional structure in as much as the upper bounds they lead to are close to the lower bound of Brown. There are also experimental data obtained [510] from neutron depolarization which show a clear transition from one to two domains in $Mn_{0.6}Zn_{0.35}Fe_{2.05}O_4$. This transition is quite close to the lower bound, and very considerably below the upper bound. Of course, the particles in this experiment had a rather irregular shape [510], and were certainly not spheres.

There is also another *analytic* upper bound, even though it has not been presented as such. It started from an attempt to approximate the configuration of the curling mode in a sphere after the nucleation stage, but was then actually used [462] to find the remanent state, and for some estimations [180] of superparamagnetism by curling. It *is* an upper bound for the energy of the remanent state, as is any calculation which restricts the minimization of the energy to any class of functions. In this particular case, the assumptions are [180]

$$m_\rho = 0, \quad m_z = g_0(r) + [1 - g_0(r)]\cos^2\theta, \quad m_\phi = \sqrt{1 - m_z^2}, \qquad (10.5.40)$$

where $g_0$ is a function of the radial spherical coordinate, with respect to which the energy is minimized, bound by the constraint $g_0(0) = 1$ to avoid difficulties at the centre. When this assumption is substituted in the expressions for the energy, a differential equation can be written [462] for $g_0(r)$

which minimizes the exchange, anisotropy and magnetostatic energies.

This differential equation has to be solved numerically, which is a much easier task than a numerical solution of the whole problem, because only a one-dimensional, ordinary differential equation is involved. Still, it is not as easy to use as the foregoing result of Brown, which can be expressed in a closed form. Computation has only been carried out for the case of uniaxial cobalt, for which the critical radius was only slightly larger than that computed [253] for all possible dependence on $r$ and $\theta$. The functional form computed for this model was also quite similar to that computed in [253]. It is made of two *cylindrically symmetric* domains, magnetized in opposite directions, and separated by a special kind of wall.

### 10.5.2 *Prolate Spheroid*

Generalizing Brown's lower bound to a prolate spheroid [511] is quite straightforward. All it actually takes is to use spheroidal wave functions, in spheroidal coordinates, instead of the spherical ones used in the previous section. The case considered here is a prolate spheroid for which the easy axis of the anisotropy is parallel to the long axis of the spheroid, which is taken as the $z$-axis. The semi-axis of this spheroid along $x$ or $y$ is denoted by $R$, and the demagnetizing factor along $z$ (which replaces $4\pi/3$ of the sphere) is denoted by $N_z$. The result is [511] that the lowest-energy state for such a spheroid, in zero applied field, is one of a uniform magnetization, whenever

$$R < R_{\text{co}} = \frac{q}{M_s}\sqrt{\frac{C}{N_z}}, \tag{10.5.41}$$

where $q$ is the same parameter defined in 9.2.2.3 for the nucleation by the curling mode in a prolate spheroid. Actually, this equation is identical to eqn (9.2.32), except for $N_x$ there which has been replaced by $N_z$ here. This difference should be obvious, because the non-uniform magnetization state here is compared with a uniform magnetization along the $z$-direction, whereas the curling mode in section 9.2.2.3 is compared with a coherent rotation in the $x$-direction.

It should be emphasized that this result applies to the remanent state only. If a particle is uniformly magnetized in zero applied field, it does not necessarily follow that it will remain so when a field is applied, and it does not necessarily follow that it will reverse its magnetization by the coherent rotation mode. In spite of the similarity in the mathematical expressions for the 'critical size' for changing over from one state to another, or from one mode to another, remanent states and magnetization reversal modes are different physical problems. In particular, the lowest-energy state in a given field may not even be reached during certain stages of the magnetization reversal process, as explained in chapter 9. It is quite possible, and examples were given there, that a lower-energy state exists, but the system cannot reach it because of an energy barrier in between, and it gets 'stuck' in a higher-energy one. Besides, the difference between $N_x$ and $N_y$ can be very significant for elongated ellipsoids.

A plot of $q/\sqrt{N_z}$ can be found in [511], which should give an idea of the values involved for not-too-elongated ellipsoids. For an aspect ratio of about 3 or larger, that plot is not necessary, because $q_1/\sqrt{N_z}$ becomes [511] a reasonably good approximation for $q/\sqrt{N_z}$, where $q_1 = 1.8412$ is the limiting value of $q$ for an infinite cylinder, as defined in section 9.2.2.1.

An upper bound has not been calculated, nor is it necessary for finding out what the situation is for typical particles used in practical recording materials. Thus, for example, $q/\sqrt{N_z} \approx 3.1$ for an aspect ratio of 8:1, which yields a 'critical' diameter of at least $2R_{c_0} \approx 150\,\text{nm}$, for magnetite with [511] $C = 1.34 \times 10^{-6}\,\text{erg/cm}$ and $M_s = 480\,\text{emu}$. Other examples are also given in [511], with the conclusion that *all* particles used in recording media are 'single domain' in zero applied field, in as much as their shape may be approximated by that of an ellipsoid. A more recent material for perpendicular recording [512] is made of an array of parallel nickel pillars, with a uniform diameter and distance. Using for nickel $M_s = 484\,\text{emu}$ and $C = 2 \times 10^{-6}\,\text{erg/cm}$, for particles with a diameter of 35 nm and a height of 120 nm, eqn (10.5.41) yields a critical *radius* of about 50 nm, so that there can be no doubt that these particles are uniformly magnetized in zero applied field. This estimate is already sufficient to make sure that the other sample [512] with a diameter of 75 nm is also made out of single domains, which would have been true even if it were the same aspect ratio, and is even more so with the larger aspect ratio for this sample.

For an ellipsoid which is not very different from a sphere, there is also an expansion [513] around the lower bound for a sphere. This expansion is not necessary for ellipsoids, now that a rigorous solution is known for prolate spheroids, and it should not be difficult to extend to oblate spheroids as well. It may be possible, however, to adopt such an expansion for shapes which are not ellipsoidal at all, as is the case in real particles.

### 10.5.3 *Cube*

Most computations on the remanent state of a cube (and other shapes), in the size range which is normally defined to be a 'fine particle', will be discussed in the next chapter. For the particular case of a cube, there is a seemingly complete study [514] of all possible functions of space, which can be expressed by cutting the cubes into slices. This computation is rigorous within its own framework, but the constraint of a one-dimensional slicing makes the critical size thus obtained only an upper bound.

There is also another problem with such computations for a cube, or actually for any non-ellipsoidal body. The demagnetizing field inside such bodies is not homogeneous when they are uniformly magnetized, and it seems that the field components perpendicular to the magnetization vector must enforce some non-uniformity. Moreover, for a uniformly magnetized cube (or a finite cylinder, or any other body which has a sharp corner) the demagnetizing field is formally *infinite* at the corners. The integrated magnetostatic energy is finite, but some still claim [431, 515] that it is wrong to use this finite energy as in [514], because the local infinite field will never allow a uniformly magnetized configuration. This claim has been controversial for many years, because it can also be said [392] that a sharp corner is only an approximation which cannot exist on an atomic scale, any more than a smooth, ellipsoidal surface can.

The calculation described in this section is restricted to an unusual case of a cube made of a small number of atoms, in an attempt to understand the transition into the atomic limit, by particularly avoiding the use of the approximation of a continuous material as introduced in chapter 7. It has already been mentioned in that chapter that this approximation must break down if the theory tries to deal with distances of the order of a unit cell, and it is always possible that even tens of unit cells may be too small for the safe usage of the approximations of micromagnetics. It may be necessary to start wondering about using this theory for cases where the whole size of the particle is only a few tens of unit cells, as are some of the very small particles which are already being studied experimentally. And it may shed some light on, or at least give some preliminary indication for, the meaning of a sharp corner in small particles.

The atomic limit in the present context does not extend as far as an attempt to start with a model which may reproduce the magnetization as depicted in Fig. 3.1. Nobody has ever approached this part of the problem. Also, it is not practical to study hundreds of lattice sites by the method which is described here, and only nine spins are used, in what is supposed to be an indication for what happens with the others. Allowing thermal fluctuations in such a small 'particle' would have made them too strong for this case, thus distorting the physical picture for somewhat larger particles. Therefore, thermal agitation is not used, nor is there any attempt to change the classical spins into quantum-mechanical ones, which should be done in the atomic limit. The spins are left to be the classical vectors used in micromagnetics, because the main purpose is study the limit of the assumption of a physically small sphere.

The model is thus made out of point spins which are localized in the lattice points of a bcc unit cell, whose cube edge is $a$. Only nine such spins are considered, which complete one unit cell. The first one of these spins is located at the body centre, which is taken as the point (0,0,0) in Cartesian coordinates. The other eight are at $\frac{1}{2}a\mathbf{p}_i$, where $\mathbf{p}_i = (\pm 1, \pm 1, \pm 1)$ is the position vector for the $i$-th spin. The numbering of the spins is according to the scheme shown in Fig. 10.2.

The exchange interaction is assumed to be non-zero only between nearest neighbours, which means between $\mathbf{S}_1$ and the other eight spins. The exchange energy is taken as the expression in eqn (2.2.25), but $J$ is replaced

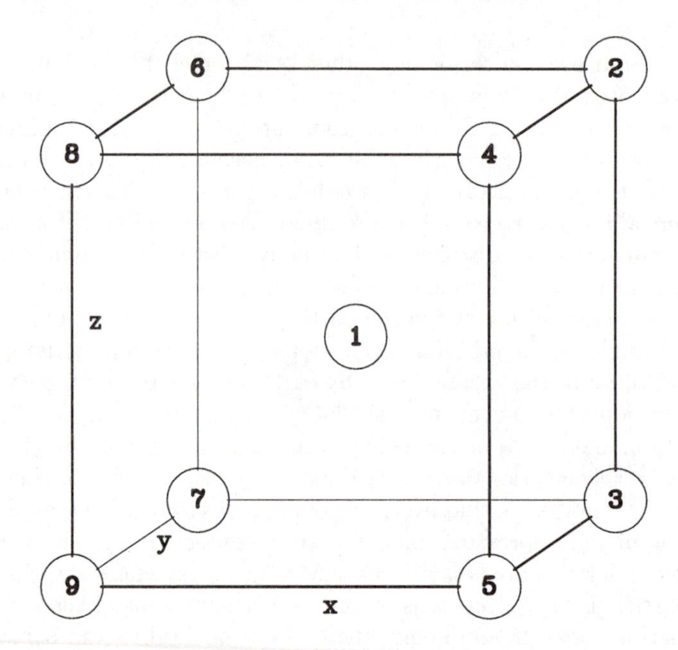


FIG. 10.2. The numbering scheme of nine spins arranged in a bcc lattice.

by the more familiar exchange constant, $C$, according to the definition in eqn (7.1.6), leading to

$$\mathcal{E}_{\mathrm{ex}} = -\frac{C}{2a^2}\sum_{i=2}^{9}\mathbf{m}_1\cdot\mathbf{m}_i, \tag{10.5.42}$$

per unit volume, where $\mathbf{m}_i$ denotes the unit vector in the direction of the $i$-th spin. A similar expression was also used in some calculation [332] which ignored the magnetostatic energy term, but did have an interaction with an applied field. That case [332] assumed a surface anisotropy, without any anisotropy in the body. Here only a volume cubic anisotropy is used, without $K_2$, which leads to the anisotropy energy per unit volume,

$$\mathcal{E}_{\mathrm{a}} = K_1\sum_{i=1}^{9}\left(m_{i_x}^2 m_{i_y}^2 + m_{i_y}^2 m_{i_z}^2 + m_{i_z}^2 m_{i_x}^2\right). \tag{10.5.43}$$

In practice [516] it turned out that this equation, as written, was useful only for positive $K_1$, namely for the easy axes along [100]. For $K_1 < 0$, which means easy axes along [111], the accuracy of computation from this relation was rather poor. A much better accuracy could be obtained by rotating the axes to $x'$, $y'$ and $z'$, with $z'$ along [111] of the original axes $x$, $y$ and $z$, and writing eqn (10.5.43) in these coordinates.

The magnetostatic energy is taken as the interaction of dipoles at the lattice sites with the dipolar field of eqn (7.3.21), before the introduction of the physically small sphere. This energy is written in terms of the vectors

$$\mathbf{p}_{i,j} = \mathbf{p}_i - \mathbf{p}_j, \tag{10.5.44}$$

where $\mathbf{p}_i = (\pm 1, \pm 1, \pm 1)$ is the position vector for the $i$-th spin, already mentioned above. With this notation, the magnetostatic energy, per unit volume, in the particular lattice assumed here is

$$\mathcal{E}_{\mathrm{M}} = \frac{8M_s^2}{81}\left\{\sum_{i=2}^{8}\sum_{j=i+1}^{9}\left[\frac{\mathbf{m}_i \cdot \mathbf{m}_j}{|\mathbf{p}_{i,j}|^3} - 3\frac{(\mathbf{m}_i \cdot \mathbf{p}_{i,j})(\mathbf{m}_j \cdot \mathbf{p}_{i,j})}{|\mathbf{p}_{i,j}|^5}\right] + \mathcal{E}_1\right\} \tag{10.5.45}$$

where the term for $i = 1$ is written separately, and is

$$\mathcal{E}_1 = \frac{1}{3\sqrt{3}}\sum_{j=2}^{9}\left[\mathbf{m}_1 \cdot \mathbf{m}_j - (\mathbf{m}_1 \cdot \mathbf{p}_j)(\mathbf{m}_j \cdot \mathbf{p}_j)\right] . \tag{10.5.46}$$

The total energy of this system was minimized numerically for the 18 directions of the nine spins for $C = 1.73 \times 10^{-6}$ erg/cm, and for either $M_s = 1700\,\mathrm{emu/cm^3}$ and $K_1 = 4.7 \times 10^5\,\mathrm{erg/cm^3}$, or $M_s = 484\,\mathrm{emu/cm^3}$ and $K_1 = -4.5 \times 10^4\,\mathrm{erg/cm^3}$. The first of these cases has the physical parameters of iron, and the second one has those of nickel, used in spite of the fact that real nickel has an fcc and not bcc structure. For either of these cases, the exchange energy is many orders of magnitude larger than the other energy terms, if a realistic value is used for the lattice constant, $a$. Such a big difference makes it very difficult to obtain any reasonable accuracy in calculating the total energy, and any attempt to minimize the energy encounters a large noise. The trick used [516] for the minimization was to start with an unphysically large value of the cube edge $a$, for which the exchange energy was only four or five orders of magnitude larger than the other energy terms. For such a case, the energy minimization could be carried out with a sufficient accuracy. The value of $a$ was then reduced, and the energy was minimized again, using as a start the values of the 18 angles obtained for the previous $a$. This procedure was repeated, and by eliminating the necessity to compute the energy for angles which were rather far from the minimum, it allowed an extrapolation of the minimum energy state to the physically significant lattice constant $a$, of several Å.

The result for positive $K_1$ was [516] that the lowest-energy state was *not* one of uniform magnetization. It was a state in which $m_{1_z} = 1$, but $m_{i_z}$ for $i > 1$ was not exactly 1, although they were all very nearly 1. The actual lowest-energy configuration could be expressed by one angle, because to a very high accuracy it had $m_{3_x} = m_{5_x} = m_{6_x} = m_{8_x} = m_{3_y} = m_{4_y} = m_{7_y} = m_{8_y}$ and $m_{2_x} = m_{4_x} = m_{7_x} = m_{9_x} = m_{2_y} = m_{5_y} = m_{6_y} = m_{9_y}$, with

$m_{2_x} = -m_{3_x}$ of approximately $10^{-4}$. For negative $K_1$, the same relations were obtained in the transformed coordinate system $x'$, $y'$ and $z'$ with $z'$ along the easy axis to the anisotropy. The angles were an order of magnitude smaller than for the $K_1 > 0$ case, which is to be expected because of the same factor in the assumed values for $M_s^2$. It leads to a difference of two orders of magnitude in the average deviation from saturation, $1 - \langle m_z \rangle$.

It has thus been proved that there is no cube which is small enough for the uniformly magnetized state to be the one with the lowest energy, in spite of the fact that over short ranges the exchange energy is several orders of magnitude larger than the other energy terms. In other words, the 'fundamental theorem' of Brown, which has been rigorously proved for a sphere and for a prolate spheroid, does not hold for a cube, and will obviously not hold for an extension of a cube to a prism. The mathematical difference between a sphere and a cube is that the demagnetizing field of a uniformly magnetized sphere is homogeneous, while that of a cube is not. In a uniformly magnetized cube, there is always a transverse demagnetizing field, which may be very small compared with the exchange field, but it is never zero. Therefore, at least when the angles change continuously and do not have certain discrete values, there is always a slight, but finite, tilt out of the uniformly magnetized state. The physical significance of this effect is not very clear, and is quite controversial, depending on the view of whether a cube or a sphere is a better approximation for the behaviour of a small magnetic particle in real life.

Of course, it is not the matter of the extremely small deviation from saturation, which is negligible for *almost* all applications. The controversy is about the paradox of Brown, described in section 9.5. The same reasoning about a non-uniform demagnetizing field applies also to a saturation by the application of a magnetic field, and leads to the conclusion that a cube can never be strictly saturated by any finite, *uniform* field. And if the crystal does not start from saturation, the whole argument of section 9.5 may not be valid. This possibility for the resolution of the paradox had been suggested [515] long before this calculation of the cube, and had already been discussed in the old review [392] of the paradox. It has been strengthened by many observations [517, 518, 519] of domains at the tip of whiskers, which are not driven away by the applied field.

These two problems are not quite the same, and there is a difference between a slightly incomplete saturation in the vicinity of the corners, and a whole reversed domain there. These reversed domains are undoubtedly real, but they only apply to *large* crystals, and their existence may only mean that some samples must be put in a larger field before measuring the nucleation, as discussed in section 9.5. The slight deviations at the edges may or may not be sufficient to start the nucleation process from, and the cube calculation does not really change this controversial issue. It is still the same old problem of which is a better approximation for real particles, and it should be remembered that micromagnetics results do agree with certain experiments on *small* particles, including particles with very odd shapes. Theoretically, the nucleation problem was also solved for infinitely long prisms, which have a square [520] or a rectangular [291, 521] cross-section. However, an infinity is always suspicious in these problems, and at any rate, the infinity in this case only evades, not solves, the problem of the saturation near the edges.

A modification of the cube calculation [522] was used to address the problem of whether the minimal energy non-uniform state is due to the high symmetry of the cube. In order to break this symmetry, spin number 9 of Fig. 10.2 was moved from the position $\mathbf{p}_9 = -(1,1,1)$ to $-\lambda(1,1,1)$, with $\lambda$ having either the value $\frac{1}{2}$ or the value 1.25. For both cases the results for large $a$ were quite different from those for the previous case, but both of them extrapolated to the same results as before for realistic values of several Å for $a$.

# 11

# NUMERICAL MICROMAGNETICS

In all the numerical computations in micromagnetics, nearly all the computer time is spent on computing the magnetostatic energy term for the different magnetization configurations which are being tried. It must be emphasized that this feature is independent of the computational method, and is a result of the inevitable fact that the magnetostatic energy is defined by a six-fold integral, as explained in section 6.3, whereas all other energy terms involve only a three-fold integration. For this reason it is important to choose an efficient and effective method for computing the magnetostatic term, while any numerical analysis method will do for the other terms. The description will, therefore, start from this term.

## 11.1 Magnetostatic Energy

Many of the numerical computations are based on the method developed by LaBonte, for computing first a one-dimensional domain wall [301], and then a two-dimensional one [310].

In two dimensions, the wall structure is assumed to be independent of the dimension $z$ of Fig. 8.1. The wall region, $|x| \leq a$ and $|y| \leq b$, of the $xy$-plane is divided into $N_x \times N_y$ square prisms, of side

$$\Delta = \frac{2b}{N_y} = \frac{2a}{N_x}\,. \tag{11.1.1}$$

The latter relation limits the parameters to those which satisfy $bN_x = aN_y$, but this limitation does not affect the generality of the method, because the width $2a$ can be extended arbitrarily into the domains. The basic assumption is that within each of the prisms, $-a + I\Delta \leq x \leq -a + (I+1)\Delta$, $-b + J\Delta \leq x \leq -b + (J+1)\Delta$, the magnetization does not vary.

For a constant magnetization in each prism, the first term in eqn (6.3.48) vanishes, while $\mathbf{n} \cdot \mathbf{M}$ in the second term is a constant, which can be moved in front of the integral. The integrand then contains an algebraic function, whose integral is known in principle. It is thus possible to obtain an analytic expression for the contribution of each prism to the magnetostatic potential, and the whole potential is the summation of these contributions over all the prisms. It should be noted that this technique transforms a volume-charge contribution, if it exists, to that of surface charge on the four surfaces of the prisms. If the magnetization does not change between one prism and its neighbour, the contribution of a positive charge on one side of the surface between them will be exactly cancelled by the contribution of the negative charge on the other side of the same surface. But if there is some change, the difference between these surface charges expresses, to a first order in small quantities, the contribution of the spatial derivative of the magnetization, which eventually converges to the contribution of the first integral in eqn (6.3.48). Once the total potential is known, it can be substituted in eqn (6.1.2), and then in eqn (6.1.7), to find the total magnetostatic energy. The latter integration can be expressed again as *another* summation of the integration over each of the prisms, which can also be carried out analytically, once $\mathbf{M}$ is moved in front of the integral sign. The result is [310] that the magnetostatic energy per unit wall length in the $z$-direction is

$$F_{\mathrm{M}} = \Delta^2 \sum_{I=1}^{N_x} \sum_{J=1}^{N_y} \Bigg\{ \pi \left[M_x^2(I,J) + M_y^2(I,J)\right] + \frac{1}{2} \sum_{I'=1}^{N_x} \sum_{J'=1}^{N_y} \Bigg[ A_m(I-I',J - J') \left[M_x(I,J)M_x(I',J') - M_y(I,J)M_y(I',J')\right] + C_m(I-I',J - J') \left[M_x(I,J)M_y(I',J') + M_y(I,J)M_x(I',J')\right] \Bigg] \Bigg\}, \quad (11.1.2)$$

where $A_m$ and $C_m$ are evaluated from the above-mentioned integrals. The expressions for these coefficients have been evaluated, and are given, in [310]. Some algebraic transformations have been implemented in order to make these coefficients more suitable for *accurate* numerical computations. They are listed in this improved fashion in [523]. It should be particularly emphasized that the accuracy of these coefficients is very important for a reliable computation. Experience shows that even a rather small inaccuracy in the coefficients can lead the whole computation astray, and end up in a completely wrong configuration. Another way to increase the accuracy is to combine together the contributions from two neighbouring prisms, thus avoiding the subtraction of nearly equal numbers. The details of this modification are described in [524].

The four-times summation in eqn (11.1.2) (which would become six times in three dimensions) is a manifestation of the long-range nature of magnetostatic interaction. Every change of the magnetization in any one prism affects the energy evaluation for all the other prisms. As has already been mentioned, this property makes the computation time much larger than it is for the other energy terms, but there is no way to avoid it. There were certain attempts, reviewed in [270], to approximate this long-range interaction by a local field. They were not successful, and only led to intolerable mistakes. It is just impossible to substitute a short-range force for a long-range one, except for a special case which will be discussed in section 11.3.4.

The most important advantage of this method, namely of writing the magnetostatic energy in the form of eqn (11.1.2), is that the coefficients $A_m$ and $C_m$ need not be evaluated over and over again with every iteration of the minimization process. They are computed, and stored, only *once*, before the actual computation starts. Any other method involves at least some computation equivalent to these coefficients, which has to be carried out again for every iteration. This repetition many times over does make a big difference. Therefore, all other methods which have ever been used, and which are reviewed in [288], either use an impossibly long computation time, or go into rough approximations, or both.

Computing $A_m$ and $C_m$ only once also means that only negligible computer time is spent on them. Therefore, it does not make any difference if they are easy or difficult to compute, nor is there any reason to try to save some time on their computation by introducing certain approximations, or other inaccuracies. In spite of that, there have been many computations, also reviewed in [288], in which the integration over the prism faces was just replaced by the field of a dipole at its centre. This presumed simplification leads to the *same* result of eqn (11.1.2) with slightly different values of $A_m$ and $C_m$. It is not a big numerical difference in the coefficients, but it can make [288] a big difference in the results. And it is completely unnecessary when the expressions for $A_m$ and $C_m$ are known and published, and it is no trouble at all to store them once and for all on the computer. Moreover, many authors are so convinced that it does not make a difference that they do not even mention if they use the correct coefficients of LaBonte, or only an approximation for them. It is thus difficult to evaluate properly many of the results in the literature, or to compare them with each other.

The same argument can also be applied to another approximation, used in many two- and three-dimensional computations. In this approximation, the integration over the square (or the cube) is replaced by the field at its centre, due to the charge on the surfaces around it. Again, this method leads to the same relation as in eqn (11.1.2), only with somewhat different values of $A_m$ and $C_m$, and again it is quite unnecessary. This approximation has been justified [299] by claiming that it is easy to change it from a cube to other geometries. But to me, at least, it seems very strange to introduce an inaccuracy into a calculation only because it is easy to introduce the same inaccuracy into another calculation. It is true, of course, that the method of LaBonte is restricted to one particular geometry, and when another geometry is needed, it is necessary to work out those integrals from the beginning. It may also be necessary to use certain approximations for geometries for which these integrals have not been evaluated. But there is no reason at all to introduce inaccuracies into cases for which the accurate coefficients are already known.

This method has been generalized to three dimensions, with prisms replaced by cubes. The necessary coefficients for writing the magnetostatic interactions among cubes, and the magnetostatic energy term, are listed in [525] enabling computation in three-dimensional bodies, such as prisms. A subdivision into cubes was also used for the study of finite circular cylinders [526], which will be further discussed in section 11.3.6. In as much as the surface details are not important, or when the cubes are small enough, such a subdivision can be used in principle for other shapes as well.

For a sphere, the full three-dimensional case has not been worked out, but two simplified cases exist, for two different physical assumptions. One is a one-dimensional approach, in which the sphere is sliced [252] into planes along the direction of one of the Cartesian coordinates. In the other, a cylindrical symmetry is assumed, and the sphere is subdivided into $N_r \times N_\theta$ *quasi-toroids*,

$$\frac{I-1}{N_\theta} \leq \frac{\theta}{\pi} \leq \frac{I}{N_\theta} \quad \text{and} \quad \frac{J-1}{N_r} \leq \frac{r}{R} \leq \frac{J}{N_r}, \tag{11.1.3}$$

where $r$ and $\theta$ are the polar coordinates, $R$ is the radius of the sphere, and $1 \leq I \leq N_\theta$ and $1 \leq J \leq N_r$ are integers. The assumption here is that the magnetization is constant in each of these quasi-toroids, so that there is no dependence on $\phi$, and the problem is essentially a two-dimensional one. For the magnetostatic energy, however, $M_r(I,J)$ and $M_\theta/\sin\theta(I,J)$ are taken as constants for the integration, which then proceeds as for the prisms of LaBonte. The result is [253]

$$\begin{aligned}
\mathcal{E}_\mathrm{M} = \frac{\pi M_s^2 R^3}{N_r} \Bigg\{ &\sum_{J=2}^{N_r} \sum_{J'=1}^{J-1} \sum_{I=1}^{N_\theta} \sum_{I'=1}^{N_\theta} \Big[ A(I,J,I',J') m_r(I,J) m_r(I',J') \\
&+ B(I,J,I',J') m_r(I,J) m_\theta(I',J') + C(I,J,I',J') m_\theta(I,J) m_r(I',J') \\
&+ D(I,J,I',J') m_\theta(I,J) m_\theta(I',J') \Big] + \sum_{J=1}^{N_r} \sum_{I=1}^{N_\theta} \sum_{I'=1}^{N_\theta} \Big[ E(I,J,I') m_r(I,J) \\
&\times m_r(I',J) + F(I,J,I') m_r(I,J) m_\theta(I',J) + G(I,J,I') m_\theta(I,J) \\
&\times m_\theta(I',J) \Big] + 2 \sin\left(\frac{\pi}{2N_\theta}\right) \sum_{J=1}^{N_r} \left(J^2 - J + \frac{1}{3}\right) \sum_{I=1}^{N_\theta} \sin\left(\frac{\pi(2I-1)}{2N_\theta}\right) \\
&\times [m_r(I,J)]^2 \Bigg\}
\end{aligned} \tag{11.1.4}$$

where the coefficients are obtained from integrations over the surfaces of the quasi-toroids, and are all listed in [253]. They are expressed as series in Legendre polynomials, whose convergence is rather slow, but then the convergence is not important for coefficients which are computed only once, before starting the time-consuming minimizations. Taking $M_\theta/\sin\theta$ as a constant in each of the quasi-toroids, instead of $M_\theta$, makes it possible to carry out all these integrations analytically, but of course it introduces a certain error by neglecting the variation of $\sin\theta$ over the range of such a toroid. Obviously, this error is negligible in principle if the subdivision is sufficiently fine. In practice it works to quite a high accuracy even for moderate $N_\theta$, as has been checked by using this method for computing the magnetostatic energy of several configurations for which the result can also be evaluated analytically.

There is no other three-dimensional body for which these coefficients have been calculated, mostly because the expressions become long and cumbersome. A beginning of a calculation for a finite circular cylinder, which has never been carried as far as yielding a practical form of the coefficients, is found in [259]. Of course, it is always possible to compute all these coefficients numerically for any given body shape, store them and then use them for minimizing the total energy. Actually, such a numerical computation has been carried out for a cube [263], claiming that it is simpler to do it this way, 'although complicated analytic forms for the interaction energy can be obtained'. For a different geometry, this method has only been used in one recent case, which will be discussed at the end of this section. Most of those who carry out such computations still prefer rougher approximations, or completely different numerical analysis methods.

As has already been mentioned, other methods are impractically time consuming. This point must be emphasized again, because computational packages are available nowadays, which can be used to compute magnetic fields, without even knowing what they contain. Also, there are special conferences on magnetic field computations, and for several years now the proceedings of each of them is a thicker book than this one is. Most of these programs, and studies of improved methods, can only be used for 'linear' magnetics, which means that they work only for cases in which eqn (1.1.3) is valid, but there are also many studies in which the magnetic field is computed for *ferromagnetic* materials. Some of them obtain the field **H** from eqn (6.1.2), with the potential obtained either by a numerical solution of the differential equations with their boundary conditions of eqns (6.1.4)–(6.1.6), or by a numerical integration of eqn (6.3.48). Others obtain **B** from the vector potential, briefly mentioned in section 6.1, which has not been used in this book. However, all these methods are devised [527] for computing the field of a given magnetization configuration only *once*. Most of them can also be extended, without difficulty, to the computation of the magnetostatic energy of a given structure, but only if it is done once. In a typical energy minimization, this energy must be computed thousands of times, from a magnetization distribution which keeps changing with every iteration. Present-day computers are too slow for this task, except for those who are ready to spend [528] 'several to many' CPU weeks on solving such a problem. Improving the technique [529, 530] did not change this time-scale sufficiently to make it more useful.

In one recent case [531] the numerical analysis technique has returned to the idea of LaBonte, of computing (and storing) all the coefficients for the magnetostatic energy term, before starting the actual iterative process of the energy minimization. It has not been presented as such, although the authors should have known about this idea, and the computations already carried out by using it in many problems. In this case, the basic discretization element is a tetrahedron, for which the LaBonte coefficients are computed numerically [531] by two different methods. It is claimed to be a very general method, but it has only been used for a certain prism, for which cubic elements, with the already-known coefficients, *could* have been used just as well.

## 11.2 Energy Minimization

Calculation of the other energy terms is straightforward. For the anisotropy energy term, integration of densities such as eqn (5.1.5) or (5.1.8) is just broken into a sum of integrals over individual prisms (or cubes). And since the magnetization is assumed to be a constant in each of these subdivisions, each integration is equal to the area of the prism, or the volume of the cube. The same applies to the energy of interaction with an applied field, if used.

The exchange energy can be obtained directly from eqn (6.2.45), after subtracting the energy of the uniform state, as is done in eqn (6.2.46). There is no contribution from the body of the subdivisions, where the neighbours are parallel to each other. Therefore, the total exchange energy is the sum over all the surfaces between neighbouring subdivisions, of an expression similar to eqn (10.5.42), which is derived directly from the theory of Chapter 2. LaBonte [310] did not do it this way. Instead, he started from the classical expression of eqn (7.1.4), and approximated it for small angles, practically working out the derivation in section 7.1 backwards. The result, however, is the same. For curved subdivisions, such as the quasi-toroids used [253] for a sphere, the procedure is essentially the same. It is also the same for a one-dimensional study [532, 533] of an infinite cylinder, in which the magnetization depends only on $\rho$. It has been presented as a new technique, and named the 'atomic layer model', in order to distinguish it from micromagnetics, that these authors understand to apply only to analytic calculations. The latter case also left out the magnetostatic energy term altogether, because they deal only with the curling 'model'.

The foregoing should be a sufficient outline for writing the total energy in a form which can be coded as a computer subroutine. There are very efficient computer programs for minimizing an expression in a subroutine which depends on several parameters, and it may seem at first sight that there should be no difficulty in minimizing the total energy. However, these programs are limited to minimization with respect to a relatively small number of parameters, usually of the order of 10. They cannot be used for minimizing the energy with respect to the magnetization vector in each of the discrete subdivisions in the present context, because their number is typically in the range of thousands or tens of thousands, and even more. There are two methods in the literature for the minimization itself, and in both of them the expression for the energy is first used to compute the effective magnetic field, defined in eqn (8.3.41), at each of the subdivisions. This field, $H_{\mathrm{eff}}$, is essentially the derivative of the energy with respect to the local magnetization vector, and can be evaluated numerically for each of the subdivisions directly from the energy expression, without going back to the definition in eqn (8.3.41).

In one method, used in *e.g.* [253, 310, 323, 326, 327, 337, 524] and others, the magnetization vector in each subdivision is rotated to the direction of this field, $\mathbf{H}_{\mathrm{eff}}$, at that position. After sweeping through all the subdivisions, the maximum angle of this rotation in any one of them is compared with a preset tolerance. The process of rotating the set of $\mathbf{m}(I, J)$ point by point throughout the grid is continued until this maximum angle is smaller than the required tolerance, at which stage eqn (8.3.40) is obviously fulfilled to within this tolerance. It can be shown that in this method the energy always decreases from one iteration to the next. This property is an advantage for relatively simple energy manifolds, and at least it can never go wrong if there is only one minimum. It may be a disadvantage if there are at least two energy minima, with a certain barrier between them, in which case a start in the vicinity of the higher minimum may converge there, without ever crossing over to the lower minimum.

The other method, used in *e.g.* [299, 324], and in many of the numerical computations which will be discussed later, is to solve numerically the dynamic equation (8.5.50), or one of its variations discussed in section 8.5. For static problems, such as the structure of a stationary domain wall, or the remanent state of a particle, a damping parameter for inclusion in that equation is chosen arbitrarily. The main advantage of this method is that it is readily adapted [327] for real dynamic problems, such as a moving wall, or the variation of the magnetization after a magnetic field is applied. For strictly static magnetization configurations, this method does not seem to have any advantage over the previous one, or at least none has been claimed in any of the publications describing it. It may also be time consuming, because intermediate stages evolved in time are of no particular interest in this kind of computation. In an improved variation of this method [531], the dynamic equation is written for every one of the subdivisions separately, and not for the whole sample, and the time-step is adjusted in each iteration to be as large as possible, as long as the energy is not allowed to *increase* in any single step. Since this purpose is automatically achieved by the other method, described in the foregoing, the advantage of this method is doubtful at best, for any static problem. Of course, it is irreplaceable when the real dynamics is sought, except for the case of a wall moving uniformly at a constant speed, for which the static techniques can be applied, as discussed in section 10.4, and as done in *e.g.* [337]. The best description of the details of integrating the time-dependent equation can be found in [534], and a comparison of the different methods for this integration is given in [535].

In either case, the boundary conditions have to be enforced by choosing special rules for the magnetization in the subdivisions on the surface. For example, when a domain wall is supposed to end in a domain on both sides, the magnetization in the first and the last row of prisms is always kept fixed in the appropriate directions [310]. The boundary condition $\partial \mathbf{M}/\partial \mathbf{n} = 0$ can be enforced by adding extra subdivisions just outside the material, in which the magnetization is [536] a mirror image of those just inside, or by other [310] methods.

The convergence of either method is quite slow. It has been noted [537] that a much faster convergence can be achieved very often by grouping subdivisions to be changed together at each iteration, instead of setting the magnetization in one subdivision at a time. Therefore, the number of iterations, and hence the computation time, can be much reduced, if a certain pattern can be identified, for a cooperative magnetic change, or a *mode*, in a large group. This method, however, has only been used in a special case [537] of a domain wall motion, for which it is rather easy to identify the parts of the wall, and make them move together. Identifying the relevant 'modes' in other cases is not that simple, and it still takes a deeper study before this method may be more generally used.

A more drastic approach of this kind is to group together several of the subdivisions *permanently* and treat them as one entity, by assuming that the magnetization is always the same in each member of the group. If done for the whole sample, this assumption only means a crude mesh, which is easier to solve than a fine mesh, but leads to a lower accuracy. However, when this technique is used selectively, it may save computations without losing the accuracy. It was actually used [323] in the computation of the domain wall in very thick films, for which it can be safely assumed that the magnetization varies much more rapidly near the surfaces than in the middle of the film. Therefore, a fine square mesh was taken [323] only near the surfaces. Farther away from the surfaces, several squares were grouped together into rather elongated rectangles. Of course, there is a big difference between the qualitative statement that the variation is more rapid near the surface, and the quantitative choice of a particular size for the rectangles away from the surface. This particular case was justified by its passing the self-consistency test of section 8.4, which makes it at least better than some wild guesses, which are also being published. But this test can only be applied after all the computations are done. There should be some way for a quantitative justification of the use of this technique, preferably before starting the main part of the iterations, but this part has never been done.

This drawback is only one of several unknown and unestablished points and assumptions which are just being used in micromagnetics computations with no justification. In the old days, when computer resources were limited and expensive, programming used to be approached much more carefully than in more recent years. It was taken for granted, for example, that a program must be checked by running at least one case which can be solved analytically, and comparing the results. Of course, it is not practical any more to do it for every problem, but some sort of checking is essential, even if the full three-dimensional problem is studied without approximations, but even more so when some 'simplifying assumptions' are introduced. The other extreme of having no checks at all is too dangerous, and its results are never reliable. The computer is a very useful and powerful *tool*, and it can do wonders if properly used. But it is certainly not a substitute for thinking, and it is too common a mistake to let the computer make all the decisions. It has become much too easy nowadays to write a program and run it, so that one often wonders if certain published results reflect some physical reality, or are merely the effect of an overlooked error in the programming, or in the logic leading to that program. It may also be just due to an approximation which the programmer does not stop to think about, or which may have been copied from another work, in which that approximation is justified, and yet it may not be justified in the new context of another particular problem. Several such difficulties in the particular case of micromagnetics computations, which still await a solution, are listed here. Some of them *may* have been addressed, or even solved, in some of the studies. But they are not mentioned in the publications, which may be because many of the results are being published in conference proceedings with strictly enforced size limits, in which the most interesting part is often omitted.

1. The size of subdivisions is chosen arbitrarily in most of the reported computations, and in many cases this size is not even *mentioned*, as if it *were* an unimportant parameter. No clear-cut criterion is known, but there is an obvious guideline between two limits. On the one hand, the mesh should not be too fine, so that the approximation of a continuous material, discussed in Chapter 7, is still valid. On the other hand, the mesh should be sufficiently fine to allow the magnetostatic energy term to develop fully the complicated structures that it usually prefers. If the subdivisions are too crude, the magnetostatic energy is too high, instead of becoming negligible as it normally is, besides the possible introduction [538, 539] of discontinuities, and converging to [540] a wrong result. A crude subdivision may be adequate for

cases such as [541, 542], when the walls and the domains are studied without the details of the walls. But it is certainly not justified when the wall details are wanted, such as in [324] where a 10 $\mu$m thick film is divided into 128 prisms, making the prism size about 78 nm. This prism size is *an order of magnitude* larger than in some of the previous works, and it is only used in order to claim some results for thicker films than ever done before. This article [324] does not even comment on this choice of the prism size, and it has other faults too, such as not even mentioning if the magnetostatic energy term is computed as in section 11.1, or by some approximation.

The safest method is to compute for a certain mesh, then subdivide it further and repeat the computations, to see the effect of the finer mesh. This procedure is almost standard in less complicated computations. Thus for example in [529] the accuracy was checked for different meshes by comparing the field of a saturated sphere with the known analytic result. This check is certainly much better than no check at all, but it is inadequate, because the field of a saturated sphere may not be sufficiently well connected with the field of the more complex configurations. For the actual energy minimization, 2639 elements were just chosen [529] in the sphere, without trying any other number. In a Ritz model for a sphere, for which only the exchange term was computed numerically, while the magnetostatic energy could be solved analytically, it was possible [539] to start with a $31\times31$ subdivision, then increase it to $33\times33$, and finally use $66\times66$. However, such a procedure is not practical for problems in which the computer resources are pushed to their maximum limit, as does often happen whenever magnetostatic computations are involved. Thus, a detailed study of the accuracy has been reported for some numerical computations [531, 543] of the LaBonte coefficients, but never of the complex minimization itself. Of course it is important to make sure that the coefficients are accurate, but it is not sufficient. Even in a case such as [531] which is compared directly with experiment, it is also necessary to check that the rest of the theory is done properly.

2. The approach to infinity can never be well-defined numerically. It can only be expected to be taken sufficiently far from the main body of the computations. Again, there is no clear-cut guideline on how far is sufficiently far, and again it is necessary to try more than one value, and yet such a check has not been reported in *any* of the published studies. In domain wall computations, for example, the domains are expected to start at $x = \pm a$ of Fig. 8.1, where $a$ should be large enough to allow a full spread of the wall. And yet, this $a$ is just chosen arbitrarily, without asking if it is large enough, and without trying to see the effect of using a larger $a$. In a numerical solution of

the differential equations (6.1.4)–(6.1.6), the behaviour at infinity is an important boundary condition, but should the infinity be taken as 5 times the radius of the ferromagnetic body, or 50 times, or what? In studying a sphere, it is mentioned that the potential need not be computed only in the sphere, but also in a 'much larger surrounding region of free space' [528], but no indication is given on how large is 'much larger'. For the actual computations, 2639 elements were used in the sphere, and 7534 outside, and if any other numbers were tried, they are not mentioned in the publication.

3. Self-consistency tests, such as those discussed in section 8.4, are very important to check that the results make sense, and are not merely the result of some mistakes, or of wrong approximations. Using them is the only known way to claim that for a certain range of the film thickness, the neglected third dimension cannot change the computed two-dimensional wall energy very considerably. No improvement of the computational accuracy by itself can reveal this information, even if those computations are 100% reliable and free from errors. The difficulty is that these tests have only been developed for static, or for uniformly moving, domain walls. It is, therefore, important to develop similar tests for other cases, *and* to use the existing tests wherever they apply. Any publication in which the self-consistency test is ignored should be suspected and not relied upon. The test is *quantitative*, and an article such as [302], which only says that the results were 'tested' this way, without specifying the resulting numbers, looks strange at best.

4. The way computational results are presented is the most difficult problem in trying to extract information from published results. The standard method for presenting two-dimensional walls is by plots such as the one in Fig. 8.3. They do not reveal all the fine details, but at least they give a good idea of the main structure. For three-dimensional structures, this method is definitely not good enough, but no better way has been developed yet. The three-dimensional pictures made out of two-dimensional arrows in [528, 530] are incomprehensible. Even relatively simple structures, such as those of [526], or of others which will be discussed in the next section, are not much clearer, although a *cut* [544] can sometimes help to see more details. Such plots are relatively easy to follow in the oral presentations in conferences, when different directions are shown in different *colours*, but these colours are usually lost in the published proceedings. The problem will be at least partially solved if these figures are published in colour, but there seem to be some technical difficulties, which also apply to the presentation of certain experimental results, such as [545]. Although there are already several publications in colour, *e.g.*

[546, 547, 548], they are still quite rare, and the whole problem needs a more drastic solution. Actually, the best way to present numerical results is to build an analytic approximation to them, as discussed in section 8.2. This method, however, is very difficult, and has hardly ever been used.

5. The convergence criterion for terminating the iterations is the least-defined parameter in micromagnetics computations. Different authors use different criteria, and their values seem to be chosen arbitrarily. There is never any attempt in the publication to justify the value used in a particular study, and in many cases this number is not even mentioned there. At a first glance it might seem that a rather large value of this criterion is adequate if only a rough estimation is wanted, but experience shows that this criterion should be *much* smaller than the required relative accuracy of structure or energy. In many cases a choice of, say, $10^{-4}$ may converge to a structure which can be changed drastically if the computation is continued with a convergence criterion of $10^{-6}$. In one (unpublished) case, the energy of the system changed by about 3% when the maximum angle was changed from $10^{-4}$ to $10^{-6}$. It may not be the same change in other cases, but the mere fact that it *can* happen should warn workers that they must be very careful. More than one value should always be tried, and it is wrong to rely on any one guess.

Because of all these difficulties and uncertainties, many of the results in the literature are doubtful. In the next section, some results that seem more reliable than others will be summarized, but even these results may change with newer research. Details which cannot be given here can be found in the cited publications.

## 11.3 Computational Results

### 11.3.1 *Domain Walls*

Most of the computational results for static, 180° walls have already been listed in section 8.2, and will only be briefly reviewed here. There is neither experimental nor theoretical information on the wall structure or energy in bulk materials. The best estimation of the wall energy in the bulk is still based on the one-dimensional calculation of the Landau and Lifshitz wall, described in section 7.2, which is most probably wrong. Computer resources are exhausted at an iron film thickness of about 3 to $4\,\mu m$ and there is no way to tell what happens in a larger thickness. At about that thickness, there *seems* to be a transition [323] from a thin film configuration (where the curve on which $M_z = 0$ is shaped like the letter 'C') to a different configuration for which that curve is shaped like the letter 'S'. If this transition exists, it may be an indication for the wall structure in the bulk. However, even in the thickest films which could be computed, there was no changeover from the C- to the S-type structure. The energies of these structures being very nearly the same at this thickness, either one of them could be obtained, depending on the symmetry of the starting configuration [323].

For an intermediate thickness, between that of a few $\mu$m and that of about 0.1 $\mu$m, or somewhat less, the situation is quite clear. There are some two-dimensional computations of the wall structure and energy, which give the results as described in section 8.2. There is also a three-dimensional computation [549] in this thickness region, namely up to a thickness of 500 nm, which is taken to be already 'bulk'. But this study uses a rough approximation for the magnetostatic energy, and very crude subdivisions. For still thinner films, a three-dimensional computation becomes essential, but none of the published results is clear cut, because they are based on an approximation for the magnetostatic energy term. In particular, there is no computation that shows a transition from the Bloch wall to the cross-tie wall at any thickness which may be compatible with the experimental observations on metallic films. A changeover was sought [550] from the two-dimensional Bloch wall to a symmetric Néel wall. Those computations did not find such a well-defined transition, and encountered instead a wide thickness region in which none of these two structures was stable. This result is not surprising in view of the fact that cross-tie walls are observed in *that* size region, and this structure cannot appear in a computation which is constrained to be two dimensional. Ignoring this experimental fact, some other explanation was sought in [550], with an attempt to enforce that transition by an application of a magnetic field.

In the case of ferrites, the periodicity along $z$ of Fig. 8.1 is usually replaced by a kink in the wall, which is less pronounced than the cross-tie structure of permalloy films. For this case, many more details have been computed, and compared with experiments [551], but only for a small *part* of the wall, in the vicinity of the change of wall chirality along $z$. Some two-dimensional computations [552, 553] point the way on how to take into account the dependence on the third dimension, $z$, by summing over the periodicity along that direction. However, these computations did not really address the problem of a cross-tie wall, and were only done for the periodic structure known as the 'strong stripe domains'. The theory of the Néel wall in very thin films is not very clear either. In particular, there is no computation of such a wall which obeys any self-consistency test.

A wall must develop an extra asymmetry when it moves, as explained in section 10.4. This asymmetry can be clearly seen in numerical solutions of eqn (8.5.52) such as [554, 555]. These computations have also been done [556] for the case of a two-dimensional wall moving through a region in which the anisotropy constant changes, as in similar, but one-dimensional, calculations discussed in section 10.2. These studies, as well as the one which addresses the effect of eddy currents on the wall motion [557], are restricted to one- or two-dimensional structures in rather thin films. The conclusions [557] about a thickness of 1 $\mu$m, for example, are *not* based on computations at this film thickness. They are actually obtained by scaling the physical parameters of the material, and computing for 0.1 $\mu$m film thickness, without paying any attention to the changing subdivision size, or convergence criterion, or any of the other points listed in the previous section. A much higher accuracy for thicker films was reported [337], but that study was only for uniformly moving walls, and could not be extended to large velocities. It could not be extended to very thick films either, because of the same limitations of the computer resources, mentioned in the foregoing for the case of a static wall. A dynamic study of a 2 $\mu$m thick film showed [558] a transition from the 'C' to the 'S' shape of the wall structure, mentioned in the foregoing, but this result was obtained with a crude subdivision. Similarly, studies of the motion [559, 560] of a *part* of a three-dimensional wall is subject to the same limitations mentioned in the foregoing for the statics of such a wall. A particularly interesting trick [561] is to apply a *periodic* field to the computed wall, to stabilize its periodic structure.

As has been mentioned in section 8.2, a very good approximation for the wall structure in intermediate film thickness can be obtained by enforcing zero magnetostatic energy. Such computations are much faster than any of the others described in this chapter, but they involve the problem of fitting $\nabla \cdot \mathbf{M} = 0$ together with the constraint $|\mathbf{M}| = M_s$. In a variation of this method [562], used for three-dimensional computations, the first constraint $\nabla \cdot \mathbf{M} = 0$ was maintained at each grid point, but the second one, $|\mathbf{M}| = M_s$, was not. Instead, $|\mathbf{M}|$ was allowed to change from one point to another, with an eventual convergence towards the same value everywhere. This technique was later ignored, and never used again.

Some computations of domain wall structures have also been extended for looking into the analysis of magnetic force microscopy (MFM) data. In one extreme [563, 564, 565] the measured magnetic configuration was taken as constant, and the resulting magnetic configuration in the MFM tip was computed. In the other extreme [566] the fine details of a two-dimensional wall were computed, taking into account the field due to the measuring tip, but the details of the tip magnetization were neglected. This magnetization was assumed to be constant within the (*spherical*) tip, and was allowed only one degree of freedom, for its direction to be rotated in such a way that the total energy of tip and sample is a minimum. Such an approximation may be justified by measurements of the magnetic field in the vicinity of a *sharp* MFM tip, which seem to be well approximated [567] by a (dipolar) field of a sphere. Of course, these measurements were made without a sample, and may not be indicative of what happens when the sample is nearby. More detailed computations of an energy minimum of the total energy, allowing a variable magnetic distribution in both the sample and the tip, are hinted at in Ref. 19 of [566], but neither the details of the computations, nor their results, were ever published.

The conclusion from [566] was that the measurement has a negligible effect on the measured pattern. An opposite conclusion, that the existence of the measuring tip has a very large effect on the measured magnetization pattern, was reached in [536]. The reason for these different conclusions is not known.

Finally, it should be emphasized that this discussion has been limited to a 180° wall, ignoring other walls, such as 90° ones. When the latter are straight lines, their computation is rather similar to those presented here. There is also a large number of works on the statics [330] and dynamics [568] of *curved* walls, which are outside the scope of this book.

### 11.3.2 *Sphere*

The theoretical remanent state of a ferromagnetic sphere has been discussed in section 10.5.1. It was proved there that below a certain 'critical' radius, the lowest energy is that of the uniformly magnetized state, but this radius was only given there as a reliable lower bound. The analytically calculated upper bounds turned out to be much too large, and their evaluation had to be done by numerical methods.

Using the method described in section 11.1, of subdividing a sphere into the quasi-toroids defined by eqn (11.1.3), the lowest-energy configuration was computed for one case [253] of a uniaxial anisotropy and two cases [509] of a cubic anisotropy. It should be emphasized that because these computations are constrained to cylindrically symmetric configurations, the result is in principle an upper bound to the energy of *all* possible configurations. Therefore, the critical radius which they imply is also an upper bound to the true radius. In the two computed cases for cubic materials, the upper bounds thus obtained were only about 10% larger than the lower bounds. Together they thus define the critical radius to within the accuracy with which the values of the physical parameters are known. For uniaxial cobalt, however, the difference is much larger. The computed upper bound [253] of 34.1 nm is smaller than all upper bounds computed before, but it is still three times the value of 11.5 nm, which eqn (10.5.34) yields for the physical parameters of this material. It is still a rough estimation, and it is not clear what happens in the case of very large anisotropies, for which the upper and lower bounds of Brown are particularly different from each other.

The magnetization configuration just above the critical radius is made essentially of two curved domains, although the word may not be proper in this context, because the 'wall' between these domains extends over an appreciable part of the sphere. These domains have a cylindrical symmetry, not only in the computations for which this symmetry is assumed, but also in a full, three-dimensional computation [528, 530] with no constraints, at least in as much as it is possible to see in the published configuration.

It seems that the whole configuration is very well approximated by the Ritz model [462], as defined in eqn (10.5.40). The latter was originally designed [462] to study the magnetization reversal by the curling mode beyond the nucleation field. But since the study of curling had already been forgotten, this configuration was presented [528] as a new type, called a 'vortex' structure. The above-mentioned two domains should be imagined as oppositely magnetized, in directions parallel to an easy anisotropy axis, and the wall between them is mostly magnetized in circles.

If the anisotropy vanishes, the inner domain is uniformly magnetized, in the direction $z$ of the previously applied field, or very nearly so. The outer domain is mostly magnetized in circles, namely $M_\phi$ is almost equal to $M_s$, with a small tilt towards $z$. In both domains, the cylindrical component $M_\rho$ is very small. The average $M_z$ in the remanent state is quite close to $M_s$ for a radius just above the critical value, and is rather large even for considerably larger radii. All this description is actually based only on those computations which assume a cylindrical symmetry [569] to start with, because the results of the full, three-dimensional computations [528, 530] are not very clear in the published figures. The latter were time-consuming computations, which approached the sphere as a limit of a polyhedron with very many faces, for which it was difficult to obtain a sufficient accuracy for parameters known from the analytic analysis, such as the nucleation field. It should be noted, though, that it is still an open question [570] if a sphere is a better approximation than a polyhedron for the real physical situation, when it comes to very small particles, which do not contain very many atoms.

The direction $z$ does not have any meaning when both the field and the anisotropy are zero. Therefore, the described two-domain structure has the same energy for any direction, and can be rotated from one direction to another without any energy barrier. Therefore, an infinitesimal magnetic field should be able to turn a large part of the magnetization into that field's direction. In other words, the initial susceptibility is infinite, at least theoretically. Such an infinity (or a very large value in a non-ideal case) could be very useful for transformers, or read heads, if it were possible to make a sufficiently good approximation for such small, and isotropic, spheres in real life. This idea is not a complete fantasy, now that very small particles, with an almost spherical shape, have been made [473, 474]. Their anisotropy is not zero, and they may not be small enough yet, but these further steps may still be possible.

For such an isotropic sphere, if it is smaller than the critical size of eqn (9.2.29), the nucleation should be by coherent rotation. The nucleation field for this mode is 0 according to eqn (9.2.10), because $K_1 = 0$ and $N_x = N_z$. In this case, the Stoner–Wohlfarth model of section 5.4 should apply, with all the arguments as presented there. The whole magnetization curve then consists of three branches. First there is the line $M_z = M_s$, of a saturation in the $+z$-direction, from a high field down to zero field. At $H = 0$ there is one jump to a saturation in the $-z$-direction, which is followed by the branch $M_z = -M_s$ for all negative fields. This curve is reversible, namely it is followed again for a field increasing from negative values. The coercivity is thus zero, the remanent magnetization equals $M_s$, and the initial susceptibility is infinite.

If the radius of the sphere is just above that critical size, the remanent state consists of the two domains described in the foregoing, for which $M_r = \langle M_z \rangle < M_s$. There is still no hysteresis, and the magnetization curve is completely reversible, according to the collective information from Ritz models [462, 539, 571] and numerical computations [528, 530, 569]. The saturation $M_z = M_s$ is followed from the high field, down to the curling nucleation field of eqn (9.2.28), which is *positive* in this case of $K_1 = 0$. Then there is a curve, which is nearly but not quite a straight line, leading from the nucleation field to a positive remanence value, $M_r$. At $H = 0$ there is one jump to $-M_r$, and the negative part of the magnetization curve is symmetric to its positive part. The coercivity is thus zero, and the initial susceptibility is still infinite, although the initial jump brings the average magnetization only to $M_r$, and not all the way to $M_s$. With increasing radius, $H_n$ increases, and $M_r$ decreases, until at some size the line from the nucleation point leads to the point $\langle M_z \rangle = 0$. For this size, and larger ones, the infinity disappears, the initial susceptibility becomes finite, tending to $4\pi/3$, and $M_r = 0$.

The theory is much less developed for $K_1 \neq 0$, and definitely needs more studies. There are very few computations [528, 530], and even for them most of the details have not been published. They are all for a sphere whose radius is such that $M_r/M_s \approx 0.2$. It seems that when a uniaxial anisotropy is added to this sphere, but so that the nucleation field is still positive, $M_r$ vanishes. It is rather difficult to understand the reason for this effect, and even more difficult to find out if this behaviour is typical, or if it applies only to a special case. There is no hysteresis around $H = 0$ in these computed curves, and the coercivity is zero. There is, however, a certain hysteresis loop for numerically large, positive and negative, applied fields, near the nucleation and the approach to saturation. A similar hysteresis near saturation has been observed in single-crystal iron *films* [572]. It was then presented as a verification of a certain theory of phase transitions, which predicted that this hysteresis should only exist for cubic anisotropy, and only for a field applied in the [111] direction. The theoretical existence of this phenomenon in uniaxial spheres definitely proves that there may be mechanisms other than this phase transition that can account for its observation, but the details have not been worked out.

All these results, as well as all of chapter 9, are limited to the case of zero surface anisotropy. The latter may play an important role in real particles, but its theory has not been sufficiently developed yet. Except for some approximations for trivial cases, there are only two articles about a sphere with a certain form of surface anisotropy, and both deal only with the nucleation field. One is an analytic evaluation [146] of $H_n$ by the curling mode. The other is a numerical computation [573] of $H_n$ for another mode, replacing the coherent rotation which is not an eigenmode once such an anisotropy is introduced. Mathematically, it is somewhat equivalent to the buckling mode in an infinite cylinder, or in elongated prolate spheroids, if there is such a mode in them. For lack of a better name, this mode may, therefore, be called buckling. Surface anisotropy has also been included in some detailed computations [348] of the whole hysteresis curve. Formally they were carried out for several two- and three-dimensional shapes, but since the magnetostatic energy was not included, the actual shape cannot really play any significant role.

### 11.3.3 *Prolate Spheroid*

Nothing equivalent to eqn (11.1.4) has been designed for any ellipsoid other than a sphere, which makes it difficult to compute any of its properties. For the prolate spheroid there is some, but very limited, guidance from analytic calculations. It is known that the uniformly magnetized state is the lowest-energy remanent state below a certain size, but only the lower bound of eqn (10.5.41) can be given for that size, and no reliable upper bound has ever been calculated. The nucleation has been proved to be by coherent rotation or by curling, but the possibility of a third mode has *not* been ruled out. It may exist for very long and very narrow prolate spheroids, in the regions marked so in Fig. 9.2.

There are only two numerical studies [530, 574] of prolate spheroids, and only for an aspect ratio of 2:1. They used time-consuming methods to compute the hysteresis curve for several particular radii, with and without anisotropy. The nucleation field increased [574] by exactly $2K_1/M_s$ from its value for $K_1 = 0$ but the coercivity increased by less than $2K_1/M_s$.

The smallest semi-major axis, $a$, tried is reported as $C/(4\pi a^2) = 0.04$, which is probably meant to be $C/(4\pi a^2 M_s^2) = 0.04$. In the notation of eqn (9.2.20), the semi-minor axis is $S = 1.0$, which is in the region of coherent rotation in Fig. 9.2. For this radius, there was one jump at nucleation from a positive to a negative saturation, and 'it was not clear visually whether curling was very slight or nonexistent' [574]. The next size they tried was 0.02 in those units, which should mean $S = 1.4$, well above the transition to the curling mode in Fig. 9.2. A well-developed curling structure was found to nucleate for this size, but no attempt was made to check the change-over size. Neither was there any attempt to compare the computed curling nucleation fields with the analytic expression, not even in order to check the accuracy of the computations.

The curling configuration at this size, which probably corresponds to $S = 1.4$, was [530, 574] quite similar to that of a sphere. The hysteresis curve consisted of a continuous change from nucleation down to a certain field, at which there was a jump all the way to the negative saturation. For a certain larger radius, this jump brought the magnetization into a curling state which was a mirror image of the structure from which the jump started. More complex behaviours were found [530, 574] for still larger particles, including some form of two domains with a complex wall between them. However, all these cases were studied for one specific radius each, with no attempt to follow the transition from one case to another. Neither was there any attempt to follow the drastic and qualitative change of the hysteresis between a sphere and the particular aspect ratio of 2:1. It is probably impossible to do any more by this method, which is very wasteful in computer time, but one would still expect such reports to give some idea of the estimated accuracy, or convergence criterion, or at least how the computer was kept from jumping from one possible solution to another for some of the radii. The published articles do not even mention the relation between the particle size and the discretization size used in this work.

### 11.3.4 *Thin Films*

Some computations try to account for the experimental hysteresis curves of thin films, and relate them to some measurable properties. One way to do it is to consider the film as a collection of non-interacting particles. By pretending that the *whole* hysteresis curve of each particle is known, a certain average is computed over the distribution of the particles. Such computations, *e.g.* [575], sometimes reproduce the measured properties, at least qualitatively. Of course, this process is not limited to thin films, and has been used for other systems of particles, including [214] models for interacting particles. In this context it is also worth noting a computational scheme [576] for finding the whole magnetization configuration in a thin film from experimental data of Lorentz microscopy, and of the measurement of the magnetic field pattern outside the film.

Other computations are concerned with the domains in thin films, and are done on a rough scale which cannot take into account the walls between the domains. For this particular purpose, it is convenient to expand the magnetization in a Fourier series [577],

$$\mathbf{M}(\mathbf{r}) = \sum_{\mathbf{k}} \boldsymbol{\mu}_{\mathbf{k}} e^{i\mathbf{k}\cdot\mathbf{r}}, \tag{11.3.5}$$

where $\mathbf{k}$ has certain discrete values in the $xy$-plane, such as an $x$-component of the form $2n\pi/L_x$ for integral values of $n$. The coefficients $\boldsymbol{\mu}_{\mathbf{k}}$ may be constants or functions of $z$, and in either case the solution of the potential problem from the differential equations of section 6.1, by expanding the potential in a similar Fourier series, is very much simplified. The difficulty is that it is usually impossible to fit the expansion in eqn (11.3.5) with the constraint of eqn (7.1.7). For this reason, this method could actually be used only when the walls were taken to be step functions [577], or in some similar applications reviewed in [288].

A similar technique has also been used in many other computations which impose a false periodicity, in order to use the fast Fourier transforms which reduce [546] the computation time by a very large factor. It has been argued [546] that this approximation is justified for a two-dimensional system, because 'the effective range of demagnetizing field is comparable to the film thickness'. This argument sounds quite convincing, but it would have been more convincing if there was any case for which a computation using the fast Fourier transform was compared quantitatively with a more rigorous computation of the same case. In a variation [541] of the method, the potential is expanded in a Fourier series, but the magnetization is not. The expansion is used to eliminate the potential outside, $U_{\text{out}}$, and formulate the whole problem within the (infinite) ferromagnetic film, with the boundary conditions expressed as an integral over the upper and lower surfaces of the film. But even this formulation is actually applicable only to a *periodic* domain structure, for which the potential is really periodic. Or at least it has only been used for such a periodic configuration, in a crude-meshed computation [541] of domains, that oversimplifies the structure of the walls separating those domains. In a certain study [578] of very thin films, it was found adequate to neglect the magnetostatic energy altogether, except for restricting the magnetization to be in the plane of the film.

Another method, which is also confined to two dimensions only, takes advantage of the upper bound to the magnetostatic energy in eqn (7.3.46). In principle, the magnetostatic energy term can be replaced by this upper bound, and the total energy can be minimized with respect to both $\mathbf{M}$ and $\mathbf{A}$. Minimization with respect to $\mathbf{A}$ makes the upper bound converge towards the true magnetostatic energy. Therefore, this minimization leads to the true minimal energy configuration, while $\mathbf{A}$ becomes the true vector potential of the problem. The advantage of this technique (first suggested in [579], but never carried out by these authors for any particular case) is that the six-fold integral for the magnetostatic energy is replaced by a three-fold one, which should reduce the computation time enormously. Increasing the number of variables from the two independent components of $\mathbf{M}$ to the five components of both $\mathbf{M}$ and $\mathbf{A}$ is a small price to pay for this localization of the problem, which does not call for evaluating interactions among different discretization points. The disadvantage, already discussed in section 7.3.4, is that eqn (7.3.46) contains an integral over the whole space, which in practice means carrying the integration over a much larger volume than the sample, thus increasing the number of grid points far beyond those used in more conventional methods. This difficulty makes this method impractical for almost any three-dimensional problem. In two dimensions, however, the integral over the outer space may be evaluated by a conformal mapping of it, making this method practical and convenient. It has thus been used, for example, in the study [580] of the effect of exchange coupling across grain boundaries of the nucleation field of a film.

A particularly popular computational method subdivides the film into either two-dimensional hexagons, or three-dimensional hexagonal columns. The hexagonal 'grains' are assumed to be somewhat separated from each other, so that the exchange coupling between them is smaller than it is in a continuous film. In practice it actually means that the expression [581] for the exchange energy between neighbouring grains has the *same* functional form as described in section 11.1 for subdividing the sample into prisms or cubes. Of course, a hexagon has more neighbours to interact with than a square. But besides this difference in the summation, the only difference is that the numerical value of the exchange constant, $C$, is taken to be somewhere between 0 and the experimental value for a continuous film. The actual value for this effective $C$ is often picked at random, although it is possible [582] to estimate it from measurements of the domain wall energy. The coefficients for the tensor describing the magnetostatic energy have been evaluated analytically [583] by integrating the surface charge on the faces of the hexagons, in a similar way to that used for the prisms discussed in section 11.1, but many computations use an approximation for those coefficients.

This model was used for two- and three-dimensional computations of both the statics and the dynamics of magnetization patterns that do not involve the fine details of the walls between the domains. These include, for example, the magnetization ripple, or the formation and rearrangement of domains and other configurations, and their effect on the full hysteresis curve for the whole film, as well as a simulation of the magnetic recording process. The details of these computations, and all their results, are fully described in [581]. This detailed description, however, does not give any information on the choice of the discretization size, or of the convergence criterion for the distribution in *space*. It does specify that the step size in *time* was chosen so that the maximum relative change of the magnetization in that step was kept at approximately $10^{-4}$. It is also mentioned that a magnetization configuration was accepted as a minimum only after trying to add to it small random perturbations, and checking that it evolved back to the initially obtained configuration. It is a nice check on the validity of the minimization process, which should be adopted by other workers as well, because it guarantees against *converging* into a saddle point in the energy manifold. It does *not* guarantee, however, against the computations converging into a high-energy minimum when a lower minimum is available, especially since these computations are not made to start from a well-defined nucleation. The use of this subdivision into hexagons continues, *e.g.* in studying [584] the effect of grain boundaries, or of [585] a random anisotropy. It was also extended [586] to elongated hexagons.

Some computations address the magnetization configuration inside such a hexagon [587], or a one-dimensional stacking [547, 588, 589, 590] of similar hexagons. There are also computations of rectangular particles made out of thin films [530, 591, 592, 593, 594, 595], or a *pair* [596, 597, 598, 599] of such rectangles, and various other two-dimensional [600] shapes. A class by itself are planar arrays of thin film particles [441, 601, 602, 603] for which the computations are being compared with experimental studies, and the agreement seems to be quite good. The published articles, however, are not adequate yet for drawing any conclusions.

### 11.3.5 *Prism*

As mentioned in section 10.5.3, Brown's 'fundamental theorem' does not hold for a cube, and there is no size under which the remanent state of a cube will be the uniformly magnetized one. Actually, this conclusion could have been drawn directly from eqns (8.3.37) and (8.3.38) which include, in principle, all the minimum energy states. If the uniform state, $m_z = 1$ and $m_x = m_y = 0$, is substituted in these equations, it is seen that they can be fulfilled only if $H_x = H_y = 0$. And these relations can only be fulfilled if either the body is an ellipsoid, or the applied field is not homogeneous. However, this property of a cube was not seriously discussed until some computations [604] revealed the equilibrium states of such a cube, and made the problem quantitative.

Before discussing these results, a semantic point needs to be clarified. When Brown [520] looked into the nucleation in an infinitely long prism, he noted that all the possible eigenfunctions could be arranged in groups, according to the symmetry class of the components $m_x$ and $m_y$. This classification was then extended [291] for the case of a rectangular prism, $-a \leq x \leq a$, $-b \leq y \leq b$, with $z$ extending all the way to infinity. In particular, the nucleation mode for which $m_x$ is an even function in $x$ and an odd function in $y$, while $m_y$ is odd in $x$ and even in $y$, was given the name 'curling', because it is basically made out of a magnetization vector which goes around the prism in quasi-circles. It is topologically the same structure as that of the curling mode in a sphere, or in other ellipsoids. The mode for which $m_x$ is odd in $x$ and even in $y$, while $m_y$ is even in $x$ and odd in $y$, looks like the vectors describing the flow out of a centre. It was given the name 'anticurling', because all the above-mentioned symmetries are opposite to those of the curling mode. For some readers it may be easier to visualize this structure by following the equations for the components of $\mathbf{m}_2$, $\mathbf{m}_3$, etc., as specified in section 10.5.3. When these same magnetization structures were rediscovered as possible minimal energy states [604] in zero applied field, it was somehow felt necessary to give them new names. One of the reasons may have been an attempt [604] to draw a distinctive line between the 'classical' micromagnetics, and the new, numerical studies. The *stated* justification was [604] that the name 'curling' was 'commonly taken to mean the reversal mode' and as such it did not fit as a name for a *state*. Therefore, the curling was renamed the 'vortex configuration' and the anticurling was renamed the 'flower' state. I never could figure out why two magnetization configurations that look the same cannot be called by the same name, but the new names have stuck in the meantime [490], and are commonly used by many.

The result of these computations was that in zero applied field the lowest-energy state was that of the anticurling, or vortex, below a certain size. This statement was formulated more cautiously in [604], because the accuracy was not really sufficient for very small cubes, but this result has been confirmed [516] in the meantime. Of course, the actual configuration in extremely small particles is of very little interest, because the system will not stay there anyway. For a very small size, the superparamagnetism of section 5.2 should take place.

In this structure, the largest deviation from the easy-axis direction $z$ is for the magnetization at the corners of the cube. The magnetization in the internal part of the cube seems to be parallel to $z$. With increasing cube size, the magnetization at the corners of the cube tilts farther in the radial direction, reaching quite large angles from the direction of $z$. However, the magnetization inside the cube is not affected, so that the decrease in the average $M_z$ is small. Above a certain size, the lowest-energy state becomes that of curling, or vortex, with a sharp decrease in the average $M_z$. Similar results were also reported [605] for different prisms, with more complex structures for large and elongated ones.

If a large field is applied in the $z$-direction, the minimal energy 'flower' configuration shrinks, namely it tilts more towards $z$, but the cube never saturates, and this structure does not completely close. Therefore, it was not considered necessary to go into the nucleation of chapter 9, and the full hysteresis curve was computed [604], by applying a large field, reducing it, then reversing it, and minimizing the total energy for each field. There was no attempt to check if the coercivity thus obtained depended on the value of the initial, 'saturating' field, or at least no such check was reported.

Of course, it is formally true that it is not absolutely necessary to look into the nucleation if there is no saturation, and a continuously evolving magnetization configuration *can* be computed. However, it is too risky at best to use this approach. The non-linear differential equations have an enormous number of solutions, not all of which are minimal energy states. These solutions belong to different branches, which are intermixed together in the non-linear case, and can only be separated and resolved when the equations are linearized. In my mind, allowing the computer to decide on how to stay on one branch, and at which point to jump to another branch, involves a too-optimistic view of the ability of the computer. I believe that even in problems in which linearization in order to find a nucleation mode can be avoided, or does not exist, it must be introduced anyway before the solution can be considered meaningful.

In the first place, there is the problem, already discussed in section 9.5.2, of whether the sharp corner of a cube (or similar bodies) has any physical meaning. I think that it is better to study ellipsoids, for which the demagnetizing field is better defined, at least as a first stage, until the basic problems are understood. But even if it is preferable for any reason to consider cubes, they need not be *perfect* cubes, which are made out of little cubes all the way to the edges. The last subunits on the surface are anyway treated differently than the rest in *all* these computations, and it is *not* necessary to use the same cubes there. It is possible for example to put more rounded bodies at the corners, so that the local infinity in the demagnetizing field will be eliminated, without complicating the rest of the calculation. For elongated prisms it is possible to taper off the edges, which will also make them look more like many of the real particles as seen in the electron microscope. There may also be other ways, but none of these modifications was ever tried. It is mainly because most people are happy to get rid of the nucleation problem, which they consider to be an unnecessary nuisance. It is not. It is an important guide on where and on which branch to start the computations. It is an essential part for those who want their computations to have a physical meaning, and to allow an insight on how to continue from there. It is a nuisance only for those who want to compute *something* fast enough for presenting at the next conference, and do not want to be bothered by the necessity to check the validity of their results.

Even in computations of a simple cube as described in the foregoing, in which the particle never saturates, magnetization configurations do not just keep evolving continuously. There are still [604] certain 'switching modes', and I do not see why they must be distinguished from the nucleation modes in the 'classical' sense of the word, as described in chapter 9. For example, in very small cubes, the basic structure of the 'flower' state is maintained during the reversal, but in order to switch, this flower has to close first. For a small field applied in the negative direction, the tendency of the flower is to *open* [604] further, which makes it more difficult to reverse. It takes a still more negative field to make the structure suddenly close, and switch into the direction of the field. For larger cubes, this flower opens until it suddenly jumps [604] into the curling configurations. In either case, there is a clear-cut energy barrier, and actually it is quite obvious that there is no hysteresis without a barrier. Giving the process a different name, such as a jump or a switch, does not change the fact that there is a well-defined nucleation process in the foregoing description, and that this process is confined to a particular *mode*. This mode must be the first one encountered when the field is changed, which means the field for which the energy barrier just flattens, as in Fig. 9.1. Evading this issue, and letting the computer decide on the jump, *may* lead to the correct result. But it is certainly not guaranteed to do so. Some of the reported results [604] for the very small particles are not much different from those obtained by the Stoner–Wohlfarth model, described in section 5.4, for *ellipsoids*. These ones make sense, and may, therefore, be taken as some sort of a check of the computer program. For the bigger particles, however, there are sometimes 'rather complicated magnetization configurations' [604], and it seems to me that more care must be exercised before determining the jump into or out of such states.

This warning of having to be very careful also applies to studies of elongated prisms made of little cubes, which have been most thoroughly reviewed in [606], as well as to the other cases mentioned in that review. It also applies to other computations of bodies with sharp corners, such as *two* interacting cubes [607, 608], and the cylinders which will be discussed in the next section. But the problem has a much more general nature, and is not restricted only to these shapes. Consider, for example, the computations of a sphere, discussed in section 11.3.2, where the beginning of curling can be checked against the analytic result for the nucleation field. After this start, the structure changes continuously, till a certain field is reached at which there is a jump, and the curling configuration is replaced by something else. There is no analytic guide for the location of the latter jump, and the computation in that vicinity needs the same care which is needed for the first jump in a cube, or a prism, or any other body shape. There must be something *analogous* to the nucleation theory, which would determine the beginning of a new mode, even when it starts from a complex configuration.

The nucleation problem is usually presented, as it was in chapter 9, as a study of the mode by which a deviation from the *saturated state* can just start. This case was indeed the only one for which the full details have been studied so far, but in principle it is not the only possible one. Actually there were *some* starts of an analytic treatment of other cases, although in a rather crude way. In one case, the hysteresis curve started [343] by rotation of the magnetization along the Stoner–Wohlfarth curve, and jumping to curling from there. The difference from the description in chapter 9 is not large, because the state before the jump was a uniformly magnetized one, even if it was not the saturated state, and there was no search for *other* modes. In another case [304], small deviations from one-dimensional magnetization structures were considered, in order to check their stability. It was found that such structures would always be unstable, and just collapse, so that it was not necessary to look for the details of that collapse.

There is no special difficulty, however, in developing a more general theory, of a 'nucleation' from a non-uniform magnetization state, say $\mathbf{M}_0(\mathbf{r})$. One way is to add a small perturbation, $\epsilon\mathbf{M}_1(\mathbf{r})$, so that both $\mathbf{M}_0(\mathbf{r})$ and $\mathbf{M}_0(\mathbf{r})+\epsilon\mathbf{M}_1(\mathbf{r})$ are solutions of Brown's static equations (8.3.40) with the appropriate boundary conditions. By substituting both $\mathbf{M}_0$ and $\mathbf{M}_0+\mathbf{M}_1$ in these equations, and leaving out every term which is higher than the first order in $\epsilon$, it is possible to obtain a set of linear differential equations with boundary conditions, for determining $\mathbf{M}_1(\mathbf{r})$. It is then possible, in principle, to look for the whole eigenvalue spectrum of these equations, in the same way as in some cases in chapter 9. The field value at which another mode will start to 'nucleate' will then be determined by the *first-encountered* eigenvalue. Another way, which has been studied [609] in some more detail, is to start from the expression for the energy, work out the first variation that gives the equilibrium states as in the derivation of Brown's equations, but proceed also to the second variation, which determines the stability of that equilibrium. A jump becomes possible when the second variation vanishes, which leads to the same differential equations. In either case the knowledge of the field at which the jump should occur can be used to guide the computer as to where to look for this jump, or transition to another configuration.

The difficulty is not in writing down the equations, but in solving them. There is no case for which the starting configuration, before the jump, is known in a closed form, except when it is the uniformly magnetized state. Therefore, the only way right now is to incorporate into the computational program the search for a possible 'nucleation' of another magnetization configuration. For this purpose it is necessary to use sufficiently small field-steps (or time-steps where applicable), and to avoid all sorts of short-cuts and approximations. Otherwise, the computation may just *skip* such a jump and continue elsewhere. Then, when a jump is encountered, the program should go sufficiently back, and restart tracing from there, using even finer steps, *and* a finer mesh. The convergence criterion has already been discussed in section 11.2, but it must be emphasized again here that starting to compute in another field before the structure in a given field is *properly* completed can lead to meaningless results. And, above all, every program should contain a special search for possible saddle points, with a check for the possibility of a jump *there*. If any of the published works contained any of these measures, they were not mentioned in the publications.

The precautions taken by [581] were already mentioned in section 11.3.4. There a magnetization configuration was accepted as a minimal energy state, only after trying to add to it small random perturbations, and checking that it evolved back to the initially obtained configuration. This method *almost* does what a numerical nucleation theory should be doing, but the way it was designed (or at least reported) just avoided *all* saddle points. Such avoidance should not do, because in the physical problem as outlined here, a saddle point may be a 'chance' for the magnetization to escape from the branch it is on, to a lower-energy one. The computer should be programmed instead to stop at each field value which does not answer the above-mentioned criterion, look around, and check if the point may be one from which it is possible to go to a lower-energy configuration. If it is, a jump to the new branch should take place. It may not be easy to formulate this 'looking around' in a programmer's language, but it can and should be done in any computation for which the nucleation is not investigated analytically. If it *is* implemented, it is essentially the old nucleation theory, or its generalization, in a numerical form, which is equally applicable for a sphere or for a cube. This addition will slow down the production of results, but it will produce only reliable ones.

It is also essential to try to obtain from such computations more than just the numerical value for one particular case. It takes a very small change in a program designed for a 'perfect' particle, to make it also applicable for finding out the effect of defects. This effect was discussed in chapter 9, but there it was based more on guesses than on facts, and a better study is needed. In a recent computation the effect of surface roughness was studied [610] by removing certain cubic elements (or making them non-magnetic), in a particular pattern along a ferromagnetic bar. In another simulation [611] (of a thin film) just one such cube was made non-magnetic, either on the surface or at the film centre. In both cases this simulation of an imperfection was found to make a significant difference in the result, but both used rather crude subdivisions. It is possible to use the same technique, of creating an inside 'void' or a 'scratch' on the surface by removing some of the little cubes, for a deeper and more detailed simulation of the De Blois experiment. In the same way, a different value of $K_1$, or of any of the other physical parameters, can be easily assigned for some *part* of the cube or prism, and so on. Actually, *all* the models for explaining the paradox of Brown, that never reached any conclusive results by analytic estimation, can be very readily studied by such a slight modification of the existing computer programs.

No serious effort has ever been put into this sort of simulation, but it seems to be the *only* way of gaining some real physical information on the true nature of the magnetization process in real particles, and a physical insight on how to proceed from there. Instead of such an attempt to solve the real problem, the literature is just getting filled up with results of computations of many different and uncorrelated cases which involve different and unspecified arbitrary assumptions. This method can lead nowhere, and the situation cannot improve as long as people hold the mistaken idea that such computations are in a different class of a 'new' micromagnetics, which should not be confused with the old, 'classical' micromagnetics. It is quite possible that such an effort to solve the problem of real particles will run out of computer resources before reaching a size for which something of interest may happen, as was the case with the attempt to compute correctly the domain wall structure in thick films. However, if this limit is reached it will at least be known how far meaningful computations may be pushed.

There are other effects which need a more serious consideration than is given to them. For example, eddy currents are known [612] to be very important for large, metallic particles. They cannot play an important role in small particles or very thin films, but a more *quantitative* estimation of the limit to which they may be neglected is still missing. Also, computations of the energy barrier for a superparamagnetic transition [196] are still done separately. In principle they should be combined with the computation of the static hysteresis, taking into account the possibility that thermal agitation may help the static jump, thus reducing the coercivity.

### 11.3.6 *Cylinder*

Practically all the discussion of the cubes and prisms in section 11.3.5 applies also to the study of a finite circular cylinder. It is listed separately here for two reasons. The first one is that for this case of a finite cylinder, there is an *analytic* proof [613] that the uniformly magnetized state can never be the lowest-energy state in zero field. At some stage I tried to prove the opposite, from some upper and lower bounds as in section 10.5.1. In the evaluation of the lower bound I used eqn (7.3.43) with a certain vector $\mathbf{H}''$ instead of the $\nabla\Phi$ as written into the equation here. I did not notice that the vector $\mathbf{H}''$ which I used could not be the gradient of a potential, because $\nabla \times \mathbf{H}''$ was not zero. This mistake was pointed out in [613], and I have already reported it in [570], but I find it necessary to emphasize it again.

The second reason is the need to mention a recent perturbation scheme [614] for calculating the deviations of the 'flower' structure from the uniform magnetization in the cylinder. Different analytic approximations are derived for a flat and for an elongated cylinder, and the plotted spatial variations in both cases turn out to be in fair agreement with those of the numerical computation [526] for a cylinder. The subdivision of the latter is into *cubes*, and it seems to be a rather crude one. It is not clear from the presentation which are supposed to be more accurate, the analytic or the numerical results. But the method is certainly unique, and the attempt to represent numerical results in an analytic form should be encouraged. A similar analytic approximation was also tried [615] for a prism.

There is also a recent experimental study [616] of an elongated, but finite cylinder. The angular dependence of its switching field has some of the features of the curling mode, but it is independent of the size of the cylinder. In spite of the discussion in [616], it should be obvious from chapter 9 that a size-independent mode and the curling mode are mutually exclusive, because the curling does work against exchange, which varies as $R^{-2}$. Details have not been given, and it is quite possible that this size independence is a result of using $R$ in a region for which the first term of eqn (9.2.18) is much larger than the second term, or similarly in eqn (9.2.31), as can be seen in Fig. 1 of [392]. If it is a *real* size independence, it is a real challenge to theorists to find, by analytic or numerical methods, what reversal mode is measured in this experiment.

# REFERENCES


1. Brown, W. F. Jr. (1962). *Magnetostatic Principles in Ferromagnetism* (North-Holland, Amsterdam).
2. Potter, H. H. (1934). The magneto-caloric effect and other magnetic phenomena in iron, *Proc. Roy. Soc. London*, Ser. A, **146**, 362–83.
3. Wagner, D. (1972). *Introduction to the Theory of Magnetism*, translated by F. Cap (Pergamon Press, Oxford).
4. Wood, D. W. and Dalton, N. W. (1966). Determination of exchange interactions in $Cu(NH_4)_2Cl_4.2H_2O$ and $CuK_2Cl_4.2H_2O$, *Proc. Phys. Soc.*, **87**, 755–65.
5. Aharoni, A. (1981). *Bull. Amer. Phys. Soc.*, **26**, 1218.
6. Néel, L. (1971). Magnetism and local molecular field, *Science*, **174**, 985–92.
7. van Vleck, J. H. (1973). $\chi = C/(T + \Delta)$, the most overworked formula in the history of paramagnetism, *Physica*, **69**, 177–92.
8. Smit, J. and Wijn, H. P. J. (1959). *Ferrites* (Wiley, New York).
9. Anderson, P. W. (1963). *Theory of magnetic exchange interactions: Exchange in insulators and semiconductors*, in *Solid State Physics* edited by F. Seitz and D. Turnbull (Academic Press, New York), Vol. 14, pp. 99–214.
10. Brown, H. A. (1971). Heisenberg ferromagnet with biquadratic exchange, *Phys. Rev. B*, **4**, 115–21.
11. Slonczewski, J. C. (1991). Fluctuation mechanism for biquadratic exchange coupling in magnetic multilayers, *Phys. Rev. Lett.*, **67**, 3172–5.
12. Demokritov, S., Tsymbal, E., Grünberg, P., Zinn, W. and Schuller, I. K. (1994). Magnetic-dipole mechanism for biquadratic interlayer coupling. *Phys. Rev. B*, **49**, 720–3.
13. Rodmacq, C., Dumesnil, K., Magin, P. and Hennion, M. (1993). Biquadratic magnetic coupling in NiFe/Ag multilayers, *Phys. Rev. B*, **48**, 3556–9.
14. Moriya, T. (1960). Anisotropic superexchange interaction and weak ferromagnetism, *Phys. Rev.*, **120**, 91–8.
15. Erdös, P. (1966). Theory of ion pairs coupled by exchange interaction, *J. Phys. Chem. Solids*, **27**, 1705–20.
16. Moriya, T. (1963). *Weak Ferromagnetism*, in *Magnetism* edited by G. T. Rado and H. Suhl (Academic Press, New York), Vol. I, pp. 85–125.
17. Treves, D. (1965). Studies on orthoferrites at the Weizmann Institute of Science, *J. Appl. Phys.*, **36**, 1033–9.

18. van de Braak, H. P. and Caspers, W. J. (1967). The local magnetization of an incomplete ferromagnet, *Z. Phys.*, **200**, 270–86.
19. Scott, G. G. (1951). A precise mechanical measurement of the gyromagnetic ratio in iron, *Phys. Rev.*, **82**, 542–7.
20. von Baeyer, H. C. (1991). Einstein at the bench, *The Sciences*, published by the New York Acad. of Sciences, November/December, 12–14.
21. Herring, C. (1963). *Direct exchange between well-separated atoms*, in *Magnetism* edited by G. T. Rado and H. Suhl (Academic Press, New York), Vol. II B, pp. 1–181.
22. Wang, C. S., Prange, R. E. and Korenman, V. (1982). Magnetism in iron and nickel, *Phys. Rev. B*, **25**, 5766–77.
23. Liechtenstein, A. I., Katsnelson, M. I., Antropov, V. P. and Gubanov, V. A. (1987). Local spin density functional approach to the theory of exchange interactions in ferromagnetic metals and alloys, *J. M. M. M.*, **67**, 65–74.
24. Mryasov, O. N., Gubanov, V. A. and Liechtenstein, A. I. (1992). Spiral-spin-density-wave states in fcc iron: Linear-muffin-tin-orbitals band-structure approach, *Phys. Rev. B*, **45**, 12330–6.
25. Garcia, N., Crespo, P., Hernando, A., Bovier, C., Serughetti, J. and Duval, E. (1993). Magnetic properties of Cu-doped porous silica gels: A possible Cu ferromagnet, *Phys. Rev. B*, **47**, 570–3.
26. Lang, P., Nordström, L., Zeller, R. and Dederichs, P. H. (1992). *Ab initio* calculations of the exchange coupling of Fe and Co monolayers in Cu, *Phys. Rev. Lett.*, **71**, 1927–30.
27. Erickson, R. P. (1994). Temperature-dependent non-Heisenberg exchange coupling of ferromagnetic layers, *J. Appl. Phys.*, **75**, 6163–8.
28. Herring, C. (1966). *Exchange interactions among itinerant electrons*, in *Magnetism* edited by G. T. Rado and H. Suhl (Academic Press, New York), Vol. IV.
29. Arajs, S. (1969). Ferromagnetic behavior of iron alloys containing molybdenum, manganese and tantalum, *Phys. stat. sol.*, **31**, 217–22.
30. Bardos, D. I., Beeby, J. L. and Aldred, A. T. (1969). Magnetic moments in body-centered cubic Fe-Ni-Al alloys, *Phys. Rev.*, **177**, 878–81.
31. Wohlfarth, E. P. and Cornwell, J. F. (1961). Critical points and ferromagnetism, *Phys. Rev. Lett.*, **7**, 342–3.
32. Tawil, R. A. and Callaway, J. (1973). Energy bands in ferromagnetic iron, *Phys. Rev. B*, **7**, 4242–52.
33. Zhang, S., Levy, P. M. and Fert, A. (1992). Conductivity and magnetoresistance of magnetic multilayered structure, *Phys. Rev. B*, **45**, 8689–702.
34. Zhang, S. and Levy, P. M. (1994). Effects of domains on giant magnetoresistance, structure, *Phys. Rev. B*, **50**, 6089–93.
35. Stearns, M. B. (1978). Why is iron ferromagnetic?, *Phys. Today*, April, 34–9.

36. Stearns, M. B. (1972). Model for the origin of ferromagnetism in Fe, *Phys. Rev. B*, **6**, 3326–31.
37. Shull, C. G. and Mook, H. A. (1966). Distribution of internal magnetization in iron, *Phys. Rev. Lett.*, **16**, 184–6.
38. Duff, K. J. and Das, T. P. (1971). Electron states in ferromagnetic iron. II. Wave-function properties, *Phys. Rev. B*, **3**, 2294–306.
39. Holstein, T. and Primakoff, H. (1940). Field dependence of the intrinsic domain magnetization of a ferromagnet, *Phys. Rev.*, **58**, 1098–113.
40. Dyson, F. J. (1956). General theory of spin-wave interactions, *and* Thermodynamic behavior of an ideal ferromagnet, *Phys. Rev.*, **102**, 1217–30 *and* 1230–44.
41. Ziman, J. M. (1965). *Principles of the Theory of Solids*, Cambridge Univ. Press.
42. Mayer, Joseph Edward and Mayer, Maria Goeppert (1950). *Statistical Mechanics* (Wiley, New York).
43. Kouvel, J. S. and Wilson, R. H. (1961). Magnetization of iron-nickel alloys under hydrostatic pressure, *J. Appl. Phys.*, **32**, 435–41.
44. Samara, G. A. and Giardini, A. A. (1969). Effect of pressure on the Néel temperature of magnetite, *Phys. Rev.*, **186**, 577–80.
45. Rayl, M., Wojtowicz, P. J., Abrahams, M. S., Harvey, R. L. and Buiocchi, C. J. (1971). Effect of lattice expansion on the Curie temperature of granular nickel films, *Phys. Lett.*, **36A**, 477–8.
46. Abd-Elmeguid, M. M. and Micklitz, H. (1982). High-pressure Mössbauer studies of amorphous and crystalline $Fe_3B$ and $(Fe_{0.25}Ni_{0.75})_3B$, *Phys. Rev. B*, **25**, 1–7.
47. Potapkov, N. A. (1974). On the calculation of the magnetization for the Heisenberg model at low temperature, *Phys. stat. sol. (b)*, **64**, 395–401.
48. Aldred, A. T. (1975). Temperature dependence of the magnetization of nickel, *Phys. Rev. B*, **11**, 2597–601.
49. Argyle, B. E., Charap, S. H. and Pugh, E. W. (1963). Deviations from $T^{3/2}$ law for magnetization of ferrometals: Ni, Fe, and Fe + 3% Si, *Phys. Rev.*, **132**, 2051–62.
50. Arrott, A. S. and Heinrich, B. (1981). Application of magnetization measurements in iron to high temperature thermometry, *J. Appl. Phys.*, **52**, 2113–15.
51. Yamada, Hideji (1974). The low temperature magnetization of ferromagnetic metals, *J. Phys. F*, **4**, 1819–31.
52. Craik, D. J. and Tebble, R. S. (1961). Magnetic domains, *Repts. Prog. Phys.*, **24**, 116–66.
53. Carey, R. and Isaac, E. D. (1966). *Magnetic domains and techniques for their observation* (Academic Press, New York).
54. Onishi, K., Tonomura, H. and Sakurai, Y. (1979). Measurement of magnetic potential distribution and wall velocity in amorphous films, *J. Appl. Phys.*, **50**, 7624–6.

55. Scheinfein, M. R., Unguris, J., Kelley, M. H., Pierce, D. T. and Celotta, R. J. (1990). Scanning electron microscopy with polarization analysis (SEMPA), *Rev. Sci. Instrum.*, **61**, 2501–26.
56. Schönenberger, C. and Alvarado, S. F. (1990). Understanding magnetic force microscopy, *Z. Phys. B*, **80**, 373–83.
57. Silva, T. J., Schultz, S. and Weller, D. (1994). Scanning near-field optical microscope for the imaging of magnetic domains in optically opaque materials, *Appl. Phys. Lett.*, **65**, 658–60.
58. De Blois, R. W. (1966). Ferromagnetic and structural properties of nearly perfect thin nickel platelets, *J. Vac. Sci. Technol.*, **3**, 146–55.
59. Griffiths, R. B. (1966). Spontaneous magnetization in idealized ferromagnets, *Phys. Rev.*, **152**, 240–6.
60. van der Woude, E. and Sawatzky, G. A. (1974). Mössbauer effect in iron and dilute iron based alloys, *Phys. Repts.*, **12C**, 335–74.
61. Feldmann, D., Kirchmayr, H. R., Schmolz, A. and Velicesku, M. (1971). Magnetic materials analysis by nuclear spectrometry: A joint approach to Mössbauer effect and nuclear magnetic resonance, *IEEE Trans. Magnetics*, **7**, 61–91.
62. Riedi, P. C. (1977). Temperature dependence of the magnetization of nickel using $^{61}$Ni NMR, *Phys. Rev. B*, **15**, 5197–203.
63. Landau, L. D. and Lifshitz, E. M. (1980). *Statistical Physics, Part 1*, 3rd Edition (Vol. 5 of *Course of Theoretical Physics*), (Pergamon Press, Oxford), Chapter XIV.
64. Heller, P. (1967). Experimental investigations of critical phenomena, *Repts. Prog. Phys.*, **30**, 731–826.
65. Noakes, J. E. and Arrott, A. (1968). Magnetization of nickel near its critical temperature, *J. Appl. Phys.*, **39**, 1235–6.
66. Arrott, A. (1971). Problem of using kink-point locus to determine the critical exponent $\beta$, *J. Appl. Phys.*, **42**, 1282–3.
67. Fisher, M. E. (1974). The renormalisation group in the theory of critical behavior, *Revs. Modern Phys.*, **46**, 597–616.
68. Pfeuty, P. and Toulouse, G. (1977). *Introduction to the renormalization group and to critical phenomena* (Wiley, New York).
69. Newell, G. F. and Montroll, E. W. (1953). On the theory of the Ising model of ferromagnetism, *Revs. Modern Phys.*, **25**, 353–89.
70. Mermin, N. D. and Wagner, H. (1966). Absence of ferromagnetism or antiferromagnetism in one- or two-dimensional isotropic Heisenberg models, *Phys. Rev. Lett.*, **17**, 1133–6.
71. Hone, D. W. and Richards, P. M. (1974). One- and two-dimensional magnetic systems, *Annual Rev. Material Sci.*, **4**, 337–63.
72. Birgeneau, R. J. and Shirane, G. (1978). Magnetism in one dimension, *Phys. Today*, December, 32–43.
73. de Jongh, L. J. and Miedema, A. R. (1974). Experiments on simple magnetic model systems, *Advan. Phys.*, **23**, 1–260.

74. Tomita, H. and Mashiyama, H. (1972). Spin-wave modes in a Heisenberg linear chain of classical spins, *Prog. Theoret. Phys.*, **48**, 1133–49.
75. Balucani, U., Pini, M. G., Tognetti, V. and Rettori, A. (1982). Classical one-dimensional ferromagnets in a magnetic field: Static and dynamic properties, *Phys. Rev. B*, **26**, 4974–86.
76. Yafet, Y., Kwo, J. and Gyorgy, E. M. (1986). Dipole-dipole interactions and two-dimensional magnetism, *Phys. Rev. B*, **33**, 6519–22.
77. Stanley, H. E. and Kaplan, T. A. (1966). Possibility of a phase transition for the two-dimensional Heisenberg model, *Phys. Rev. Lett.*, **17**, 913–15.
78. Falicov, L. M. (1992). Metallic magnetic superlattices, *Phys. Today*, October, 46–51.
79. Heinrich, B. and Cochran, J. F. (1993). Ultrathin metallic magnetic films: magnetic anisotropies and exchange interactions, *Advan. Phys.*, **42**, 523–639.
80. Freeman, A. J. and Wu, R.-Q. (1992). Magnetism in man made materials, *J. M. M. M.*, **104–7**, 1–6.
81. Neugebauer, C. A. (1959). Saturation magnetization of nickel films of thickness less than 100 Å, *Phys. Rev.*, **116**, 1441–6.
82. Lee, E. L., Bolduc, P. E. and Violet, C. E. (1964). Magnetic ordering and critical thickness in ultrathin iron films, *Phys. Rev. Lett.*, **13**, 800–2.
83. Reale, C. (1967). Magnetization of thin nickel films versus their thickness, *Phys. Lett.*, **25A**, 358–9.
84. Zinn, W. (1971). Mössbauer effect studies on magnetic thin films, *Czech. J. Phys. B*, **21**, 391–406.
85. Zuppero, A. C. and Hoffman, R. W. (1970). Mössbauer spectra of monolayer iron films, *J. Vac. Sci. Technol.*, **7**, 118–21.
86. Varma, M. N. and Hoffman, R. W. (1971). UHV Mössbauer emission spectra of thin iron films, *J. Appl. Phys.*, **42**, 1727–9.
87. Hellenthal, W. (1968). Superparamagnetic effects in thin films, *IEEE Trans. Magnetics*, **4**, 11–14.
88. Pomerantz, M. (1978). Experimental evidence for magnetic ordering in a literally two-dimensional magnet, *Solid St. Comm.*, **27**, 1413–16. See also *Phys. Today*, January 1981, pp. 20–1.
89. Kouvel, J. S. (1957). Methods for determining the Curie temperature of a ferromagnet, *General Electric Research Lab. Report* No. 57-RL-1799.
90. Arrott, A. (1957). Criterion for ferromagnetism from observations of magnetic isotherms, *Phys. Rev.*, **108**, 1394–6.
91. Arrott, A. and Noakes, J. E. (1967). Approximate equation of state for nickel near its critical temperature, *Phys. Rev. Lett.*, **14**, 786–9.
92. Aharoni, A. (1985). Use of high-field data for determining the Curie temperature, *J. Appl. Phys.*, **57**, 648–9.
93. Wohlfarth, E. P. (1971). Homogeneous and heterogeneous ferromagnetic alloys, *J. de Phys.* (Paris) Colloq. C1, **32**, 636–7.

94. Brommer, P. E. (1982). Magnetic properties of inhomogeneous weakly ferromagnetic materials, *Physica*, **113B**, 391–9.
95. Aharoni, A. (1984). Amorphicity, heterogeneity, and the Arrott plots, *J. Appl. Phys.*, **56**, 3479–84.
96. Yeung, I., Roshko, R. M. and Williams, G. (1986). Arrott-plot criterion for ferromagnetism in disordered systems, *Phys. Rev. B*, **34**, 3456–7.
97. Aharoni, A. (1986). A possible interpretation of non-linear Arrott plots, *J. M. M. M.*, **58**, 297–302.
98. Callen, E. R. and Callen, H. B. (1960). Anisotropic magnetization, *J. Phys. Chem. Solids*, **16**, 310–28.
99. Callen, E. R. (1961). Anisotropic Curie temperature, *Phys. Rev.*, **124**, 1373–9.
100. Mori, N. (1969). Calculation of ferromagnetic anisotropy energies for Ni and Fe metals, *J. Phys. Soc. Japan*, **27**, 307–312.
101. Kleman, M. (1969). Influence of internal magnetostriction on the formation of periodic magnetization configurations, *Philos. Mag.*, **19**, 285–303.
102. Jirák, Z. and Zelený, M. (1969). Influence of magnetostriction on the domain structure of cobalt, *Czech. J. Phys. B*, **19**, 44–7.
103. Brown, W. F. Jr. (1966). *Magnetoelastic Interactions* (Springer-Verlag, Berlin).
104. Maugin, G. A. (1976). A continuum theory of deformable ferrimagnetic bodies. I. Field equations, & II. Thermodynamics, constitutive theory, *J. Math. Phys.*, **17**, 1727–51.
105. Bartel, L. C. (1969). Theory of strain-induced anisotropy and the rotation of the magnetization in cubic single crystals, *J. Appl. Phys.*, **40**, 661–9.
106. Bartel, L. C. (1969). Approximate solution for the strain-induced deviation of the magnetization from saturation in polycrystalline cubic ferrites, *J. Appl. Phys.*, **40**, 3988–94.
107. Wayne, R. C., Samara, G. A. and LeFever, R. A. (1970). Effects of pressure on the magnetization of ferrites: Anomalies due to strain-induced anisotropy in porous samples, *J. Appl. Phys.*, **41**, 633–40.
108. Kronmüller, H. (1967). The contribution of dislocations to the magnetocrystalline energy and their effect on rotational hysteresis processes, *J. Appl. Phys.*, **38**, 1314–15.
109. Gessinger, H., Köster, E. and Kronmüller, H. (1968). Magnetostrictive crystalline energy of dislocations in Ni single crystals and anisotropy of the approach to saturation, *J. Appl. Phys.*, **39**, 986–8.
110. Aharoni, A. (1970). Approach to magnetic saturation in the vicinity of impurity atoms, *Phys. Rev. B*, **2**, 3794–805.
111. Malozemoff, A. P. (1983). Laws of approach to magnetic saturation for interacting and isolated spherical and cylindrical defects in isotropic magnetostrictive media, *IEEE Trans. Magnetics*, **19**, 1520–3.

112. Richter, H. G. and Dietrich, H. E. (1968). On the magnetic properties of fine-milled barium and strontium ferrite, *IEEE Trans. Magnetics*, **4**, 263–7.
113. Hoselitz, K. and Nolan, R. D. (1969). Anisotropy-field distribution in barium ferrite micropowders, *J. Phys. D*, Ser. 2, **2**, 1625–33.
114. Haneda, K. and Kojima, H. (1974). Effect of milling on the intrinsic coercivity of barium ferrite powders, *J. Amer. Ceramic Soc.*, **57**, 68–71.
115. Seeger, A. (1966). The effect of dislocations on the magnetization curves of ferromagnetic crystals, *J. de Phys.* (Paris) Colloq. C3, **27**, 68–77.
116. Hirsch, A. A., Ahilea, E. and Friedman, N. (1969). Magnetic anisotropy induced by an electric field, *Phys. Lett.*, **28A**, 763–4.
117. Lewis, B. (1964). The permalloy problem and magnetic annealing in bulk nickel-iron alloys, *Brit. J. Appl. Phys.*, **15**, 407–12.
118. Sambongi, T. and Mitui, T. (1963). Magnetic annealing effect in cobalt, *J. Phys. Soc. Japan*, **18**, 1253–60.
119. Takahashi, M. (1962). Induced magnetic anisotropy of evaporated films formed in a magnetic field, *J. Appl. Phys.*, **33**, 1101–6.
120. Lewis, B. (1964). The permalloy problem and anisotropy in nickel-iron magnetic films, *Brit. J. Appl. Phys.*, **15**, 531–42.
121. Hibino, M. and Maruse, S. (1969). Rotatable magnetic anisotropy in electron-microscope specimens of permalloy, *Japan. J. Appl. Phys.*, **8**, 366–73.
122. Cohen, M. S., Huber, E. E. Jr., Weiss, G. P. and Smith, D. O. (1960). Investigations into the origin of anisotropy in oblique-incidence films, *J. Appl. Phys.*, **31**, 291 S–2 S.
123. Torok, E. J. (1965). Origin and effects of local regions of complex biaxial anisotropy in thin ferromagnetic films with uniaxial anisotropy, *J. Appl. Phys.*, **36**, 952–60.
124. Kench, J. R. and Schuldt, S. B. (1970). Concerning the origin of uniaxial magnetic anisotropy in electrodeposited permalloy films, *J. Appl. Phys.*, **41**, 3338–46.
125. Cohen, M. S. (1967). Oblique-incidence magnetic anisotropy in co-deposited alloy films, *J. Appl. Phys.*, **38**, 860–9.
126. Williams, C. M. (1968). Effects of $^3$He irradiation on anisotropy-field inhomogeneity and coercive force in thin permalloy films, *J. Appl. Phys.*, **39**, 4741–4.
127. Yaegashi, S., Kurihara. T. and Segawa, H. (1993). Epitaxial growth and magnetic properties of Fe(211), *J. Appl. Phys.*, **74**, 4506–12.
128. Ono, H., Ishida, M., Fujinaga, M., Shishido, H. and Inaba, H. (1993). Texture, microstructure, and magnetic properties of Fe-Co alloy films formed by sputtering at an oblique angle of incidence, *J. Appl. Phys.*, **74**, 5124–8.
129. Johnson, K. E., Mirzamaani, M. and Doerner, M. F. (1995). In-plane

anisotropy in thin-film media: Physical origins of orientation ratio, *IEEE Trans. Magnetics*, **31**, 2721–7.
130. Spain, R. J. and Puchalska, I. B. (1964). Mechanism of ripple formation in thin films, *J. Appl. Phys.*, **35**, 824–5.
131. Maruse, S. and Hibino, M. (1969). The ripple contrast in Lorentz images of magnetic thin films, *Z. angew. Phys.*, **27**, 183–7.
132. Callen, H. B., Coren, R. L. and Doyle, W. D. (1965). Magnetization ripple and arctic foxes, *J. Appl. Phys.*, **36**, 1064–6.
133. Hoper, J. H. (1970). Magnetization-ripple internal fields in thin magnetic films measured by ferromagnetic resonance, *J. Appl. Phys.*, **41**, 1331–3.
134. Richards, P. M. (1974). Field dependence of magnetization in random fcc ferromagnets, *AIP Conference Proceedings*, No. 18, Part 1, 600–4.
135. Hayashi, N. (1966). Effective triaxial anisotropy in triple-layered thin magnetic films, *Japan. J. Appl. Phys.*, **5**, 1148–56.
136. Viscrian, I. (1970). Magnetic properties of cobalt thin films deposited on a scratched substrate, *Thin Solid Films*, **6**, R20–4.
137. Hoffmann, F. (1970). Dynamic pinning induced by nickel layers on permalloy films, *Phys. stat. sol.*, **41**, 807–13.
138. Engel, B. N., Wiedmann, M. H., Van Leeuwen, R. A. and Falco, C. M. (1993). Anomalous magnetic anisotropy in ultrafine transition metals, *Phys. Rev. B*, **48**, 9894–7.
139. Engel, B. N., Wiedmann, M. H. and Falco, C. M. (1994). Overlayer-induced perpendicular anisotropy in ultrathin Co films, *J. Appl. Phys.*, **75**, 6401–5.
140. Barnaś, J. (1991). On the Hoffmann boundary conditions at the interface between two ferromagnets, *J. M. M. M.*, **102**, 319–22.
141. Bennett, A. J. and Cooper, B. R. (1971). Origin of magnetic "surface anisotropy" of thin ferromagnetic films, *Phys. Rev. B*, **3**, 1642–9.
142. Cinal, M., Edwards, D. M. and Mathon, J. (1994). Magnetocrystalline anisotropy in ferromagnetic films, *Phys. Rev. B*, **50**, 3754–60.
143. Schulz, B. and Baberschke, K. (1994). Crossover from in-plane to perpendicular magnetization in ultrathin Ni/Cu(001) films, *Phys. Rev. B*, **50**, 13467–71.
144. Wulfhekel, W., Knappmann, S., Gehring, B. and Oepen, H. P. (1994). Temperature-induced magnetic anisotropies in Co/Cu (1117), *Phys. Rev. B*, **50**, 16074–7.
145. Brown, W. F. Jr. (1963). *Micromagnetics* (Interscience, New York).
146. Aharoni, A. (1987). Surface anisotropy in micromagnetics, *J. Appl. Phys.*, **61**, 3302–4.
147. Baron, R. B. and Hoffman, R. W. (1970). Saturation magnetisation and perpendicular anisotropy of nickel films, *J. Appl. Phys.*, **41**, 1623–32.
148. Sievert, J. D. and Voigt, C. (1976). Influence of domains on the magnetic torque in cubic crystals, *Phys. stat. sol. (a)*, **37**, 205–9.

149. Veerman, J., Franse, J. J. M. and Rathenau, G. W. (1963). Dependence of magnetocrystalline anisotropy on field strength, *J. Phys. Chem. Solids*, **24**, 947–51.
150. Rado, G. T. and Ferrari, J. M. (1975). Electric field dependence of the magnetic anisotropy energy in magnetite, *Phys. Rev. B*, **12**, 5166–74.
151. Abelmann, L., Kamberský, V., Lodder, C. and Popma, T. J. A. (1993). Analysis of torque measurements on films with oblique anisotropy, *IEEE Trans. Magnetics*, **29**, 3022–4.
152. Flanders, P. J. and Shtrikman, S. (1962). Experimental determination of the anisotropy distribution in ferromagnetic powders, *J. Appl. Phys.*, **33**, 216–19.
153. Gradmann, U., Bergholz, R. and Bergter, E. (1984). Magnetic surface anisotropies of clean Ni (111) surfaces and of Ni (111)/metal interfaces, *IEEE Trans. Magnetics*, **20**, 1840–5.
154. Gradmann, U. (1986). Magnetic surface anisotropies, *J. M. M. M.*, **54–7**, 733–6.
155. Sakurai, M. (1994). Magnetic anisotropy of epitaxial Fe/Pt(001), *Phys. Rev. B*, **50**, 3761–6.
156. Ho, K.-Y., Xiong, X.-Y., Zhi, J. and Cheng, L.-Z. (1993). Measurement of effective magnetic anisotropy of nanocrystalline Fe-Cu-Nb-Si-B soft magnetic alloys, *J. Appl. Phys.*, **74**, 6788–90.
157. Dionne, G. F. (1969). Determination of magnetic anisotropy and porosity from the approach to saturation of polycrystalline ferrites, *J. Appl. Phys.*, **40**, 1839–48.
158. Quach, H. T., Friedmann, A., Wu, C. Y. and Yelon, A. (1978). Surface anisotropy of Ni-Co from ferromagnetic resonance, *Phys. Rev. B*, **17**, 312–17.
159. Heinrich, B., Celinski, Z., Cochran, J. F., Arrott, A. S. and Myrtle, K. (1991). Magnetic anisotropies in single and multilayered structures, *J. Appl. Phys.*, **70**, 5769–74.
160. Hicken, R. J. and Rado, G. T. (1992). Magnetic surface anisotropy in ultrathin amorphous $Fe_{70}B_{30}$ and $Co_{80}B_{20}$ multilayer films, *Phys. Rev. B*, **46**, 11688–96.
161. Aharoni, A., Frei, E. H., Shtrikman, S. and Treves, D. (1957). The reversible susceptibility tensor of the Stoner-Wohlfarth model, *Bull. Res. Counc. Israel*, **6A**, 215–38.
162. Pareti, L. and Turilli, G. (1987). Detection of singularities in the reversible transverse susceptibility of an uniaxial ferromagnet, *J. Appl. Phys.*, **61**, 5098–100.
163. Sollis, P. M., Hoare, A., Peters, A., Orth, Th., Bissell, P. R., Chantrell, R. W. and Pelzl, J. (1992). Experimental and theoretical studies of transverse susceptibility in recording media, *IEEE Trans. Magnetics*, **28**, 2695–7.

164. Asti, G. (1994). Singularities in the magnetization processes of high anisotropy materials, *IEEE Trans. Magnetics*, **30**, 991–6.
165. Zimmermann, G. (1993). Determination of magnetic anisotropy by transverse susceptibility measurements – an application to NdFeB, *J. Appl. Phys.*, **73**, 8436–40.
166. Zimmermann, G. and Hempel, K. A. (1994). Transverse susceptibility and magnetic anisotropy of CoTi-doped barium hexaferrite single crystals, *J. Appl. Phys.*, **76**, 6062–4.
167. Asti, G. and Bolzoni, F. (1985). Singular point detection of discontinuous magnetization processes, *J. Appl. Phys.*, **58**, 1924–34.
168. Asti, G., Bolzoni, F. and Cabassi, R. (1993). Singular point detection in multidomain samples, *J. Appl. Phys.*, **73**, 323–33.
169. García-Arribas, A., Barandiarán, J. M. and Herzer, G. (1992). Anisotropy field distribution in amorphous ferromagnetic alloys from second harmonic response, *J. Appl. Phys.*, **71**, 3047–9.
170. Aharoni, A. (1994). Elongated superparamagnetic particles, *J. Appl. Phys.*, **75**, 5891–3.
171. Brown, W. F. Jr. (1963). Thermal fluctuations of a single-domain particle, *Phys. Rev.*, **130**, 1677–86.
172. Brown, W. F. Jr. (1977). Time constants of superparamagnetic particles, *Physica*, **86-88B**, 1423–4.
173. Aharoni, A. (1964). Thermal agitation of single domain particles, *Phys. Rev.*, **135**, A447–9.
174. Aharoni, A. (1969). Effect of a magnetic field on the superparamagnetic relaxation time, *Phys. Rev.*, **177**, 793–6.
175. Coffey, W. T., Crothers, D. S. F., Kalmykov, Yu. P., Massawe, E. S. and Waldron, J. T. (1994). Exact analytic formula for the correlation time of a single-domain ferromagnetic particle, *Phys. Rev. B*, **49**, 1869–82.
176. Coffey, W. T., Cregg, P. J., Crothers, D. S. F., Waldron, J. T. and Wickstead, A. W. (1994). Simple approximate formulae for the magnetic relaxation time of single domain ferromagnetic particles with uniaxial anisotropy, *J. M. M. M.*, **131**, L301–3.
177. Aharoni, A. (1973). Relaxation time of superparamagnetic particles with cubic anisotropy, *Phys. Rev. B*, **7**, 1103–7.
178. Aharoni, A. and Eisenstein, I. (1975). Theoretical relaxation times of large superparamagnetic particles with cubic anisotropy, *Phys. Rev. B*, **11**, 514–19.
179. Krop, K., Korecki, J., Żukrowski, J. and Karaś, W. (1974). The relaxation time of superparamagnetic particles as determined from Mössbauer spectra, *Int. J. Magnetism*, **6**, 19–23.
180. Eisenstein, I. and Aharoni, A. (1976). Magnetization curling in superparamagnetic spheres, *Phys. Rev. B*, **14**, 2078–95.

181. Dunlop, D. J. (1976). Thermal fluctuation analysis: A new technique in rock magnetism, *J. Geophys. Res.*, **81**, 3511–17.
182. Dunlop, D. J. (1977). Rocks as high-fidelity tape recorders, *IEEE Trans. Magnetics*, **13**, 1267–71.
183. Yatsuya, S., Hayashi, T., Akoh, H., Nakamura, E. and Akira, T. (1978). Magnetic properties of extremely fine particles of iron prepared by vacuum evaporation on running oil substrate, *Japan. J. Appl. Phys.*, **17**, 355–9.
184. Giessen, A. A. v. d. (1967). Magnetic properties of ultra-fine iron (III) oxide-hydrate particles made from iron (III) oxide-hydrate gels, *J. Phys. Chem. Solids*, **28**, 343–6.
185. Goldfarb, R. B. and Patton, C. E. (1981). Superparamagnetism and spin-glass freezing in nickel-manganese alloys, *Phys. Rev. B*, **24**, 1360–73.
186. Wirtz, G. P. and Fine, M. E. (1967). Superparamagnetic magnesioferrite precipitates from dilute solutions of iron in MgO, *J. Appl. Phys.*, **38**, 3729–37.
187. Mørup, S. (1990). Mössbauer effect in small particles, *Hyperfine Int.*, **60**, 959–74.
188. Shinjo, T. (1990). Mössbauer effect in antiferromagnetic fine particles, *J. Phys. Soc. Japan*, **21**, 917–22.
189. Mørup, S. (1983). Magnetic hyperfine splitting in Mössbauer spectra of microcrystals, *J. M. M. M.*, **37**, 39–50.
190. Kneller, E. F. and Luborsky, F. E. (1963). Particle size dependence of coercivity and remanence of single-domain particles, *J. Appl. Phys.*, **34**, 656–8.
191. Gaunt, P. (1968). The temperature dependence of single domain particle properties, *Philos. Mag.*, **17**, 263–6.
192. Joffe, I. (1969). The temperature dependence of the coercivity of a random array of uniaxial single domain particles, *J. Phys. C*, **2**, 1537–41.
193. Mizia, J., Figiel, H. and Krop, K. (1971). The temperature dependence of the magnetization of a single-domain particle assembly, *Acta Phys. Polon.*, **A39**, 71–81.
194. Victora, R. H. (1989). Predicted time dependence of the switching field for magnetic materials, *Phys. Rev. Lett.*, **63**, 457–60.
195. Heuberger, R. and Joffe, I. (1971). Magnetic properties of assemblies of single domain elliptical platelets, *J. Phys. C*, **4**, 805–10.
196. Lyberatos, A. and Chantrell, R. W. (1990). Calculation of the size dependence of the coercive force in fine particles, *IEEE Trans. Magnetics*, **26**, 2119–21.
197. Gaunt, P. (1973). Thermally activated domain wall motion, *IEEE Trans. Magnetics*, **9**, 171–3.
198. Jacobs, I. S. and Bean, C. P. (1963). *Fine particles, thin films and exchange anisotropy*, in *Magnetism* edited by G. T. Rado and H. Suhl

(Academic Press, New York), Vol. III, pp. 271–350.
199. de Biasi, R. S. and Fernandes, A. A. R. (1990). Ferromagnetic resonance evidence for superparamagnetism in a partially crystallized metallic glass, *Phys. Rev. B*, **42**, 527–9.
200. Coey, J. M. D. (1971). Noncollinear spin arrangement in ultrafine ferrimagnetic crystallites, *Phys. Rev. Lett.*, **27**, 1140–2.
201. Morrish, A. H. and Haneda, K. (1981). Magnetic structure of small $NiFe_2O_4$ particles, *J. Appl. Phys.*, **52**, 2496–8.
202. Haneda, K. and Morrish, A. H. (1988). Noncollinear magnetic structure of $CoFe_2O_4$ small particles, *J. Appl. Phys.*, **63**, 4258–60.
203. Linderoth, S., Hendriksen, P. V., Bødker, F., Wells, S., Davies, K., Charles, S. W. and Mørup, S. (1994). On spin-canting in maghemite particles, *J. Appl. Phys.*, **75**, 6583–5.
204. Gangopadhyay, S., Hadjipanayis, G. C., Dale, B., Sorensen, C. M., Klabunde, K. J., Papaefthymiou, V. and Kostikas, A. (1992). Magnetic properties of ultrafine iron particles, *Phys. Rev. B*, **45**, 9778–87.
205. Gangopadhyay, S., Hadjipanayis, G.C., Sorensen, C. M. and Klabunde, K. J. (1993). Magnetism in ultrafine Fe and Co particles, *IEEE Trans. Magnetics*, **29**, 2602–7.
206. Berkowitz, A. E., Lahut, J. A. and VanBuren, C. E. (1980). Properties of magnetic fluid particles, *IEEE Trans. Magnetics*, **16**, 184–90.
207. Tasaki, A., Sakutaro, T., Iida, S., Wada, N. and Uyeda, R. (1965). Magnetic properties of ferromagnetic metal fine particles prepared by evaporation in argon gas, *Japan. J. Appl. Phys.*, **4**, 707–11.
208. Tanaka, T. and Tamagawa, N. (1967). Magnetic properties of Fe-Co alloys fine particles, *Japan. J. Appl. Phys.*, **6**, 1096–100.
209. Waring, R. K. (1967). Magnetic interactions in assemblies of single-domain particles. The effect of aggregation, *J. Appl. Phys.*, **38**, 1005–6.
210. Rodé, D., Bertram, H. N. and Fredkin, D. R. (1987). Effective volume of interacting particles, *IEEE Trans. Magnetics*, **23**, 2224–6.
211. Lyberatos, A. and Chantrell, R. W. (1993). Thermal fluctuations in a pair of magnetostatically coupled particles, *J. Appl. Phys.*, **73**, 6501–3.
212. Lyberatos, A. (1994). Activation volume of a pair of magnetostatically coupled particles, *J. Appl. Phys.*, **75**, 5704–6.
213. O'Grady, K., El-Hilo, M. and Chantrell, R. W. (1993). The characterisation of interaction effects in fine particle systems, *IEEE Trans. Magnetics*, **29**, 2608–13.
214. Mørup, S. and Tronc, E. (1994). Superparamagnetic relaxation of weakly interacting particles, *Phys. Rev. Lett.*, **67**, 3172–5.
215. Rancourt, D. G. and Daniels, J. M. (1984). Influence of unequal magnetization direction probabilities on the Mössbauer spectra of superparamagnetic particles, *Phys. Rev. B*, **29**, 2410–14.
216. Nunes, A. C. and Yu, Z.-C. (1987). Fractionation of a water-based ferrofluid, *J. M. M. M.*, **65**, 265–8.

217. Chen, J. P., Sorensen, C. M., Klabunde, K. J. and Hadjipanayis, G. C. (1994). Magnetic properties of nanophase cobalt particles synthesized in inversed micelles, *J. Appl. Phys.*, **76**, 6316–18.
218. Abeledo, C. R. and Selwood, P. W. (1961). Temperature dependence of spontaneous magnetization in superparamagnetic nickel, *J. Appl. Phys.*, **32**, 229S–30S.
219. Billas, I. M. L., Becker, J. A., Châtelain, A. and de Heer, W. A. (1993). Magnetic moments of iron clusters with 25 to 700 atoms and their dependence on temperature, *Phys. Rev. Lett.*, **71**, 4067–70.
220. Bean, C. P., Livingston, J. D. and Rodbell, D. S. (1959). The anisotropy of very small cobalt particles, *J. Phys. Radium*, **20**, 298–301.
221. Du, Y.-w., Xu, M.-x., Wu, J., Shi, Y.-b., Lu, H.-x. and Xue, R.-h. (1991). Magnetic properties of ultrafine nickel particles, *J. Appl. Phys.*, **70**, 5903–5.
222. Kulkarni, G. U., Kannan, K. R., Arunarkavalli, T. and Rao, C. N. R. (1994). Particle-size effects on the value of $T_c$ of $MnFe_2O_4$: Evidence for finite-size scaling, *Phys. Rev. B*, **49**, 724–7.
223. Levinson, L. M., Luban, M. and Shtrikman, S. (1969). Mössbauer studies on $^{57}$Fe near the Curie temperature, *Phys. Rev.*, **177**, 864–71.
224. Gaunt, P. (1960). Correction to a single domain calculation, *Proc. Roy. Soc. London*, **75**, 625–6.
225. Chantrell, R. W., Lyberatos, A., El-Hilo, M. and O'Grady, K. (1994). Models of slow relaxation in particulate and thin film materials, *J. Appl. Phys.*, **76**, 6407–12.
226. Street, R. and Brown, S. D. (1994). Magnetic viscosity, fluctuation fields, and activation energies, *J. Appl. Phys.*, **76**, 6386–90.
227. Aharoni, A. (1992). Susceptibility resonance and magnetic viscosity, *Phys. Rev. B*, **46**, 5434–41.
228. Berkov, D. V. (1992). On the concept of the magnetic viscosity: analytic expression for the time dependent magnetization, *J. M. M. M.*, **111**, 327–9.
229. Dahlberg, E. D., Lottis, D. K., White, R. M., Matson, M. and Engle, E. (1994). Ubiquitous nonexponential decay: The effect of long-range couplings?, *J. Appl. Phys.*, **76**, 6396–400.
230. Figiel, H. (1975). The temperature growth of the coercive force for an assembly of single domain particles, *Phys. Lett.*, **53A**, 25–6.
231. Charap, S. H. (1988). Magnetic viscosity in recording media, *J. Appl. Phys.*, **63**, 2054–7.
232. Chamberlin, R. V. (1994). Mesoscopic model for the primary response of magnetic materials, *J. Appl. Phys.*, **76**, 6401–6.
233. Labarta, A., Iglesias, O., Balcells, Ll. and Badia, F. (1993). Magnetic relaxation in small-particle systems: $T\ln(t/\tau_0)$ scaling, *Phys. Rev. B*, **48**, 10240–6.

234. Stoner, E. C. and Wohlfarth, E. P. (1948). A mechanism of magnetic hysteresis in heterogeneous alloys, *Philos. Trans. Roy. Soc. London*, **A240**, 599–642. Reprinted (1991) in *IEEE Trans. Magnetics*, **27**, 3475–518.
235. Lee, E. W. and Bishop, J. E. L. (1966). Magnetic behaviour of single-domain particles, *Proc. Phys. Soc.*, **89**, 661–75.
236. Hagedorn, B. F. (1967). Effect of a randomly oriented cubic crystalline anisotropy on Stoner-Wohlfarth magnetic hysteresis loops, *J. Appl. Phys.*, **38**, 263–71.
237. Zimmermann, G. (1995). Transverse susceptibility of particulate recording media in different remanence states, *J. Appl. Phys.*, **77**, 2097–101.
238. Vos, M. J., Brott, R. L., Zhu, J.-G. and Carlson, L. W. (1995). Computed hysteresis behavior and interaction effects in spheroidal particle assemblies, *IEEE Trans. Magnetics*, **29**, 3652–7.
239. Mallinson, J. C. (1981). On the properties of two-dimensional dipoles and magnetized bodies, *IEEE Trans. Magnetics*, **17**, 2453–60.
240. Oda, H., Hirai, H., Kondo, K., and Sato, T. (1994). Magnetic properties of shock-compacted high-coercivity magnets with a nanometer-sized microstructure, *J. Appl. Phys.*, **76**, 3381–6.
241. Kronmüller, H., Durst, K.-D. and Martinek, G. (1987). Angular dependence of the coercive field in sintered $Fe_{77}Nd_{15}B_8$ magnets, *J. M. M. M.*, **69**, 149–57.
242. Osborn, J. A. (1945). Demagnetizing factors for the general ellipsoid, *Phys. Rev.*, **67**, 351–7.
243. Cronemeyer, D. C. (1991). Demagnetization factors for general ellipsoids, *J. Appl. Phys.*, **70**, 2911–14.
244. Aharoni, A. (1980). Magnetostatic energy of a ferromagnetic sphere, *J. Appl. Phys.*, **51**, 5906–8.
245. Leoni, F. and Natoli, C. (1971). Ferromagnetic stability and elementary excitations in a $S = \frac{1}{2}$ Heisenberg ferromagnet with dipolar interaction by the Green's function method, *Phys. Rev. B*, **4**, 2243–8.
246. Bryant, P. and Suhl, H. (1989). Micromagnetics below saturation, *J. Appl. Phys.*, **66**, 4329–37.
247. DeSimone, A. (1994). Magnetization and magnetostriction curves from micromagnetics, *J. Appl. Phys.*, **76**, 7018–20.
248. Hubert, A. (1969). Stray-field-free magnetization configurations, *Phys. stat. sol.*, **32**, 519–34.
249. Hubert, A. (1970). Stray-field-free and related domain wall configurations in thin magnetic films, *Phys. stat. sol.*, **38**, 699–713.
250. van den Berg, H. A. M. (1987). Domain structure in soft ferromagnetic thin film objects, *J. Appl. Phys.*, **61**, 4194–9.
251. Hahn, A. (1970). Theory of the Heisenberg superparamagnet, *Phys. Rev. B*, **1**, 3133–42.

252. Stapper, C. H. Jr. (1969). Micromagnetic solutions for ferromagnetic spheres, *J. Appl. Phys.*, **40**, 798–802.
253. Aharoni, A. and Jakubovics, J. P. (1986). Cylindrical magnetic domains in small ferromagnetic spheres with uniaxial anisotropy, *Philos. Mag. B*, **53**, 133–45.
254. Moyssides, P. G. (1989). Calculation of the sixfold integrals of the Biot-Savart-Lorentz force law in a close circuit, *and* Calculation of the sixfold integrals of the Ampere force law in a close circuit, *and* Experimental verification of the Biot-Savart-Lorentz and Ampere force laws in a close circuit, revisited, *IEEE Trans. Magnetics*, **25**, 4298–321.
255. Brown, W. F. Jr. and Morrish, A. H. (1957). Effect of a cavity on a single-domain magnetic particle, *Phys. Rev.*, **105**, 1198–201.
256. Joseph, R. I. (1976). Demagnetizing factors in nonellipsoidal samples – a review, *Geophysics*, **41**, 1052–4.
257. Joseph, R. I. (1966). Ballistic demagnetizing factor in uniformly magnetized cylinders, *J. Appl. Phys.*, **37**, 4639–43.
258. Chen, Du-Xing, Brug, J. A. and Goldfarb, R. B. (1991). Demagnetizing factors for cylinders, *IEEE Trans. Magnetics*, **27**, 3601–19.
259. Hegedus, C. J., Kadar, G. and Della Torre, E. (1979). Demagnetizing matrices for cylindrical bodies, *J. Inst. Math. Applics*, **24**, 279–91.
260. Coren, R. L. (1966). Shape demagnetizing effects in permalloy films, *J. Appl. Phys.*, **37**, 230–3.
261. Dove, D. B. (1967). Demagnetizing fields in thin magnetic films, *Bell Sys. Tech. J.*, **46**, 1527–59.
262. Moskowitz, R. and Della Torre, E. (1966). Theoretical aspects of demagnetization tensors, *IEEE Trans. Magnetics*, **2**, 739–44.
263. Yan, Y. D. and Della Torre, E. (1989). On the computation of particle demagnetizing fields, *IEEE Trans. Magnetics*, **25**, 2919–21.
264. Joseph, R. I. and Schlömann, E. (1965). Demagnetizing field in nonellipsoidal bodies, *J. Appl. Phys.*, **36**, 1579–93.
265. Brug, J. A. and Wolf, W. P. (1985). Demagnetizing fields in magnetic measurements. I. Thin discs, *and* II. Bulk and surface imperfections, *J. Appl. Phys.*, **57**, 4685–701.
266. Brown, W. F. Jr. (1984). Tutorial paper on dimensions and units, *IEEE Trans. Magnetics*, **20**, 112–17.
267. Cohen, E. R. and Giacomo, P. (1987). Symbols, units, nomenclature and fundamental constants in physics, Document I.U.P.A.P.-25, *Physica*, **146A**, 1–68. In particular: Appendix. Non-SI systems of quantities and units, pp. 62–8.
268. Brown, W. F. Jr. (1978). Domains, micromagnetics, and beyond, reminiscences and assessments, *J. Appl. Phys.*, **49**, 1937–42.
269. Minnaja, N. (1970). Micromagnetics at high temperature, *Phys. Rev. B*, **1**, 1151–9.

270. Aharoni, A. (1971). Applications of micromagnetics, *CRC Crit. Revs. Solid State Sci.*, **1**, 121–80.
271. Aharoni, A. (1982). Domain walls at high temperatures, *J. Appl. Phys.*, **53**, 7861–3.
272. Baltensperger, W. and Helman, J. S. (1988). Phenomenological theory of ferromagnets without anisotropy, *Phys. Rev. B*, **38**, 8954–7.
273. Aharoni, A. (1989). Micromagnetics and the phenomenological theory of ferromagnets, *Phys. Rev. B*, **40**, 4607–8.
274. Voltairas, P. A., Massalas, C. V. and Lagaris, I. E. (1992). Nucleation field of the infinite ferromagnetic circular cylinder at high temperature, *Mathl. Comput. Modeling*, **16**, 59–75.
275. Döring, W. (1968). Point singularities in micromagnetism, *J. Appl. Phys.*, **39**, 1006–7.
276. Aharoni, A. (1980). Exchange energy near singular points or lines, *J. Appl. Phys.*, **51**, 3330–2.
277. Arrott, A. S., Heinrich, B. and Bloomberg, D. S. (1981). Micromagnetics of magnetization processes in toroidal geometries, *IEEE Trans. Magnetics*, **10**, 950–3.
278. Jakubovics, J. P. (1978). Comments on the definition of ferromagnetic domain wall width, *Philos. Mag. B*, **38**, 401–6.
279. Kittel, C. (1949). Physical theory of ferromagnetic domains, *Revs. Modern Phys.*, **21**, 541–83.
280. Hauser, H. (1994). Energetic model of ferromagnetic hysteresis, *J. Appl. Phys.*, **75**, 2584–97.
281. Hauser, H. (1995). Energetic model of ferromagnetic hysteresis 2: Magnetization calculations of (110)[001] FeSi sheets by statistic domain behavior, *J. Appl. Phys.*, **77**, 2625–33.
282. Sato, M., Ishii, Y. and Nakae, H. (1982). Magnetic domain structures and domain walls in iron fine particles, *J. Appl. Phys.*, **53**, 6331–4.
283. Sato, M. and Ishii, Y. (1983). Critical sizes of cobalt fine particles with uniaxial magnetic anisotropy, *J. Appl. Phys.*, **54**, 1018–20.
284. Egami, T. and Graham, C. D. Jr. (1971). Domain walls in ferromagnetic Dy and Tb, *J. Appl. Phys.*, **42**, 1299–300.
285. Adam, Gh. and Corciovei, A. (1971). New approach to Lorentz approximation in micromagnetism, *J. Appl. Phys.*, **43**, 4763–7.
286. O'Dell, T. H. (1974). *Magnetic Bubbles* (Macmillan, London), section 2.1.3.
287. Kotiuga, P. R. (1988). Variational principles for three-dimensional magnetostatics based on helicity, *J. Appl. Phys.*, **63**, 3360–2.
288. Aharoni, A. (1991). Magnetostatic energy calculations, *IEEE Trans. Magnetics*, **27**, 3539–47.
289. Brown, W. F. Jr. (1962). Approximate calculation of magnetostatic energies, *J. Phys. Soc. Japan*, **17**, Suppl. B-I, 540–2.

290. Aharoni, A. (1991). Useful upper and lower bounds to the magnetostatic self-energy, *IEEE Trans. Magnetics*, **27**, 4793–5.
291. Aharoni, A. (1963). Upper and lower bounds for the nucleation field in an infinite rectangular ferromagnetic cylinder, *J. Appl. Phys.*, **34**, 2434–41.
292. Aharoni, A. (1966). Energy of one-dimensional domain walls in ferromagnetic films, *J. Appl. Phys.*, **37**, 3271–9.
293. Torok, E. J., Olson, A. L. and Oredson, H. N. (1966). Transition between Bloch and Néel walls, *J. Appl. Phys.*, **36**, 1394–9.
294. Middelhoek, S. (1963). Domain walls in thin Ni-Fe films, *J. Appl. Phys.*, **34**, 1054–9.
295. Aharoni, A. (1971). Domain walls and micromagnetics, *J. de Phys.* (Paris) Colloq. C1, **32**, 966–71.
296. Minnaja, N. (1971). Evaluation of the energy per unit surface in a cross-tie wall, *J. de Phys.* (Paris) Colloq. C1, **32**, 406–7.
297. Schwee, L. J. and Watson, J. K. (1973). A new model for cross-tie walls using parabolic coordinates, *IEEE Trans. Magnetics*, **9**, 551–4.
298. Kosiński, R. (1977). The structure and energy of cross-tie domain walls, *Acta Phys. Polon.*, **A51**, 647–57.
299. Nakatani, Y., Uesaka, Y. and Hayashi, N. (1989). Direct solution of the Landau-Lifshitz-Gilbert equation for micromagnetics, *Japan. J. Appl. Phys.*, **28**, 2485–507.
300. Ploessl, R., Chapman, J. N., Thompson, A. M., Zweck, J. and Hoffman, H. (1993). Investigation of the micromagnetic structure of cross-tie walls in permalloy, *J. Appl. Phys.*, **73**, 2447–52.
301. Brown, W. F. Jr. and LaBonte, A. E. (1965). Structure and energy of one-dimensional walls in ferromagnetic thin films, *J. Appl. Phys.*, **36**, 1380–6.
302. Berger, A. and Oepen, H. P. (1992). Magnetic domain walls in ultrathin fcc cobalt films, *Phys. Rev. B*, **45**, 12596–9.
303. Riedel, H. and Seeger, A. (1971). Micromagnetic treatment of Néel walls, *Phys. stat. sol. (b)*, **46**, 377–84.
304. Brown, W. F. Jr. and Shtrikman, S. (1962). Stability of one-dimensional ferromagnetic microstructure, *Phys. Rev.*, **125**, 825–8.
305. Shtrikman, S. and Treves, D. (1960). Fine structure of Bloch walls, *J. Appl. Phys.*, **31**, 1304.
306. Hartmann, U. and Mende, H. H. (1986). Observation of subdivided 180° Bloch wall configurations on iron whiskers, *J. Appl. Phys.*, **59**, 4123–8.
307. Hartmann, U. (1987). Néel regions in 180° Bloch walls, *Phys. Rev. B*, **36**, 2328–30.
308. Janak, J. F. (1967). Structure and energy of the periodic Bloch wall, *J. Appl. Phys.*, **38**, 1789–93.

309. Aharoni, A. (1967). Two-dimensional model for a domain wall, *J. Appl. Phys.*, **38**, 3196–9.
310. LaBonte, A. E. (1969). Two-dimensional Bloch-type domain walls in ferromagnetic films, *J. Appl. Phys.*, **40**, 2450–8.
311. Harrison, C. G. and Leaver, K. D. (1973). The analysis of two-dimensional domain wall structures by Lorentz microscopy, *Phys. stat. sol. (a)*, **15**, 415–29.
312. Schwellinger, P. (1976). The analysis of magnetic domain wall structures in the transition region of Néel and Bloch walls by Lorentz microscopy, *Phys. stat. sol. (a)*, **36**, 335–44.
313. Hothersall, D. C. (1969). The investigation of domain walls in thin sections of iron by the electron interference method, *Philos. Mag.*, **20**, 89–112.
314. Hothersall, D. C. (1972). Electron images of two-dimensional domain walls, *Phys. stat. sol. (b)*, **51**, 529–36.
315. Tsukahara, S. and Kawakatsu, H. (1972). Asymmetric 180° domain walls in single crystal iron films, *J. Phys. Soc. Japan*, **32**, 1493–9.
316. Suzuki, S. and Suzuki, K. (1977). Domain wall structures in single crystal Fe films, *IEEE Trans. Magnetics*, **13**, 1505–7.
317. Tsukahara, S. and Kawakatsu, H. (1972). Magnetic contrast and inclined 90° domain walls in thick iron films, *J. Phys. Soc. Japan*, **32**, 72–8.
318. Harrison, C. G. (1972). Lorentz electron images of tilted 180° magnetic domain walls, *Phys. Lett.*, **41A**, 53–4.
319. Green, A. and Leaver, K. D. (1975). Evidence for asymmetrical Néel walls observed by Lorentz microscopy, *Phys. stat. sol. (a)*, **27**, 69–74.
320. Tsukahara, S. (1984). Asymmetric wall structure observation by deflection pattern in transmission Lorentz microscopy, *IEEE Trans. Magnetics*, **20**, 1876–8.
321. Jakubovics, J. P. (1978). Application of the analytic representation of Bloch walls in thin ferromagnetic films to calculations of changes of wall structures with increasing anisotropy, *Philos. Mag. B*, **37**, 761–71.
322. Aharoni, A. (1975). Two-dimensional domain walls in ferromagnetic films. I. General theory. II. Cubic anisotropy. III. Uniaxial anisotropy, *J. Appl. Phys.*, **46**, 908–16 and 1783–6.
323. Aharoni, A. and Jakubovics, J. P. (1991). Magnetic domain walls in thick iron films, *Phys. Rev. B*, **43**, 1290–3.
324. Humphrey, F. B. and Redjdal, M. (1994). Domain wall structure in bulk magnetic materia, *J. M. M. M.*, **133**, 11–15.
325. Oepen, H. P. and Kirschner, J. (1989). Magnetization distribution of 180° domain walls at Fe(100) single-crystal surfaces, *Phys. Rev. Lett.*, **62**, 819–22.
326. Scheinfein, M. R., Unguris, J., Celotta, R. J. and Pierce, D. T. (1989). Influence of the surface on magnetic domain-wall microstructure, *Phys.*

*Rev. Lett.*, **63**, 668–71.

327. Scheinfein, M. R., Unguris, J., Blue, J. L., Coakley, K. J., Pierce, D. T., Celotta, R. J. and Ryan, P. J. (1991). Micromagnetics of domain walls at surfaces, *Phys. Rev. B*, **43**, 3395–422.
328. Craik, D. J. and Cooper, P. V. (1972). Derivation of magnetic parameters from domain theory and observation, *J. Phys. D*, **5**, L37–9.
329. Cooper, P. V. and Craik, D. J. (1973). Simple approximations to the magnetostatic energy of domains in uniaxial platelets, *J. Phys. D*, **6**, 1393–402.
330. Ploessl, R., Chapman, J. N., Scheinfein, M. R., Blue, J. L., Mansuripur, M. and Hoffman, H. (1993). Micromagnetic structure of domains in Co/Pt multilayers. I. Investigations of wall structure, *J. Appl. Phys.*, **74**, 7431–7.
331. Bloomberg, D. S. and Arrott, A. S. (1975). Micromagnetics and magnetostatics of an iron single crystal whisker, *Canad. J. Phys.*, **53**, 1454–71.
332. Dimitrov, D. A. and Wysin, G. M. (1995). Magnetic properties of spherical fcc clusters with radial surface anisotropy, *Phys. Rev. B*, **51**, 11947–50.
333. Aharoni, A. (1968). Measure of self-consistency in 180° domain wall models, *J. Appl. Phys.*, **39**, 861–2.
334. Aharoni, A. and Jakubovics, J. P. (1991). 90° walls in bulk ferromagnetic materials, *J. Appl. Phys.*, **69**, 4587–9.
335. Aharoni, A. and Jakubovics, J. P. (1991). Self-consistency of magnetic domain wall calculation, *Appl. Phys. Lett.*, **59**, 369–71.
336. Aharoni, A. and Jakubovics, J. P. (1993). Generalized self-consistency test of wall computations, *J. Appl. Phys.*, **73**, 3433–40.
337. Aharoni, A. and Jakubovics, J. P. (1993). Moving two-dimensional domain walls, *IEEE Trans. Magnetics*, **29**, 2527–9.
338. Brown, W. F. Jr. (1979). Thermal fluctuations of fine ferromagnetic particles, *IEEE Trans. Magnetics*, **15**, 1196–208.
339. Callen, H. B. (1958). A ferromagnetic dynamical equation, *J. Phys. Chem. Solids*, **4**, 256–70.
340. Mallinson, J. C. (1987). On damped gyromagnetic precession, *IEEE Trans. Magnetics*, **23**, 2003–4.
341. Brown, W. F. Jr. (1959). Micromagnetics, domains and resonance, *J. Appl. Phys.*, **30**, 62 S–9 S.
342. Chang, C.-R. (1991). Micromagnetic studies of coherent rotation with quartic crystalline anisotropy, *J. Appl. Phys.*, **69**, 2431–9.
343. Shtrikman, S. and Treves, D. (1959). The coercive force and rotational hysteresis of elongated ferromagnetic particles, *J. Phys. Radium*, **20**, 286–9.
344. Aharoni, A. (1969). Nucleation of magnetization reversal in ESD magnets, *IEEE Trans. Magnetics*, **5**, 207–10.

345. Aharoni, A. (1986). Angular dependence of the nucleation field in magnetic recording particles, *IEEE Trans. Magnetics*, **22**, 149–50.
346. Ishii, Y. (1991). Magnetization curling in an infinite cylinder with a uniaxial magnetocrystalline anisotropy, *J. Appl. Phys.*, **70**, 3765–9.
347. Soohoo, R. F. (1963). General exchange boundary conditions and surface anisotropy energy of a ferromagnet, *Phys. Rev.*, **131**, 594–601.
348. Dimitrov, D. A. and Wysin, G. M. (1994). Effects of surface anisotropy on hysteresis in fine magnetic particles, *Phys. Rev. B*, **43**, 3395–422.
349. Aharoni, A. (1987). Magnetization curling in coated particles, *J. Appl. Phys.*, **62**, 2576–7.
350. Aharoni, A. (1988). Magnetization buckling in elongated particles of coated iron oxides, *J. Appl. Phys.*, **63**, 4605–8.
351. Skomski, R., Müller, K.-H., Wendhausen, P. A. P. and Coey, J. M. D. (1993). Magnetic reversal in $Sm_2Fe_{17}N_y$ permanent magnets, *J. Appl. Phys.*, **73**, 6047–9.
352. Yang, J.-S. and Chang, C.-R. (1994). Magnetization curling in elongated heterostructure particles, *Phys. Rev. B*, **49**, 11877–85.
353. Brown, W. F. Jr. (1940). Theory of approach to magnetic saturation, *Phys. Rev.*, **58**, 736–43.
354. Brown, W. F. Jr. (1941). The effect of dislocations on magnetization near saturation, *Phys. Rev.*, **60**, 139–47.
355. Brown, W. F. Jr. (1957). Criterion for uniform micromagnetization, *Phys. Rev.*, **105**, 1479–82.
356. Frei, E. H., Shtrikman, S. and Treves, D. (1957). Critical size and nucleation field of ideal ferromagnetic particles, *Phys. Rev.*, **106**, 446–55.
357. Braun, H.-B. and Bertram, H. N. (1994). Nonuniform switching of single domain particles at finite temperatures, *J. Appl. Phys.*, **75**, 4609–16.
358. Aharoni, A. (1959). Some recent developments in micromagnetics at the Weizmann Institute of Science, *J. Appl. Phys.*, **30**, 70 S–8 S.
359. Aharoni, A. (1966). Magnetization curling, *Phys. stat. sol.*, **16**, 1–42.
360. Aharoni, A. and Shtrikman, S. (1958). Magnetization curve of the infinite cylinder, *Phys. Rev.*, **109**, 1522–8.
361. Forlani, F., Minnaja, N. and Sacchi, G. (1968). Application of micromagnetics to a boundless plate, *IEEE Trans. Magnetics*, **4**, 70–4.
362. Aharoni, A. (1968). Remarks on the application of micromagnetics to a boundless plate, *IEEE Trans. Magnetics*, **4**, 720–1.
363. Aharoni, A. (1986). Magnetization buckling in a prolate spheroid, *J. Appl. Phys.*, **60**, 1118–23.
364. Braun, H.-B. (1993). Thermally activated magnetization reversal in elongated ferromagnetic particles, *Phys. Rev. Lett.*, **71**, 3557–60.
365. Aharoni, A. (1995). Magnetostatic energy of the soliton in a cylinder, *J. M. M. M.*, **140–4**, 1819–20.

366. Luborsky, F. E. (1961). Development of elongated particle magnets, *J. Appl. Phys.*, **32**, 171 S–83 S.
367. Jacobs, I. S. and Bean, C. P. (1955). An approach to elongated fine-particle magnets, *Phys. Rev.*, **100**, 1060–7.
368. Kubo, O., Ido, T. and Yokoyama, H. (1987). Magnetization reversal for barium ferrite particulate media, *IEEE Trans. Magnetics*, **23**, 3140–2.
369. Knowles, J. E. (1981). Magnetic properties of individual acicular particles, *IEEE Trans. Magnetics*, **17**, 3008–13.
370. Knowles, J. E. (1984). The measurement of the anisotropy field of single "tape" particles, *IEEE Trans. Magnetics*, **20**, 84–6.
371. Kaneko, M. (1981). Magnetization reversal mechanism of nickel alumite films, *IEEE Trans. Magnetics*, **17**, 1468–71.
372. Knowles, J. E. (1986). Magnetization reversal by flipping, in acicular particles of $\gamma$-$Fe_2O_3$, *J. M. M. M.*, **61**, 121–8.
373. Aharoni, A. (1986). Perfect and imperfect particles, *IEEE Trans. Magnetics*, **22**, 478–83.
374. Knowles, J. E. (1988). A reply to "perfect and imperfect particles", *IEEE Trans. Magnetics*, **24**, 2263–5.
375. Kuo, P. C. (1988). Chain-of-spheres calculation on the coercivities of elongated fine particles with both magnetocrystalline and shape anisotropy, *J. Appl. Phys.*, **76**, 6561–3.
376. Ishii, Y. and Sato, M. (1986). Magnetic behaviors of elongated single domain particles by chain of spheres model, *J. Appl. Phys.*, **59**, 880–7.
377. Ishii, Y., Anbo, H., Nishida, K. and Mizuno, T. (1992). Magnetization behaviors in a chain of discs, *J. Appl. Phys.*, **71**, 829–35.
378. Han, D. and Yang, Z. (1994). Magnetization reversal mechanism of the chain of two oblate ellipsoids, *J. Appl. Phys.*, **75**, 4599–604.
379. Fulmek, P. F. and Hauser, H. (1994). Coercivity and switching field of single domain $\gamma$-$Fe_2O_3$ particles under consideration of the demagnetizing field, *J. Appl. Phys.*, **76**, 6561–3.
380. Huang, M. and Judy, J. H. (1991). Effects of demagnetization fields on the angular dependence of coercivity of longitudinal thin film media, *IEEE Trans. Magnetics*, **27**, 5049–51.
381. Stephenson, A. and Shao, J. C. (1993). The angular dependence of the remanent coercivity of gamma ferric oxide 'tape' particles, *IEEE Trans. Magnetics*, **29**, 7–10.
382. Aharoni, A. (1995). Agreement between theory and experiment, *Phys. Today*, June, 33–7.
383. Brown, W. F. Jr. (1945). Virtues and weaknesses of the domain concept, *Revs. Modern Phys.*, **17**, 15–19.
384. Smith, A. F. (1970). Domain wall interactions with non-magnetic inclusions observed by Lorentz microscopy, *J. Phys. D*, **3**, 1044–8.
385. Davis, P. F. (1969). A theory of the shape of spike-like magnetic domains, *J. Phys. D*, **2**, 515–21.

386. Saka, C., Shiiki, K. and Shinagawa, K. (1990). Simulation of domain structure for magnetic thin films in an applied field, *J. Appl. Phys.*, **68**, 263–8.
387. Kooy, C. and Enz, U. (1960). Experimental and theoretical study of the domain configuration in thin layers of $BaFe_{12}O_{19}$, *Philips Res. Repts.*, **15**, 7–29.
388. Kojima, H. and Goto, K. (1962). New remanent structure of magnetic domains in $BaFe_{12}O_{19}$, *J. Phys. Soc. Japan*, **17**, 584.
389. Kojima, H. and Goto, K. (1965). Remanent domain structures of $BaFe_{12}O_{19}$, *J. Appl. Phys.*, **36**, 538–43.
390. Kusunda, T. and Honda, S. (1974). Nucleation and demagnetization by needle pricking in MnBi films, *Appl. Phys. Lett.*, **24**, 516–19.
391. Shimada, Y. and Kojima, H. (1973). Bubble lattice formation in a magnetic uniaxial single-crystal thin plate, *J. Appl. Phys.*, **44**, 5125–9.
392. Aharoni, A. (1962). Theoretical search for domain nucleation, *Revs. Modern Phys.*, **34**, 227–38.
393. Middleton, B. K. (1969). The nucleation and reversal of magnetization in high fields to form cylindrical domains in thin magnetic films, *IEEE Trans. Magnetics*, **10**, 931–4.
394. Bachman, K. (1971). Nucleation and growth of magnetic domains in small $RCo_5$ particles, *IEEE Trans. Magnetics*, **7**, 647–50.
395. Onoprienko, L. G. (1973). Field of nucleation in a ferromagnetic plate with a local variation in the magnetic anisotropy constant, *Sov. Phys. Solid State*, **15**, 375–8.
396. Déportes, J., Givord, D., Lemaire, R. and Nagai, H. (1975). On the coercivity of $RCo_5$ compounds, *IEEE Trans. Magnetics*, **11**, 1414–16.
397. Ratnam, D. V. and Buessem, W. R. (1970). On the nature of defects in barium ferrite platelets, *IEEE Trans. Magnetics*, **6**, 610–14.
398. Haneda, K. and Kojima, H. (1973). Magnetization reversal process in chemically precipitated and ordinary prepared $BaFe_{12}O_{19}$, *J. Appl. Phys.*, **44**, 3760–2.
399. Seeger, A., Kronmüller, H., Rieger, H. and Träuble, H. (1964). Effect of lattice defects on the magnetization curve of ferromagnets, *J. Appl. Phys.*, **35**, 740–8.
400. Kronmüller, H. and Hilzinger, H. R. (1976). Incoherent nucleation of reversed domains in $Co_5Sm$ permanent magnets, *J. M. M. M.*, **2**, 3–10.
401. Kronmüller, H. (1978). Micromagnetism in hard magnetic materials, *J. M. M. M.*, **7**, 341–50.
402. Levinstein, H. J., Guggenheim, H. J. and Capio, C. D. (1969). Domain wall dislocation interaction in $RbFeF_3$, *J. Appl. Phys.*, **40**, 1080–1.
403. Zijlstra, H. (1970). Domain-wall processes in $SmCo_5$ powders, *J. Appl. Phys.*, **41**, 4881–5.
404. Fidler, J. and Kronmüller, H. (1978). Nucleation and pinning of magnetic domains at $Co_7Sm_2$ precipitates in $Co_5Sm$ crystals, *Phys. stat. sol.*

*(a)*, **56**, 545–56.

405. Becker, J. J. (1972). Temperature dependence of coercive force and nucleating fields in $Co_5Sm$, *IEEE Trans. Magnetics*, **8**, 520–2.
406. McCurrie, R. A. and Mitchell, R. K. (1975). Nucleation and pinning mechanisms in sintered $Sm\,Co_5$ magnets, *IEEE Trans. Magnetics*, **11**, 1408–13.
407. Livingston, J. D. (1987). Nucleation fields of permanent magnets, *IEEE Trans. Magnetics*, **23**, 2109–13.
408. Shur, Y. S., Shtoltz, E. V. and Margolina, V. I. (1960). Magnetic structure of small monocrystalline particles of MnBi alloys, *Sov. Phys. JETP*, **11**, 33–7.
409. Honda, S., Hosokawa, Y., Konishi, S. and Kusunda, T. (1973). Magnetic properties of single crystal MnBi platelet, *Japan. J. Appl. Phys.*, **12**, 1028–35.
410. Searle, C. W. (1973). Influence of surface conditions on the coercive force of $SmCo_5$ particles, *IEEE Trans. Magnetics*, **9**, 164–7.
411. Katayama, T. and Ohkoshi, M. (1976). Magnetization reversal in $GdCo_5$ single crystals, *Appl. Phys. Lett.*, **28**, 635–7.
412. Kishimoto, M. and Wakai, K. (1977). Magnetic properties of MnBi particles, *J. Appl. Phys.*, **48**, 4640–2.
413. De Blois, R. W. and Bean, C. P. (1959). Nucleation of ferromagnetic domains in iron whisker, *J. Appl. Phys.*, **30**, 225 S–6 S.
414. De Blois, R. W. (1962). Ferromagnetic nucleation sources on iron whiskers, *J. Appl. Phys.*, **32**, 1561–3.
415. Aharoni, A. and Neeman, E. (1963). Nucleation of magnetisation reversal in iron whiskers, *Phys. Lett.*, **6**, 241–2.
416. Wade, R. H. (1964). Some factors in the easy axis magnetization of permalloy films, *Philos. Mag.*, **10**, 49–66.
417. Bostanjoglo, O., Liedtke, R. and Oelmann, A. (1974). Microscopical test of current theories of magnetic wall motion, *Phys. stat. sol. (a)*, **24**, 109–13.
418. Aharoni, A. (1960). Reduction in coercive force caused by a certain type of imperfection, *Phys. Rev.*, **119**, 127–31.
419. Trueba, A. (1971). Iron whiskers under the effect of time-dependent fields, *Phys. stat. sol. (b)*, **43**, 157–62.
420. Schuler, F. (1962). Magnetic films: Nucleation, wall motion, and domain morphology, *J. Appl. Phys.*, **33**, 1845–50.
421. Self, W. B. and Edwards, P. L. (1972). Effect of tensile stress on the domain-nucleation field of iron whiskers, *J. Appl. Phys.*, **43**, 199–202.
422. Aharoni, A. (1968). Theory of the De Blois experiment. I. The effect of the small coil, II. Surface imperfection in an infinite coil, & III. Effect of neighboring imperfections, *J. Appl. Phys.*, **30**, 5846–54, & **41**, 2484–90.
423. Brown, W. F. Jr. (1962). Statistical aspects of ferromagnetic nucleation-field theory, *J. Appl. Phys.*, **33**, 3022–5.

424. Aharoni, A. (1985). Domain wall pinning at planar defects, *J. Appl. Phys.*, **58**, 2677–80.
425. Dietze, H. D. (1962). Statistical theory of coercive field, *J. Phys. Soc. Japan*, **17**, Suppl. B-I, 663–5.
426. Hilzinger, H. R. and Kronmüller, H. (1976). Statistical theory of the pinning of Bloch walls by randomly distributed defects, *J. M. M. M.*, **2**, 11–17.
427. Dietze, H. D. (1964). Theory of coercive force for randomly distributed lattice defects and precipitations, *kondens. Materie*, **2**, 117–32.
428. Baldwin, J. A. Jr. and Culler, G. J. (1969). Wall-pinning model of magnetic hysteresis, *J. Appl. Phys.*, **40**, 2828–35.
429. Baldwin, J. A. Jr. (1971). Magnetic hysteresis in simple materials. I. Theory & II. Experiment. *J. Appl. Phys.*, **42**, 1063–76.
430. Baldwin, J. A. Jr. (1974). Do ferromagnets have a true coercive force?, *J. Appl. Phys.*, **45**, 4006–12.
431. Holz, A. (1970). Formation of reversed domains in plate-shaped ferrite particles, *J. Appl. Phys.*, **41**, 1095–6.
432. Nembach, E., Chow, C. K. and Stöckel, D. (1976). Coercivity of single-domain nickel wires, *J. M. M. M.*, **3**, 281–7.
433. Huysmans, G. T. A., Lodder, J. C. and Wakui, J. (1988). Magnetization curling in perpendicular iron particle arrays (alumite media), *J. Appl. Phys.*, **64**, 2016–21.
434. Eagle, D. F. and Mallinson, J. C. (1967). On the coercivity of $\gamma Fe_2O_3$ particles, *J. Appl. Phys.*, **38**, 995–7.
435. Luborsky, F. E. and Morelock, C. R. (1964). Magnetization reversal in almost perfect whiskers, *J. Appl. Phys.*, **35**, 2055–66.
436. Ouchi, K. and Iwasaki, S.-I. (1987). Studies of magnetization reversal mechanism of perpendicular recording media by hysteresis loss measurement, *IEEE Trans. Magnetics*, **23**, 180–2.
437. Cherkaoui, R., Nogués, M., Dormann, J. L., Prené, P., Tronc, E., Jolivet, J. P., Fiorani, D. and Testa, A. M. (1994). Static magnetic properties at low and medium field of gamma-$Fe_2O_3$ particles with controlled dispersion, *IEEE Trans. Magnetics*, **30**, 1098–100.
438. Lam, J. (1992). Magnetic hysteresis of a rectangular lattice of interacting single-domain ferromagnetic spheres, *J. Appl. Phys.*, **72**, 5792–8.
439. Wirth, S. (1995). Magnetization reversal in systems of interacting magnetically-hard particles, *J. Appl. Phys.*, **77**, 3960–4.
440. Smyth, J. F., Schultz, S., Kern, D., Schmid, H. and Yee, D. (1988). Hysteresis of submicron permalloy particulate arrays, *J. Appl. Phys.*, **63**, 4237–9.
441. Smyth, J. F., Schultz, S., Fredkin, D. R., Kern, D. P., Rishton, S. A., Schmid, H., Cali, M. and Koehler, T. R. (1991). Hysteresis in lithographic arrays of permalloy particles: Experiment and theory, *J. Appl. Phys.*, **69**, 5262–9.

442. Brown, W. F. Jr. (1962). Failure of the local-field concept for hysteresis calculations, *J. Appl. Phys.*, **33**, 1308–9.
443. Bertram, H. N. and Mallinson, J. C. (1969). Theoretical coercive field for an interacting anisotropic dipole pair of arbitrary bond angle, *J. Appl. Phys.*, **40**, 1301–2.
444. Bertram, H. N. and Mallinson, J. C. (1970). Switching dynamics for an interacting dipole pair of arbitrary bond angle, *J. Appl. Phys.*, **41**, 1102–4.
445. Berkowitz, A. E., Hall, E. L. and Flanders, P. J. (1987). Microstructure of $\gamma$-$Fe_2O_3$ particles: A response to Andress *et al*, *IEEE Trans. Magnetics*, **23**, 3816–19.
446. Zijlstra, H. (1971). Hysteresis measurements of $RCo_5$ micro-particles, *J. de Phys.* (Paris) Colloq. C1, **32**, 1039–40.
447. Becker, J. J. (1971). Magnetization discontinuities in cobalt-rare-earth particles, *J. Appl. Phys.*, **42**, 1537–8.
448. McCurrie, R. A. and Willmore, L. E. (1979). Barkhausen discontinuities, nucleation, and pinning of domain walls in etched microparticles of $SmCo_5$, *J. Appl. Phys.*, **50**, 3560–4.
449. Roos, R., Voigt, C., Dederichs, H. and Hempel, K. A. (1980). Magnetization reversal in microparticles of barium ferrite, *J. M. M. M.*, **15-18**, 1455–6.
450. Salling, C., Schultz, S., McFadyen, I. and Ozaki, M. (1991). Measuring the coercivity of individual sub-micron ferromagnetic particles by Lorentz microscopy, *IEEE Trans. Magnetics*, **27**, 5184–6.
451. Chang, T., Zhu, J.-G. and Judy, J. H. (1993). Method for investigating the reversal properties of isolated barium ferrite fine particles utilizing magnetic force microscopy (MFM), *J. Appl. Phys.*, **73**, 6716–18.
452. Prené, P., Tronc, E., Jolivet, J.-P., Livage, J., Cherkaoui, R., Noguès, M., Dormann, J. L. and Fiorani, D. (1993). Magnetic properties of isolated $\gamma$-$Fe_2O_3$ particles, *IEEE Trans. Magnetics*, **29**, 2658–60.
453. Lederman, M., Gibson, G. A. and Schultz, S. (1993). Observation of thermal switching of a single ferromagnetic particle, *J. Appl. Phys.*, **73**, 6961–3.
454. Lederman, M., Schultz, S. and Ozaki, M. (1994). Measurement of the dynamics of the magnetization reversal in individual single-domain ferromagnetic particles, *Phys. Rev. Lett.*, **73**, 1986–9.
455. Lederman, M., Fredkin, D. R., O'Barr, R., Schultz, S. and Ozaki, M. (1994). Measurement of thermal switching of the magnetization of single domain particles, *J. Appl. Phys.*, **75**, 6217–22.
456. Salling, C., O'Barr, R., Schultz, S., McFadyen, I. and Ozaki, M. (1994). Investigation of the magnetization reversal mode for individual ellipsoidal single domain particles of $\gamma$-$Fe_2O_3$, *J. Appl. Phys.*, **75**, 7989–92.
457. Dionne, G. F., Weiss, J. A. and Allen, G. A. (1987). Hysteresis loops

modeled from coercivity, anisotropy and microstructure parameters, *J. Appl. Phys.*, **61**, 3862–4.
458. Arrott, A. S. (1987). Soft polycrystalline magnetization curves, *J. Appl. Phys.*, **61**, 4219–21.
459. Chang, C.-R., Lee, C. M. and Yang, J.-S. (1994). Magnetization curling reversal for an infinite hollow cylinder, *Phys. Rev. B*, **50**, 6461–4.
460. Broz, J. S., Braun, H. B., Brodbeck, O., Baltensperger, W. and Helman, J. S. (1990). Nucleation of magnetization reversal via creation of pairs of Bloch walls, *Phys. Rev. Lett.*, **65**, 787–9.
461. Aharoni, A. and Baltensperger, W. (1992). Spherical and cylindrical nucleation centers in a bulk ferromagnet, *Phys. Rev. B*, **45**, 9842–9.
462. Eisenstein, I. and Aharoni, A. (1976). Magnetization curling in a sphere, *J. Appl. Phys.*, **47**, 321–8.
463. Arrott, A. S., Heinrich, B. and Aharoni, A. (1979). Point singularities and magnetization reversal in ideally soft ferromagnetic cylinders, *IEEE Trans. Magnetics*, **15**, 1228–35.
464. Ishii, Y. and Sato, M. (1989). Magnetization curling in a finite cylinder, *J. Appl. Phys.*, **65**, 3146–50.
465. Muller, M. W. and Yang, M. H. (1971). Domain nucleation stability, *IEEE Trans. Magnetics*, **7**, 705–10.
466. Frait, Z. (1977). FMR in thin permalloy films with small surface anisotropy, *Physica*, **86-88B**, 1241–2.
467. Frait, Z. and Fraitová, D. (1980). Ferromagnetic resonance and surface anisotropy in iron single crystals, *J. M. M. M.*, **15-18**, 1081–2.
468. Rado, G. T. (1958). Effect of electronic mean free path on spin-wave resonance in ferromagnetic metals, *J. Appl. Phys.*, **29**, 330–2.
469. De Wames, R. E. and Wolfram, T. (1970). Dipole-exchange spin waves in ferromagnetic films, *J. Appl. Phys.*, **41**, 987-93.
470. Ament, W. S. and Rado, G. T. (1955). Electromagnetic effects of spin wave resonance in ferromagnetic metals, *Phys. Rev.*, 1558–66.
471. Walker, L. R. (1958). Resonant modes of ferromagnetic spheroids, *J. Appl. Phys.*, **29**, 318–23.
472. Aharoni, A. (1991). Exchange resonance modes in a ferromagnetic sphere, *J. Appl. Phys.*, **69**, 7762–4.
473. Viau, G., Ravel, F., Acher, O., Fiévet-Vincent, F. and Fiévet, F. (1994). Preparation and microwave characterization of spherical and monodisperse $Co_{20}Ni_{80}$ particles, *J. Appl. Phys.*, **76**, 6570–2.
474. Viau, G., Ravel, F., Acher, O., Fiévet-Vincent, F. and Fiévet, F. (1995). Preparation and microwave characterization of spherical and monodisperse Co–Ni particles, *J. M. M. M.*, **140–4**, 377–8.
475. Barnaś, J. (1992). Exchange modes in ferromagnetic sublattices, *Phys. Rev. B*, **45**, 10427–37.
476. Voltairas, P. A. and Massalas, C. V. (1993). Size-dependent resonance modes in ferromagnetic spheres, *J. M. M. M.*, **124**, 20–6.

477. Abraham, C. and Aharoni, A. (1960). Linear decrease in the magnetocrystalline anisotropy, *Phys. Rev.*, **120**, 1576–9.
478. Abraham, C. (1964). Model for lowering the nucleation field of ferromagnetic materials, *Phys. Rev.*, A1269–72.
479. Richter, H. J. (1989). Model calculation of the angular dependence of the switching field of imperfect ferromagnetic particles with special reference to barium ferrite, *Appl. Phys.*, **65**, 3597–601.
480. Hilzinger, H. R. (1977). The influence of planar defects on the coercive field of hard magnetic materials, *Appl. Phys.*, **12**, 253–60.
481. Jatau, J. A. and Della Torre, E. (1993). One-dimensional energy barrier model for coercivity, *Appl. Phys.*, **73**, 6829–31.
482. Della Torre, E. and Perlov, C. M. (1991). A one-dimensional model for wall motion coercivity in magneto-optic media, *J. Appl. Phys.*, **69**, 4596–8.
483. Hsieh, Y.-C. and Mansuripur, M. (1995). Coercivity of magnetic domain wall motion near the edge of a terrace, *J. Appl. Phys.*, **73**, 6829–31.
484. Mergel, D. (1993). Magnetic reversal processes in exchange-coupled double layers, *J. Appl. Phys.*, **74**, 4072–80.
485. Smith, N. and Cain, W. C. (1991). Micromagnetic model of an exchange coupled NiFe-TbCo bilayer, *J. Appl. Phys.*, **69**, 2471–9.
486. Barnaś, J. and Grünberg, P. (1991). On the static magnetization of double ferromagnetic layers with antiferromagnetic inter-layer coupling in an external magnetic field, *J. M. M. M.*, **98**, 57–9.
487. Aharoni, A. (1994). Exchange anisotropy in films, and the problem of inverted hysteresis loops, *J. Appl. Phys.*, **76**, 6977–9.
488. Chang, C.-R. (1992). Influence of roughness on magnetic surface anisotropy in ultrathin films, *J. Appl. Phys.*, **72**, 596–600.
489. Aharoni, A. (1961). Remanent state in one-dimensional micromagnetics, *Phys. Rev.*, **123**, 732–6.
490. Aharoni, A. (1993). Analytic solution to the problem of magnetic films with a perpendicular surface anisotropy, *Phys. Rev. B*, **47**, 8296–7.
491. Aharoni, A. and Jakubovics, J. P. (1992). One-dimensional domain walls in bulk magnetic materials, *J. M. M. M.*, **104-7**, 353–4.
492. Goedsche, F. (1970). Micromagnetic boundary conditions in inhomogeneous alloys, *Acta Phys. Polon.*, **A37**, 515–19.
493. Skomski, R. (1994). Aligned two-phase magnets: Permanent magnetism of the future?, *J. Appl. Phys.*, **76**, 7059–64.
494. Yang, J.-S. and Chang, C.-R. (1995). Grain size effects in nanostructured two-phase magnets, *IEEE Trans. Magnetics*, **31**, 3602–4.
495. Hubert, A. (1975). Statics and dynamics of domain walls in bubble materials, *J. Appl. Phys.*, **46**, 2276–87.
496. Rado, G. T. (1951). On the inertia of oscillating ferromagnetic domain walls, *Phys. Rev.*, **83**, 821–6.

497. Schlömann, E. (1971). Structure of moving domain walls in magnetic materials, *Appl. Phys. Lett.*, **19**, 274–6.
498. Slonczewski, J. C. (1972). Dynamics of magnetic domain walls, *Int. J. Magnetism*, **2**, 85–97.
499. Schlömann, E. (1972). Mass and critical velocity of domain walls in thin magnetic films, *J. Appl. Phys.*, **43**, 3834–42.
500. Slonczewski, J. C. (1973). Theory of domain-wall motion in magnetic films and platelets, *J. Appl. Phys.*, **44**, 1759–70.
501. Aharoni, A. (1974). Critique on the theories of domain wall motion in ferromagnets, *Phys. Lett.*, **50A**, 253–4.
502. Aharoni, A. and Jakubovics, J. P. (1979). Moving domain walls in ferromagnetic films with parallel anisotropy, *Philos. Mag. B*, **40**, 223–31.
503. Aharoni, A. and Jakubovics, J. P. (1979). Theoretical wall mobility in soft ferromagnetic films, *IEEE Trans. Magnetics*, **15**, 1818–20.
504. Konishi, S., Ueda, M. and Nakata, H. (1975). Domain wall mass in permalloy films, *IEEE Trans. Magnetics*, **11**, 1376–8.
505. Stankiewicz, A., Maziewski, A., Ivanov, B. A. and Safaryan, K. A. (1994). On the calculation of magnetic domain wall mass, *IEEE Trans. Magnetics*, **30**, 878–80.
506. Aharoni, A. (1976). Two-dimensional domain walls in ferromagnetic films. IV. Wall motion, *J. Appl. Phys.*, **47**, 3329–36.
507. Brown, W. F. Jr. (1968). The fundamental theorem of fine-ferromagnetic-particle theory, *J. Appl. Phys.*, **39**, 993–4.
508. Brown, W. F. Jr. (1969). The fundamental theorem of the theory of fine ferromagnetic particles, *Ann. NY Acad. Sci.*, **147**, 461–88.
509. Aharoni, A. and Jakubovics, J. P. (1988). Cylindrical domains in small ferromagnetic spheres with cubic anisotropy, *IEEE Trans. Magnetics*, **24**, 1892–4.
510. van der Zaag, P. J., Ruigrok, J. J. M., Noordermeer, A., van Delden, M. H. W. M., Por, P. T., Rekveldt, M. Th., Donnet, D. M. and Chapman, J. N. (1993). The initial permeability of polycrystalline MnZn ferrites: The influence of domain and microstructure, *J. Appl. Phys.*, **74**, 4085–95.
511. Aharoni, A. (1988). Elongated single-domain ferromagnetic particles, *J. Appl. Phys.*, **63**, 5879–82; [Erratum: **64**, 3330].
512. Chou, S. Y., Wei, M. S., Kraus, P. R. and Fischer, P. B. (1994). Single-domain magnetic pillar array of 35 nm diameter and 65 Gbits/in.$^2$ density for ultrahigh density quantum magnetic storage, *J. Appl. Phys.*, **76**, 6673–5.
513. Afanas'ev, A. M., Manykin, E. A. and Onishchenko, E. V. (1973). Magnetic structure of small weakly nonspherical ferromagnetic particles, *Sov. Phys. Solid State*, **14**, 2175–80.
514. Enkin, R. J. and Dunlop, D. J. (1987). A micromagnetic study of pseudo single-domain remanence in magnetite, *J. Geophys. Res.*, **92**,

12726-40.
515. Shtrikman, S. and Treves, D. (1960). On the resolution of Brown's paradox, *J. Appl. Phys.*, **31**, 72 S–3 S.
516. Aharoni, A. (1991). The concept of a single-domain particle, *IEEE Trans. Magnetics*, **27**, 4775–7.
517. Fowler, C. A. Jr., Fryer, E. M. and Treves, D. (1961). Domain structures in iron whiskers as observed by the Kerr method, *J. Appl. Phys.*, **32**, 296 S–7 S.
518. Trueba, A. (1971). Some magnetic properties of iron whiskers, *Phys. stat. sol. (a)*, **5**, 115–20.
519. Hartmann, U. (1987). Origin of Brown's coercive paradox in perfect ferromagnetic crystals, *Phys. Rev. B*, **36**, 2331–2.
520. Brown, W. F. Jr. (1962). Nucleation field of an infinitely long square ferromagnetic prism, *J. Appl. Phys.*, **33**, 3026–31; [Erratum: **34**, 1004].
521. Brown, W. F. Jr. (1964). Some magnetostatic and micromagnetic properties of the infinite rectangular bar, *J. Appl. Phys.*, **35**, 2102–6.
522. Aharoni, A. (1992). Modification of the saturated magnetization state, *Phys. Rev. B*, **45**, 1030–3.
523. Jakubovics, J. P. (1991). Interaction of Bloch-wall pairs in thin ferromagnetic films, *J. Appl. Phys.*, **69**, 4029–39.
524. Miltat, J., Thiaville, A. and Trouilloud, P. (1989). Néel line structures and energies in uniaxial ferromagnets with quality factor Q>1, *J. M. M. M.*, **82**, 297–308.
525. Schabes, M. E. and Aharoni, A. (1987). Magnetostatic interaction fields for a three-dimensional array of ferromagnetic cubes, *IEEE Trans. Magnetics*, **23**, 3882–8.
526. Usov, N. A. and Peschany, S. E. (1993). Magnetization curling in a fine cylindrical particle, *J. M. M. M.*, **118**, L290–4.
527. Cendes, Z. J. (1989). Unlocking the magic of Maxwell's equations, *IEEE Spectrum*, April, 29–33.
528. Fredkin, D. R. and Koehler, T. R. (1988). Numerical micromagnetics of small particles, *IEEE Trans. Magnetics*, **24**, 2362–7.
529. Fredkin, D. R. and Koehler, T. R. (1990). Hybrid method for computing demagnetizing fields, *IEEE Trans. Magnetics*, **26**, 415–17.
530. Fredkin, D. R. and Koehler, T. R. (1990). *Ab initio* micromagnetic calculations for particles, *J. Appl. Phys.*, **67**, 5544–8.
531. Chen, W., Fredkin, D. R. and Koehler, T. R. (1993). A new finite element method in micromagnetics, *IEEE Trans. Magnetics*, **29**, 2124–8.
532. Andrä, W. and Danan, H. (1987). Magnetization reversal by curling in terms of the atomic layer model, *Phys. stat. sol. (a)*, **102**, 367–73.
533. Andrä, W., Appel, W. and Danan, H. (1990). Magnetization reversal in cobalt-surface-coated iron-oxide particles, *IEEE Trans. Magnetics*, **26**, 231–4.

534. Hayashi, N., Nakatani, Y. and Inoue, T. (1988). Numerical solution of Landau-Lifshitz-Gilbert equation with two space variables for vertical Bloch lines, *Japan. J. Appl. Phys.*, **27**, 366–78.
535. Hayashi, N., Inoue, T., Nakatani, Y. and Fukushima, H. (1988). Direct solution of Landau-Lifshitz-Gilbert equation for domain walls in thin permalloy films, *IEEE Trans. Magnetics*, **24**, 3111–13.
536. Müller-Pfeiffer, S., Schneider, M. and Zinn, W. (1994). Imaging of magnetic domain walls in iron with a magnetic force microscope: A numerical study, *Phys. Rev. B*, **49**, 15745–52.
537. Jatau, J. A. and Della Torre, E. (1993). A methodology for mode pushing in coercivity calculations, *IEEE Trans. Magnetics*, **29**, 2374–6.
538. Ratnajeevan, S., Hoole, H., Weeber, K. and Subramaniam, S. (1991). Fictitious minima of object functions, finite element meshes, and edge elements in electromagnetic device synthesis, *IEEE Trans. Magnetics*, **27**, 5214–16.
539. Aharoni, A. (1984). Magnetization distribution in an ideally soft sphere, *J. Appl. Phys.*, **55**, 1049–51.
540. Shyamkumar, B. B. and Cendes, Z. J. (1988). Convergence of iterative methods for nonlinear magnetic field problems, *IEEE Trans. Magnetics*, **24**, 2585–7.
541. Viallix, A., Boileau, F., Klein, R., Niez, J. J. and Baras, P. (1988). A new method for finite element calculation of micromagnetic problems, *IEEE Trans. Magnetics*, **24**, 2371–4.
542. Del Vecchio, R. M., Hebbert, R. S. and Schwee, L. J. (1989). Micromagnetics calculation for two-dimensional thin-film geometries using a finite-element formulation, *IEEE Trans. Magnetics*, **29**, 4322–9.
543. Blue, J. L. and Scheinfein, M. R. (1991). Using multipoles decreases computation time for magnetostatic self-energy, *IEEE Trans. Magnetics*, **27**, 4778–80.
544. Schrefl, T., Fischer, R., Fidler, J. and Kronmüller, H. (1994). Two- and three-dimensional calculations of remanence enhancement of rare-earth based composite magnets, *J. Appl. Phys.*, **76**, 7053–8.
545. Tonomura, A. (1993). Observation of flux lines by electron holography, *IEEE Trans. Magnetics*, **29**, 2488–93.
546. Giles, R., Alexopoulos, P. S. and Mansuripur, M. (1992). Micromagnetics of thin film cobalt-based media for magnetic recording, *Comput. Phys.*, **6**, 53–70.
547. Uesaka, Y., Nakatani, Y. and Hayashi, N. (1993). Micromagnetic computation of damping constant effect on switching mechanism of a hexagonal platelet particle, *Japan. J. Appl. Phys.*, **32**, 1101-78.
548. Gomez, R. D., Adly, A. A., Mayergoyz, I. D. and Burke, E. R. (1993). Magnetic force scanning tunneling microscopy: Theory and experiment, *IEEE Trans. Magnetics*, **29**, 2494–9.
549. Guo, G. and Della Torre, E. (1994). 3-D micromagnetic modeling of

domain configurations in soft magnetic materials, *J. Appl. Phys.*, **75**, 5710–12.

550. Guo, Y.-M. and Zhu, J.-G. (1992). Transitions between intra-wall structures in permalloy thin films, *IEEE Trans. Magnetics*, **28**, 2919–21.
551. Trouilloud, P. and Miltat, J. (1987). Néel lines in ferrimagnetic garnet epilayers with orthorhombic anisotropy and canted magnetization, *J. M. M. M.*, **66**, 194–212.
552. Labrune, M. and Miltat, J. (1990). Micromagnetics of strong stripe domains in NiCo thin films, *IEEE Trans. Magnetics*, **22**, 1521–3.
553. Labrune, M. and Miltat, J. (1994). Strong stripes as a paradigm of quasi-topological hysteresis, *J. Appl. Phys.*, **75**, 2156–68.
554. Yuan, S. W. and Bertram, H. N. (1991). Domain wall structure and dynamics in thin films, *IEEE Trans. Magnetics*, **24**, 5511–13.
555. Yuan, S. W. and Bertram, H. N. (1991). Domain-wall dynamic transitions in thin films, *Phys. Rev. B*, **44**, 12395–405.
556. Yuan, S. W. and Bertram, H. N. (1992). Inhomogeneities and coercivity of soft permalloy thin films, *IEEE Trans. Magnetics*, **28**, 2916–18.
557. Yuan, S. W. and Bertram, H. N. (1993). Eddy current damping of thin film domain walls, *IEEE Trans. Magnetics*, **29**, 2515–17.
558. Yuan, S. W. and Bertram, H. N. (1993). Domain-wall dynamics in thick permalloy films, *J. Appl. Phys.*, **73**, 5992–4.
559. Miltat, J., Laska, V., Thiaville, A. and Boileau, F. (1988). Direct studies of Néel (or Bloch) line dynamics, *J. de Phys.* (Paris) Colloq. C8, **49**, 1871–5.
560. Bagnérés, A. and Humphrey, F. B. (1992). Dynamics of magnetic domain walls with loosely spaced vertical Bloch lines, *IEEE Trans. Magnetics*, **28**, 2344–6.
561. Theile, J., Kosinski, R. A. and Engemann, J. (1986). Numerical computations of vertical Bloch lines motion in the presence of a periodic in-plane magnetic field, *J. M. M. M.*, **62**, 139–42.
562. Leaver, K. D. (1975). The synthesis of three-dimensional stray-field-free magnetization distributions, *Phys. stat. sol. (a)*, **27**, 153–63.
563. Mansuripur, M. (1989). Computation of fields and forces on magnetic force microscopy, *IEEE Trans. Magnetics*, **25**, 3467–9.
564. Oti, J. O. and Rice, P. (1993). Micromagnetic simulations of tunneling stabilized magnetic force microscopy, *J. Appl. Phys.*, **73**, 5802–4.
565. Tomlinson, S. L., Hoon, S. R., Farley, A. N. and Valera, M. S. (1995). Flux closure in magnetic force microscope tips, *IEEE Trans. Magnetics*, **31**, 3352–4.
566. Aharoni, A. and Jakubovics, J. P. (1993). Effect of the MFM tip on the measured magnetic structure, *J. Appl. Phys.*, **73**, 6498–500.
567. Matteucci, G., Muccini, M. and Hartmann, U. (1993). Electron holography in the study of the leakage field of magnetic force microscopy sensor

tips, *Appl. Phys. Lett.*, **62**, 1839–41.

568. Hayashi, N. and Abe, K. (1976). Computer simulation of magnetic bubble domain wall motion, *Japan. J. Appl. Phys.*, **15**, 1683–94.
569. Aharoni, A. and Jakubovics, J. P. (1990). Approach to saturation in small isotropic spheres, *J. M. M. M.*, **83**, 451–2.
570. Aharoni, A. (1993). The remanent state of fine particles, *IEEE Trans. Magnetics*, **29**, 2596–601.
571. Aharoni, A. (1981). Magnetization curve of zero-anisotropy sphere, *J. Appl. Phys.*, **52**, 933–5.
572. Rachford, F. J., Prinz, G. A., Krebs, J. J. and Hathaway, K. B. (1982). Verification of first-order magnetic phase transition in single crystal iron films, *J. Appl. Phys.*, **53**, 7966–8.
573. Aharoni, A. (1988). Nucleation in a ferromagnetic sphere with surface anisotropy, *J. Appl. Phys.*, **64**, 6434–8.
574. Fredkin, D. R. and Koehler, T. R. (1989). Numerical micromagnetics: Prolate spheroids, *IEEE Trans. Magnetics*, **25**, 3473–7.
575. Nakamura, Y. and Iwasaki, S.-i. (1987). Magnetization models of Co-Cr film for the computer simulation of perpendicular magnetic recording process, *IEEE Trans. Magnetics*, **23**, 153–5.
576. Beardsley, I. A. (1989). Reconstruction of the magnetization in a thin film by a combination of Lorentz microscopy and external field measurement, *IEEE Trans. Magnetics*, **25**, 671–7.
577. Corciovei, A. and Adam, Gh. (1971). A general approach to the calculation of the magnetostatic energy in thin ferromagnetic films, *Rev. Roum. Phys.*, **16**, 275–89.
578. Arrott, A. S. (1990). Micromagnetics of ultrathin films and surfaces, *J. Appl. Phys.*, **69**, 5212–14.
579. Asselin, P. and Thiele, A. A. (1986). On the field Lagrangians in micromagnetics, *IEEE Trans. Magnetics*, **22**, 1876–80.
580. Schrefl, T., Schmidts, H. F., Fidler, J. and Kronmüller, H. (1993). Nucleation fields and grain boundaries in hard magnetic materials, *IEEE Trans. Magnetics*, **29**, 2878–80.
581. Bertram, H. N. and Zhu, J.-G. (1992). *Fundamental magnetization processes in thin film recording media*, in *Solid State Physics* edited by H. Ehrenreich and D. Turnbull (Academic Press, New York), Vol. 46, pp. 271–371.
582. Muller, M. W. and Indeck, R. S. (1994). Intergranular exchange coupling, *J. Appl. Phys.*, **75**, 2289–90.
583. Hughes, G. F. (1983). Magnetization reversal in cobalt-phosphorus films, *J. Appl. Phys.*, **54**, 5306–13.
584. Schrefl, T., Schmidts, H. F., Fidler, J. and Kronmüller, H. (1993). Nucleation of reversed domains at grain boundaries, *J. Appl. Phys.*, **73**, 6510–12.

585. Murata, O., Saito, K., Nakatani, Y. and Hayashi, N. (1993). Computer simulation of magnetization states in a magnetic thin film consisting of magnetostatically coupled grains with randomly oriented cubic anisotropy, *J. Appl. Phys.*, **73**, 6513–15.
586. Zhao, Y. and Bertram, H. N. (1995). Micromagnetic modeling of magnetic anisotropy in textured thin film media, *J. Appl. Phys.*, **77**, 6411–15.
587. Chang, T. and Zhu, J.-G. (1994). Angular dependence measurement of individual barium ferrite recording particles near the single domain size, *J. Appl. Phys.*, **75**, 5553–5.
588. Uesaka, Y., Nakatani, Y. and Hayashi, N. (1991). Micromagnetic calculation of applied field effect on switching mechanism of a hexagonal platelet particle, *Japan. J. Appl. Phys.*, **30**, 2489–502.
589. Nakatani, Y., Hayashi, N. and Uesaka, Y. (1991). Computer simulation of magnetization reversal mechanisms of hexagonal platelet particle: Effect of material parameters, *Japan. J. Appl. Phys.*, **30**, 2503–12.
590. Uesaka, Y., Nakatani, Y. and Hayashi, N. (1993). Computation of switching fields of stacked magnetic hexagonal particles with different heights, *J. M. M. M.*, **124**, 341–6.
591. Fredkin, D. R., Koehler, T. R., Smyth, J. F. and Schultz, S. (1991). Magnetization reversal in permalloy particles: Micromagnetic computations, *J. Appl. Phys.*, **69**, 5276–8.
592. Fredkin, D. R. and Koehler, T. R. (1990). Numerical micromagnetics: Rectangular parallelepipeds, *IEEE Trans. Magnetics*, **26**, 1518–20.
593. Koehler, T. R., Yang, B., Chen, W. and Fredkin, D. R. (1993). Simulation of magnetoresistive response in a small permalloy strip, *J. Appl. Phys.*, **73**, 6504–6.
594. Koehler, T. R. and Williams, M. L. (1995). Micromagnetic modeling of a single element MR head, *IEEE Trans. Magnetics*, **31**, 2639–41.
595. Champion, E. and Bertram, H. N. (1995). The effect of interface dispersion on noise and hysteresis in permanent magnet stabilized MR elements, *IEEE Trans. Magnetics*, **31**, 2642–4.
596. Yuan, S. W. and Bertram, H. N. (1994). Micromagnetics of GMR spin-valve heads, *J. Appl. Phys.*, **75**, 6385–7.
597. Lu, D. and Zhu, J.-G. (1995). Micromagnetic analysis of permanent magnet biased narrow track spin-valve heads, *IEEE Trans. Magnetics*, **31**, 2615–17.
598. Beech, R. S., Pohm, A. V. and Daughton, J. M. (1995). Simulation of sub-micron GMR memory cells, *IEEE Trans. Magnetics*, **31**, 3203–5.
599. Russek, S. E., Cross, R. W., Sanders, S. C. and Oti, J. O. (1995). Size effects in submicron NiFe/Ag GMR devices, *IEEE Trans. Magnetics*, **31**, 3939–42.
600. Fredkin, D. R. and Koehler, T. R. (1987). Numerical micromagnetics by the finite element method, *IEEE Trans. Magnetics*, **23**, 3385–7.
601. Koehler, T. R. and Fredkin, D. R. (1991). Micromagnetic modeling

of permalloy particles: Thickness effects, *IEEE Trans. Magnetics*, **27**, 4763–5.

602. Chen, W., Fredkin, D. R. and Koehler, T. R. (1992). Micromagnetic studies of interacting permalloy particles, *IEEE Trans. Magnetics*, **28**, 3168–70.
603. Yuan, S. W., Bertram, H. N., Smyth, J. F. and Schultz, S. (1992). Size effects of switching fields of thin permalloy particles, *IEEE Trans. Magnetics*, **28**, 3171–3.
604. Schabes, M. E. and Bertram, H. N. (1988). Magnetization processes in ferromagnetic cubes, *J. Appl. Phys.*, **64**, 1347–57.
605. Usov, N. A. and Peschany, S. E. (1992). Modeling of equilibrium magnetization structures in fine ferromagnetic particles with uniaxial anisotropy, *J. M. M. M.*, **110**, L1–5.
606. Schabes, M. E. (1991). Micromagnetic theory of non-uniform magnetization processes in magnetic recording particles, *J. M. M. M.*, **95**, 249–88.
607. Spratt, G. W. D., Uesaka, Y., Nakatani, Y. and Hayashi, N. (1991). Two interacting cubic particles: Effect of placement on switching field and magnetization reversal mechanism, *IEEE Trans. Magnetics*, **27**, 4790–2.
608. Uesaka, Y., Nakatani, Y. and Hayashi, N. (1993). Computer simulation of switching fields and magnetization states of interacting cubic particles: Cases with fields applied parallel to the easy axes, *and* Cases with fields applied parallel to the hard axes, *J. M. M. M.*, **123**, 209–18 *and* 337–58.
609. Usov, N. A., Grebenschikov, Yu. B. and Peschany, S. E. (1993). Criterion for stability of a nonuniform micromagnetic state, *Z. Phys. B*, **87**, 183–9.
610. Gadbois, J. and Zhu, J.-G. (1995). Effect of edge roughness in nanoscale magnetic bar switching, *IEEE Trans. Magnetics*, **31**, 3802–4.
611. Komine, T., Mitsui, Y. and Shiiki, K. (1995). Micromagnetics of soft magnetic thin films in presence of defects, *J. Appl. Phys.*, **78**, 7220–5.
612. Jiles, D. C. (1994). Frequency dependence of hysteresis curves in conducting magnetic material, *J. Appl. Phys.*, **76**, 5849–55.
613. Usov, N. A. (1993). On the concept of a single-domain nonellipsoidal particle, *J. M. M. M.*, **125**, L7–13.
614. Usov, N. A. and Peschany, S. E. (1994). Flower state micromagnetic structure in fine cylindrical particles, *J. M. M. M.*, **130**, 275–87.
615. Usov, N. A. and Peschany, S. E. (1994). Flower state micromagnetic structures in a fine parallelepiped and a flat cylinder, *J. M. M. M.*, **135**, 111–28.
616. Lederman, M., O'Barr, R. and Schultz, S. (1995). Experimental study of individual ferromagnetic sub-micron cylinders, *IEEE Trans. Magnetics*, **31**, 3793–5.

# AUTHOR INDEX

The nunbers in square brackets are the reference numbers. They are followed by the page numbers.

# SUBJECT INDEX